Bioinspired Nanomaterials
Synthesis and Emerging Applications

Edited by

Alagarsamy Pandikumar[1], Perumal Rameshkumar[2]

[1]Electrochemical Materials Science, Functional Materials Division, CSIR-Central Electrochemical Research Institute, Karaikudi-630003, Tamil Nadu, India

[2] Department of Chemistry, Kalasalingam Academy of Research and Education, Krishnankoil, 626 126, Tamil Nadu, India

Published by **Materials Research Forum LLC**
Millersville, PA 17551, USA

Published as part of the book series
Materials Research Foundations
Volume 111 (2021)
ISSN 2471-8890 (Print)
ISSN 2471-8904 (Online)

Print ISBN 978-1-64490-156-4
eBook ISBN 978-1-64490-157-1

This book contains information obtained from authentic and highly regarded sources. Reasonable efforts have been made to publish reliable data and information, but the author and publisher cannot assume responsibility for the validity of all materials or the consequences of their use. The authors and publishers have attempted to trace the copyright holders of all material reproduced in this publication and apologize to copyright holders if permission to publish in this form has not been obtained. If any copyright material has not been acknowledged please write and let us know so we may rectify this in any future reprints.

Distributed worldwide by

Materials Research Forum LLC
105 Springdale Lane
Millersville, PA 17551
USA
https://www.mrforum.com

Manufactured in the United States of America
10 9 8 7 6 5 4 3 2 1

Table of Contents

Preface

Biological synthesis employing microorganisms, fungi or plants is an alternative method to produce nanoparticles in a low-cost, simple and eco-friendly way. This route provides a nontoxic and reliable way for nanomaterials synthesis with diversity in size, shape, composition, and physicochemical properties. The biocomponents are capable of serving as the templates for the synthesis and organising the nanorange particles into well-defined structures. These bio-mediated nanomaterials have significantly contributed towards a variety of applications. Thus, detailed information on the bioinspired nanomaterials and their multi-font applications must be obtained in order to realize their high potential to the maximum. With this aspect, the present book aims to discuss elaborately the different biomaterials used for nanoparticles synthesis and their potential applications in multi-disciplinary fields. The synthesis part will cover the extraction process of biocomponents, synthesis of metal nanoparticles, metal oxide nanostructures and nanocomposite materials, mechanism of reducing and stabilizing process, stability of nanomaterials and the characterization of the bioinspired nanomaterials. The application part includes microbicidal property, optical and electrochemical sensors, packaging, SERS and drug delivery applications. Individual chapters are allocated for every process to elaborately discuss the functions of bioinspired nanomaterials towards particular application. The recent progress in developing bioinspired composites and the perspectives on future opportunities in various applications are also discussed in this book.

This book will cover the importance of bioinspired synthesis, extraction of biocomponents, mechanism of action in synthesis, morphology controlled structures and nanomaterials property dependent applications. Audience can easily understand the fundamentals of bioinspired synthesis, mechanism, characteristics and role of the nanocomposite materials in the current scenario of application-oriented research and development. The book will be helpful for researchers to establish their own research in the area of bioinspired nanomaterials.

We are very much grateful to all the authors who contributed their chapters to make a valuable book and for the successful completion of the process. We are thankful to the editor, Thomas Wohlbier, Materials Research Forum LLC for accepting our proposal and giving an opportunity to edit this book and his help towards the successful completion of the work is greatly acknowledged.

Dr. Alagarsamy Pandikumar (Leading Editor)
Scientist
Electro Organic and Materials Electrochemisry Division

CSIR-Central Electrochemical Research Institute
Karaikudi-630003, Tamil Nadu, India
&

Dr. Perumal Rameshkumar (Associate Editor)
Assistant Professor
Department of Chemistry
School of Advanced Sciences
Kalasalingam Academy of Research and Education
Krishnankoil, 626 126, Tamil Nadu, India

Bioinspired Nanomaterials
Materials Research Foundations 111 (2021) 1-35

Materials Research Forum LLC
https://doi.org/10.21741/9781644901571-1

Chapter 1

Introduction to Bioinspired Nanomaterials

Sangeetha Kumaravel[1,2], Prabaharan Thiruvengetam[3] and Subrata Kundu[1,2,*]

[1]Electrochemical Process Engineering (EPE) division, CSIR-Central Electrochemical Research Institute (CECRI), Karaikudi-630003, Tamil Nadu, India

[2]Academy of Scientific and Innovative Research (AcSIR), Ghaziabad-201002, India

[3]Department of Chemistry, Indian Institute of Technology Madras, Chennai-600036, India

* skundu@cecri.res.in and kundu.subrata@gmail.com

Abstract

Nanomaterials (NMs) developed using biomolecules display numerous advantages which attract the science community to explore them for a wide range of applications. In this line, bio-scaffolds are studied as templates to form nano-bio heterojunctions in the nano confined materials. With the high flexibility of bio-mediated NMs, it is possible to develop desired size and shape selective NMs. Such bio-based NMs have great benefits in wide areas including catalysis, sensors and energy related applications particularly, electrocatalysis, supercapacitor, batteries etc. The viability of bio-scaffolds in developing metal superstructures makes them better choice in the medicinal fields. This book chapter mainly focused on the advantageous and challenges of bioinspired NMs in the medicinal field, particularly in drug delivery systems. Moreover, the synthetic methods such as enzyme catalyzed wet-chemical route, photo-irradiation and incubation methods were also discussed in detail. Also, this chapter gives a better understanding to the readers about the development of new nano-bio heterojunctions for medicine, energy and environmental fields. Moreover, the morphological features of nano-bio interactions at nanoscale level show predominant activity particularly in Surface Enhanced Raman Scattering (SERS) and sensor applications. With the knowledge gained from this chapter, in futuristic, one can go for the development of new metal nanostructures with different bio-scaffolds such as microorganisms, viruses, DNA and protein to mainstream applications for the medicinal fields.

Bioinspired Nanomaterials   Materials Research Forum LLC
Materials Research Foundations **111** (2021) 1-35   https://doi.org/10.21741/9781644901571-1

Keywords

Nanomaterials, Biomolecules, Enzymatic, Non-Enzymatic, Bacteria DNA, Proteins, Viruses, Cell Wall, Wet-Chemical Method, Incubation, Photoreduction, Catalysis, Sensor, Disinfection, Electrocatalysis, Supercapacitor

Contents

1.   Introduction

Owing to the augmentation of nanoscience and nanotechnology, numerous interesting advances and possibilities have been emerged in recent years [1–3]. Nowadays, nanomaterials (NMs) have been progressively integrated with industries as well as in consumer products [4–6]. The NMs we studied are mainly of two categories such as (i) Organic nanomaterials-which consist of carbon nanoparticles (e.g., fullerenes, carbon nanotubes and graphene) and (ii) Inorganic nanomaterials-which include noble metals, magnetic nanoparticles and semiconductors such as (oxides of titanium, zinc, cadmium and silicon) [7,8]. Inorganic nanomaterials are known to be useful in day to day life such as agriculture, food safety, oil refining, medicine, pharmacy, cosmetics, textile, healthcare, transport, electronics and communication industries [7]. A recent inventory documented that more than 1800 consumer products contains NMs and amongst that many more are in non-commodity products such as industrial catalysts, etc. In traditional ways, NMs are manufactured by either top-down or bottom-up methods. In the top-down method, the nanoscale dimensions can be achieved by grinding or successive cutting of a bulk material (e.g., lithography, milling and etching) and in the bottom-up methods, the assemble of atoms/molecules and clusters occurs through the techniques such as sol-gel or epitaxy methods, etc. (Figure 1) [9,10]. Even though the top-down process is predominating, the bottom-up approach is more advantageous for sustainable manufacturing of NMs due to the production of materials *via* molecular levels and hence, reducing the unwanted wastes. In spite of the commonly used methodologies for NMs production, certain issues are need to be addressed regarding their sustainability [10].

At first, the production of ultrapure chemicals, reagents and solvents in the traditional methods needs high energy, high temperature and high vacuum for purification, which is a major issue in view of cost and energy consumption [11]. Also, the traditional methods contain several reaction steps that produce more unwanted waste to the atmosphere. Over the last 15 years, the life-cycle analysis and risk assessment evaluated the environmental and safety impact during the use of some extensive NMs such as silver, silica and titania nanoparticles [12,13].

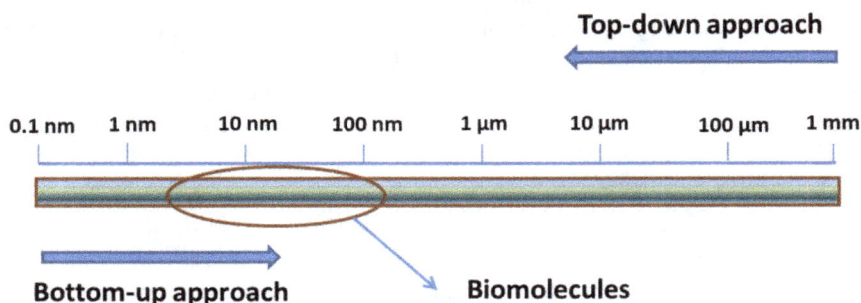

Figure 1. *Top-down and bottom-up approach in the nanomaterials preparation.*

The analysis results strongly imply that the wasteful preparation methods as well as the toxicity of NMs are being ignored [14]. The environmental impact and sustainability analysis by E-factor showing that the existing processes of manufacturing NMs produces nearly 10000 kg waste in 1 kg product, which is 1000 times more wasteful when compared to the pharmaceuticals and fine chemical industries [15]. These issues enforce the researchers to develop better manufacturing methods which will be environmental friendly as well as sustainable towards various applications. On account of this, bio-mediated methods of preparing NMs were found to be more sustainable and greener. Mostly in the biomedicine applications for treating the organisms, the NMs prepared with various bio-systems such as microorganisms, amino acids, proteins, peptides and plant extracts were found to be more suitable [16,17]. The following are some of the alternative greener approaches in the NMs synthesis which are discussed detail.

2.   Emerging strategies for greener routes

For synthesizing NMs, some new preparation techniques have evolved by considering environmental pertained routes. Such methodologies involve self-assemblies, nano-bio interactions and template mediated designing with biomolecules. Such methods can effectively control the sizes and shapes of the NMs [15,18,19]. This new method acquires high selectivity with the molecular building blocks assembly at different scales of NMs synthesis. Moreover, this bio kind of approach can largely reduce the production of undesirable wastes in the NMs synthesis. Besides, it could also minimizes the multiple processing steps and increases the efficiency in the NM preparation with higher rates **(Figure 2)** [20]. Precisely, such bio approaches avoiding the toxic solvents in an elevated temperature. The designing of NMs in an aqueous medium at a lower temperature might reducing the environmental risks in the bottom-up approach [21]. This bottom-up approach of bioconjucation greatly helps to achieve better reaction rates in many applications. After the evolution of NMs, industries people started to eradicate the traditional/old existing manufacturing processes with cost effective ones [22].

Also, the advantages of comparable size of nanoparticles (NPs) with the cellular component attracted the researchers to study more in the field of biomediated NMs [23,24]. The physico-chemical properties of the nanoparticles such as size, shape, surface charge and surface chemical composition were largely dictating their entry into the biomolecules. Simply, the biological system is an effective alternative owing to its low cost and green approach for the production of nanoparticles. Before involving into the designing of biobased NMs, researchers have to gain basic knowledge about the emulating natural designing and eco-designing principles of a bio system [18]. The use of biomolecules can play many roles such as catalytic agent and a structure directing agent for NMs design with its self-assembling of molecules at different nano scales.

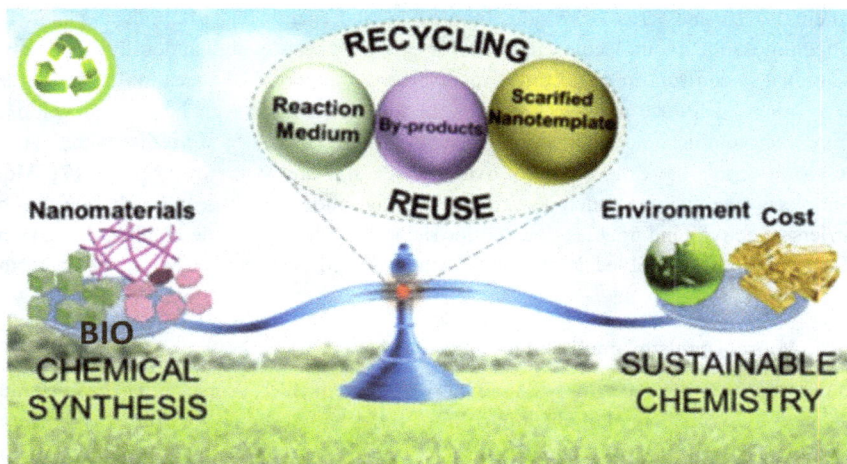

Figure 2. The environmental and economic concern for a greener way of fabricating nanomaterials. Reprinted from ref. 20 with permission. Copyright 2019. American Chemical Society.

These non-toxic and environment friendly methods of synthesizing NMs satisfies most of the principles of green chemistry [18]. Thus, these advantages of a biological approach were considered as the best alternative to address the current issues of synthesizing NMs [25,26]. Recent studies were giving more emphasis on synthesizing bioinspired inorganic NMs from micro-organisms, plants, proteins, enzymes, viruses, DNA and some specified biomolecules [27]. Despite of their eco-friendly behavior and greener routes of applying biomolecules their usage were limited on the lab scales. To date bioinspired inorganic NMs have been reported with metals, alloys, ceramics, carbonates, sulfides, selenides, arsenide and zeolites [25,26]. However, the minimum supply of these biomolecules makes them difficult to implement on an industrial scale production [27]. Therefore, it is necessary to focus on finding the supplies of biomolecules with minimal cost for industrial uptake of NMs manufacturing.

3.   Bioinspired route of nanomaterial synthesis

With the detailed preceding about biomediated NMs synthesis, the following are the discussions about some examples of bio based NMs synthesis including microorganisms, proteins, DNA and virus for various biomedical and energy related applications. At the initial times, the biomolecules utilized for NM synthesis were unmodified and more traditional to confront in a real time practice. Over the period, the vast molecular library for peptides, cellulose and DNA sequences were studied. Which helps to develop or tuning the biomolecular sequences, as results multiple structures were discovered. Also, the bio materials have more control over varying size and shape of NMs [16,17]. Most biomolecules containing functional groups such as amides, sulfides and carbonyls are the carriers of active binding sites devoted for rising the NMs [28]. The detailed discussions of methodologies involved in the biomolecules based NMs synthesis are given in the following.

3.1   Microorganisms mediated nanomaterials

The greener way of preparing NMs *via* bio-reduction of microorganism has been studied for many bio medicinal applications. In general, microorganisms such as bacteria, actinomycetes, yeast and fungi's were widely used in the NMs synthesis [29]. Among this, bacteria and fungi are the most favorable microorganisms compared to yeast and actinomycetes [30]. Considering bacteria, it is a prokaryotic type microorganism and it is an active candidate in the NMs synthesis. There are two mechanisms studied for the NMs assembly over bacteria; the primary mechanism is such that the metal ions get trapped on the cell walls (inside/outside) of the bacteria. Then the enzymatic reactions carryout to reduce the metal ions and forms corresponding NMs structures [28,31–33].

In detail, bacteria having the specialty of reducing metal ions (detoxification process) to metallic NMs. Another process where nonliving microorganisms having various functional groups such as thiols, amino, imino and carboxyl has been used to reduce metal ions to form metal NMs. Here, initially, the metal ions get attached on the surface of the microorganisms, get reduced, start nucleate over the bacteria and finally form the NMs. Less toxic chemical usage is the major advantage with bacteria mediated NMs synthesis. For example, Joerger *et al.* studied a bacterial strain such as *Pseudomonas stutzeri* AG259 for making ceramic metal composites [34]. The composite has been prepared with previously developed bacteria over the Lennox L substrate was mixed with silver nitrate solution (30 °C for 48 h). Further, the mixture was washed and spread over aluminum foil to make it a thin film (μm) and heat treated at 300 – 400 °C. The prepared Ag bacterial composite were cost-effective and good in optical properties [35]. Also, the Ag, Au and their alloys were studied with the bacteria *Lactobacillus strains*. Such,

Materials Research Forum LLC
https://doi.org/10.21741/9781644901571-1

bacterial synthesis are highly stable and produces various shapes of NMs in the Au-Ag alloy preparation [36]. Figure 3 a and b are TEM images showing Au NPs were developed over the rod-shape bacterium with the particle size of 20-50 nm, approximately. Nonetheless, there are some disadvantages with the bacteria towards controlling the perfect size and shape of the NMs. It is effective, but less selectivity towards the growth of NPs which results non uniform size and shapes. Other than, Pseudomonas stutzeri and Lactobacillus strains, there are a number of other bacteria's used in NMs preparation such as escherichia coli, pseudomonas, bacillus licheniformis, ureibacillus and rhodopseudomonas etc. [37,38].

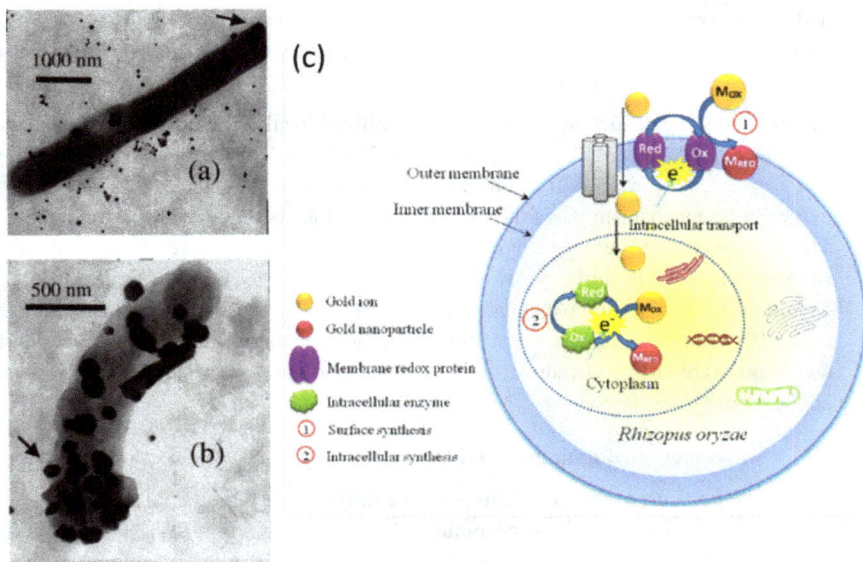

Figure 3. *TEM images of gold NPs developed on Lactobacillus strains bacteria (a and b) Reprinted from ref. 35 with permission. Copyright 2002. American Chemical Society and the proposed mechanism of gold biomineralization in R. oryzae (c). Reprinted from ref. 34 with permission. Copyright 2012. American Chemical Society.*

Similar to bacteria, the fungi's were also studied for the NMs preparation. Here the metal ions attached with the intracellular or extracellular parts of fungi gets reduced as metal

NPs [34]. Mukherjee *et al.* have studied Au NPs developed over the fungi zygomycete fungi R. oryzae [39]. They found that the reductions were carried out both in the cytoplasmic region and the cell wall of the fungi. Compared to other type of fungi, R. oryzae is plentiful in availability of mycelia (vegetative part of fungus) and nonpathogenic in nature. Therefore, handling of R. oryzae is simpler than other fungus. The process of Au NP formation was carried out using R. oryzae mycelia with $HAuCl_4$ solution was kept for 48 h of incubation and the formed Au NPs were confirmed using UV-Vis analysis. From the morphological analysis, particle size of Au NPs formed was identified as 15.9 ± 2.8 nm. There are two possible mechanisms proposed, one is the interaction of metal ions on the cell walls of the fungi and it gets reduced by enzymes to form Au NPs [40,41]. Secondly, transportation of Au ions into the cytoplasm of fungi and it gets reduced by the enzymes present in it and forms Au NPs (Figure 3 c). The fungi such as Verticillium, Fusarium oxysporum and Colletotrichum were also used in various NMs syntheses [17].

The microorganism such as yeast were also utilized for the NMs synthesis. Meenal *et al.* have developed an Ag NPs using yeast strain MKY3 [42]. Like fungi and bacteria, yeast can also process the detoxification of heavy elements [43]. In the Ag NPs synthesis, the extracellular mechanism was followed and the Ag ions get reduced to form metallic NPs. The yeast such as S. cerevisiae and C. albicans were also studied for NMs synthesis [29]. In comparison to bacteria, yeast was more advantageous in the fabrication of diode cadmium. Similarly, zinc phosphate was prepared using yeast and the resulted NPs were in the 10–80 nm size ranges. As summarized, the microorganisms such as bacteria, fungi and yeast were better to reduce the metal ions and form a structural NMs for various applications.

3.2   Virus mediated nanomaterials

Similar to the microorganisms, viruses can be utilized as a biomaterial for NMs synthesis. In general, viruses are made of genomic and proteomic moieties [44]. Study of virus is a developing field in the nanomaterial fabrication, and it is more advantageous for many biomedical fields. Particularly, it has more advantages in imaging, drug delivery and tissue engineering applications. It is believed by the researchers, that the use of virus creates a strong bond between material science and the biotechnology in near future [45]. Moreover, the major availability of genomic sequence of virus is highly desirable to produce most stable NMs. It is even possible to provide stable monodispersed size and shape controlled NPs. The morphological tuning nature and the highly organized functionality of virus were also used in the energy storage and memory device fabrications [46]. In practice, different viruses such as tobacco mosaic virus (TMV), filamentous bacteriophage fd, Cowpea Mosaic Virus (CPMV) and T4 bacteriophages provide different types of morphological outcomes in the NMs fabrication. [47–49].

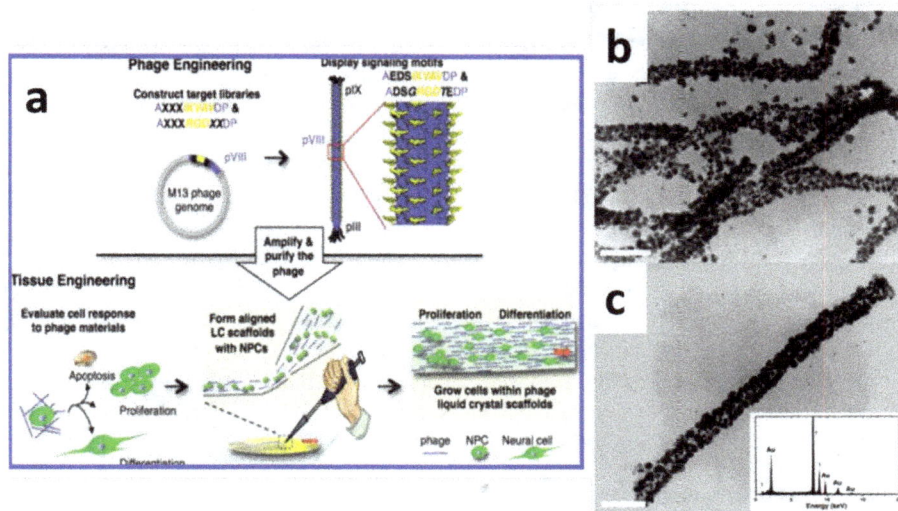

Figure 4. The tissue engineering process using the M13 phage virus (a) Reprinted from ref. 46 with permission. Copyright 2009. American Chemical Society. and TEM image of Au nanorods produced in wild-type TMV virus. Reprinted from ref. 44 with permission. Copyright 2003. American Chemical Society.

In the wild type viruses, there are more proteins sequence in the coatings which reduces the metal ion binding ability of the viruses. Therefore, the virus needed to be surface modified before it was used in the NMs synthesis. There are more methods to activate the surface of the virus and subsequently used to reduce the metal ions with or without the use of external reducing agents. Amongst the various viruses, the M13 bacteriophage is the most studied one. Lee *et al.* have prepared the nano-fiber like virus for biomedicine application such as tissue regeneration process (Figure 4a) [46]. Dujardin *et al.* have performed the deposition and organization of Pt, Au, and Ag NPs using the TMV virus. The elements such as Pt and Au salts were reduced at a maintained pH for better yield. Here, the use of wild-type TMV virus is more control over the particle size. The morphological study shows the particle size was below 10 nm and the corresponding images are shown in Figure 4b and c [44].

The surface modification with the wild type TMV virus was carried out *via* immersing the Wild-Type TMV and E95Q/D109N mutant in a buffer solution (pH-7). Similarly, the varying noble metal precursor was suspended along with the wild-type TMV virus was reduced with hydrazine hydrate solution. This kind of modification is simply known as charge modification process and this provides more affinity of metal ions with the virus. The virus template mediated synthesis results in various size and shape of NMs. It may be varied as external or internal attachment of metal ions and bring control over the NM growth. In 2011, Manocchi *et al.* studied the Pd NPs growth on TMV virus (TMV1cys) template [50]. The presence of virus template results in a perfect growth of Pd NPs with controlled morphology. This study detailed about the role of varying concentration of Pd precursor and the use of different types of reducing agents. The optimization study confirms that the high Pd precursor concentration simultaneously increases the Pd NPs formation on TMV virus. So, it is well understood that the biomolecule such as a virus can be useful and effective in the NMs synthesis by producing controlled sizes and shapes for various applications.

3.3 Deoxyribonucleic acid (DNA) mediated nanomaterials

The DNA is known as a genetic carrier in the biological system which gives instructions for various functions of the organism [51,52]. Other than that, it has a wide usage in the bottom-up approach towards NMs synthesis. DNA is a polymeric structure with the double-helix structure containing repeating nucleotides such as adenine (A), guanine (G), Thymine (T) and cytosine (C), those repeating units were linked by a hydrogen bonding **(Figure 5)** [53]. Also, the backbone of the double helix structure of DNA has phosphate groups and sugar moieties which will act as a binding sites in the NMs fabrication [54]. In addition, the diameter of DNA is about 2 nm and the length is 0.34 nm and it has the advantage of meeting the same line with material science. The presence of these aromatic groups and phosphate groups in DNA can easily accommodate more metal ions by means of electrostatic interaction which results a perfect one dimensional chain like assemblies of NMs [53,55]. The researchers did lot more in the DNA sequences/tiles to assemble different tiles such as bundles, double and triple crossover tile, cross linked with multiple junctions [56].

These programmed DNA sequence were later developed by many software to use them in a NMs synthesis and provides several 1D, 2D and 3D structures [57]. The DNA plays multiple roles of serving as a template or scaffold, stabilizer, binder, mild reducing agent and also acts as a stimulator to carry organic reactions in aqueous medium. There are

Materials Research Forum LLC
https://doi.org/10.21741/9781644901571-1

several advantages, such as the resulted 1D structure were stable for a long period, quiet easy to characterize, morphological bearing capability, easy to handle, high selectivity and control over NMs growth [52,58]. The detailed information and styles of DNA were discussed in *Chapter-5*. There are several methods involved in the DNA based NMs synthesis such as wet chemical or chemical reduction, photochemical reduction, incubation, seed-mediated synthesis and microwave heating methods [41,59]. Following are discussions about the preparations and their mechanism of DNA based NMs.

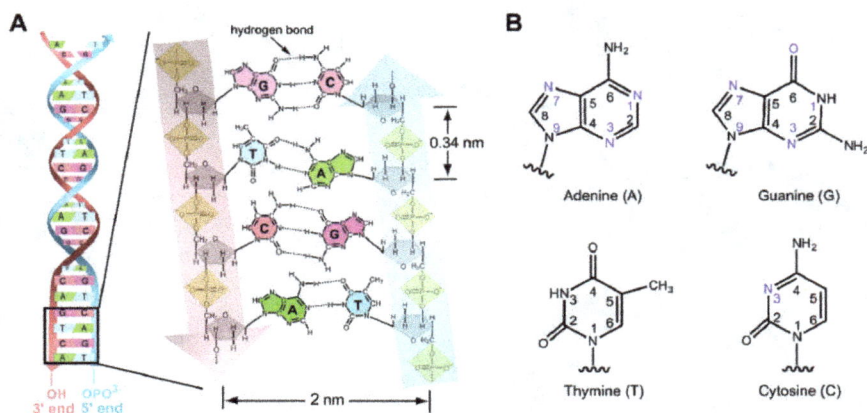

Figure 5. Watson-Crick Structure of DNA (a) and the aromatic group's presence in the center of the double helix structure (b). Reprinted from ref. 54 with permission. Copyright 2018. Royal Society of Chemistry.

In the wet-chemical method, the DNA based NMs synthesis involves three stages. At first the metal ions electrostatically interact with the double-helix of DNA and then the reduction is carried out using reducing agents such as sodium borohydride or tetrabutyl ammonium borohydride. Following, the pre-formed NMs gets nucleate on the surface and the inner core of the reduced metal act as a metal sites to assembling the metal and form nano clusters [58]. The wet chemical method of NMs synthesis was carried out both in aqueous and organosol medium and was stabilized using DNA. The DNA metallization has been done for almost all the elements including Au, Ag, Cu, Pd, Pt, Rh, Re, Pd, Pt, Os, and Ir [59–62].

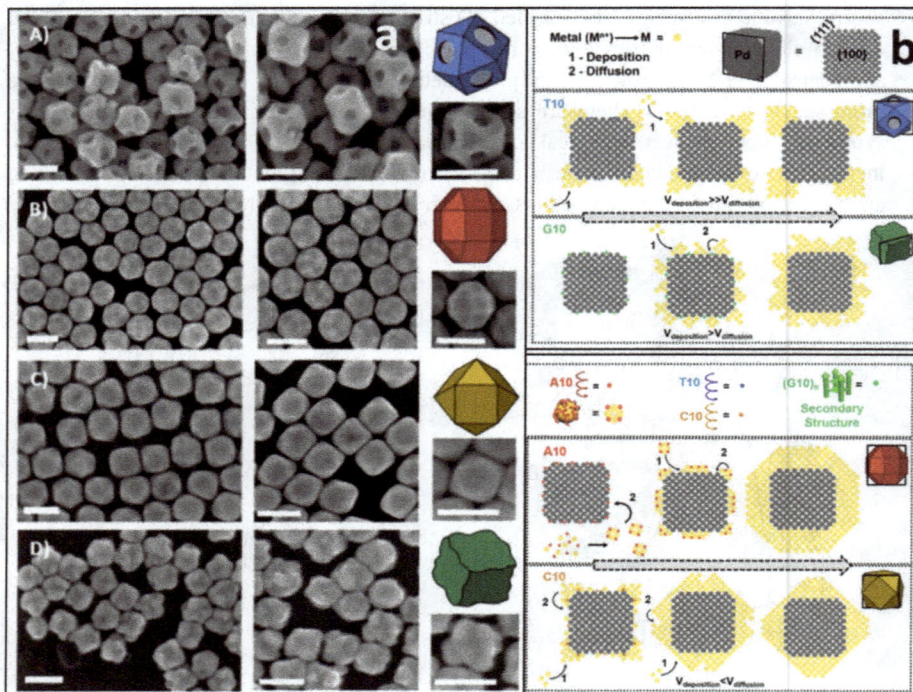

Figure 6. *Morphological analysis with SEM shows the Pd-Au nanoparticles prepared with A) T10, B) A10, C) C10 and D) G10 (a) and growth mechanism of Pd-Au bimetallic nanostructures with different DNA sequences (b). Reprinted from ref. 65 with permission. Copyright 2016. American Chemical Society.*

Also, several metal oxides and metal tungstates were prepared using DNA as template. The synthesized DNA based materials were used for wide range of applications in energy related fields such as supercapacitor, electrocatalysis, photocatalysis, sensing, optoelectronic device fabrication and in biomass conversions [59–62]. Our group has done various works in the synthesis and stabilization of metal NPs in the DNA nano-assembly in both aqueous and organic medium [63,64]. In the incubation method, the metal ions were incubated with DNA for several hours for interaction and the positive metal ions were reduced by using citrates that acts as a reducing agent and structure directing agent. Maintaining pH (around 7) using phosphate buffers is an important factor to concern in this method for successful NMs fabrication. The dimer formation with Au or Ag is more favorable and simple to fabricate *via* the incubation method. The next method is the seed-mediated NMs growth on DNA which is different than above methods. In the seed mediated growth method, the initially developed metal particles can serve as seed to grow another metal over the surface of the seeds. Lu *et al.* have studied the varying DNA sequence with the seed mediated NMs synthesis [65]. Their synthesis detailed the formation of Pd–Au bimetallic nanoparticles from palladium NC seeds. At first, the Pd NPs were developed in CTAB and reduced with ascorbic acid. Then it was incubated with Au precursor solution contains DNA oligomeric sequences and studied with different binding cites of DNA. The DNA sequences produce more binding sites and the diffusion of the NMs growth on the facets over the Pd seed. Among the four different base pairs present in DNA sequences such as adenine (A), guanine (G), Thymine (T) and cytosine (C), the adenine (A10) shows aggregated growth due to the high affinity with the Au NPs. The morphological analysis from SEM for the Pd–Au bimetallic NPs mediated by various DNA sequence shows the possible binding sites mechanism of different sequence were given in Figure 6. The results clearly explain the significance and the control over size/shape NMs with DNA.

3.4   Protein mediated nanomaterials synthesis

Small templates such as proteins were also viable to prepare NMs in a simple manner. Protein based synthesis is a bottom-up approach in the NMs synthesis [66]. It is a natural source of biomaterial for tuning the NMs into a controlled and defined shape. In comparison with other biomolecules, the complex protein has a large number of functional groups to scaffold the metal ions which easily reduces the metal ions to form NMs. Also, the method of protein based NMs synthesis is fast and safer (in room temperature) to handle than other methods. They are more particular to design size and shape of the NM structure by multiple varieties of protein fabrication. Because of the availability and low cost, the protein based NMs have been more facile to carry large scale applications. The more advantage, such as the protein template having vast binding site to accommodate almost every elements in the transition metal series, therefore it would be suitable to apply in a wide range of applications. Many researches have attempt with the proteins for biomineralization, most importantly the fabrication of metal sulfide NMs are more favorable.

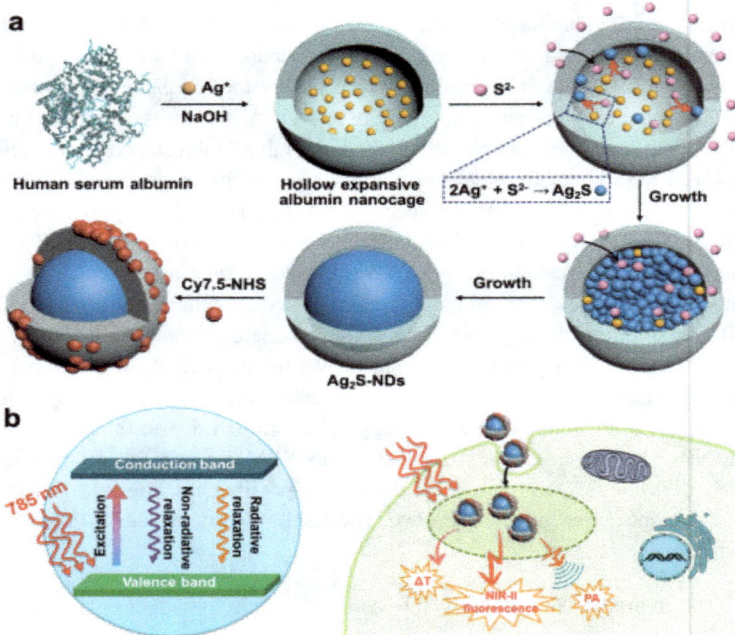

Figure 7. Synthesis steps of Theranostic Ag_2S Nanodots in human serum albumin (HSA). Reprinted from ref. 68 with permission. Copyright 2017. American Chemical Society.

Sheng *et al.* have developed a series of metal sulfides such as Ag_2S, Bi_2S_3, CdS, and CuS NPs using protein as template [67]. Following the incubation method, the mixing of bovine serum albumin (BSA) with $Cu(NO_3)_2$ at 60 °C in alkaline condition in first step. The sulfur anion from the protein interacts with metal ions and it starts to nucleate. The prepared protein based NMs were highly preferable in the biomedical field particularly in photothermal therapy. Similarly, Yang et al. have prepared an Ag_2S nanodots for photothermal therapy **(Figure 7)** [68]. The protein used in this work is human serum albumin (HSA) which is highly desirable to prepare nano-caged Ag_2S nanodots. The mechanism follows a diffusion controlled process; the formation is mainly depending on the ratio/concentration of the reactant. Collagen type proteins were also used for NMs synthesis, it is mainly found in mammals. The single collagen molecule is a complex of three polypeptide chains of helical structures. Also, such proteins having a higher number of functional groups such as carboyl group, amide group, and hydroxyl group which are more favorable to fabricate NMs [68]. The following are some of the major applications studied with the biomediated NMs.

4. Applications of bioinspired nanomaterials

Bioinspired synthesis of nanomaterials with defined size and shapes are of wide interest to apply for various applications. It is more advantage with high stability, dispersity, cost effectivity and superior in drug delivery systems. In catalysis and electrocatalysis reactions, the presence of biomolecules increases the rate of the reaction to cross the activation barriers easily. The following are some of the applications of bioinspired NMs for medicinal, energy and environmental fields **(Figure 8)**.

4.1 Medicinal applications

Developing biomolecules such as enzymes, DNA and proteins were showing promising potential applications in drug delivery systems (DDS). For years, several drugs have been used to eliminate microbes including bacteria, but these microbes tend to become resistant over a period of time due to the gene mutations [69]. Various bio based NPs have been in practice as a drug with multiple mechanisms concurrently to fight microbes. Some of the examples are metal NPs (Au, Ag, Mg, Ti, Cu and Zn NPs) contains multiple mechanisms to combat microbes *via* inhibiting the growth of antibiotic-resistant bacteria.

Some of them are E. faecalis, K. pneumoniae, E. coli, and P. aeruginosa [70]. Nowadays, bioinspired NPs are being synthesized using plant extracts became more eco-friendly and non-toxic and effective over microbes [71]. The photocatalytic degradation process with bioinspired NMs has gained a lot of recognition in the area of wastewater treatment [72,73]. It is known that organic and inorganic substances can be effectively degraded using photocatalysis [74], but the pollutants are generally of organic compounds, hence the degradation of such organic substances form an intermediate as OH radical. The reactions taking place on the surface were mainly redox reactions [75]. The photoactive nanomaterials such as Ti, Bi, W and Si with biomolecules were used in the catalyst preparation for toxic degradation. There are certain factors that influence the degradation process. Some of them are: effect of light intensity and wavelength; the kind of light source used; effect of anions and cations; pH; effect of adsorption and surface area [74]; the nature of the catalyst used. One of the very good examples of photocatalytic degradation of azo dyes using transition metal doped titanium dioxide (TiO_2) [76].

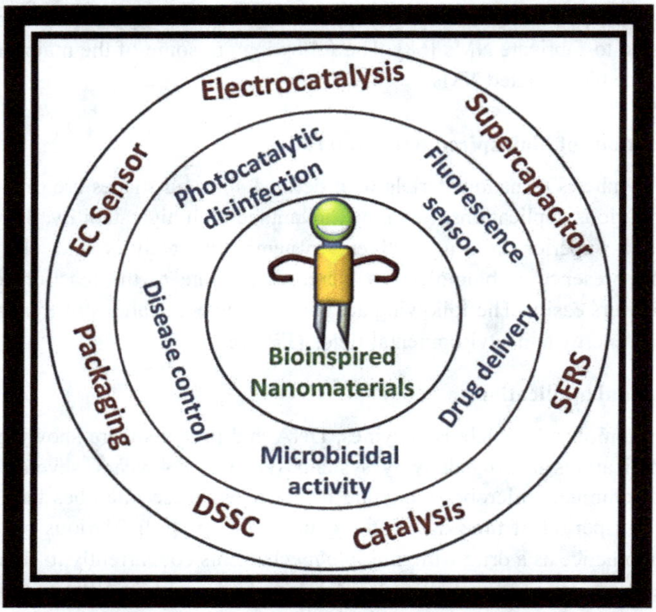

Figure 8. *Various applications with bioinspired nanomaterials.*

For the disinfection process [77], the traditional way of disinfection by chlorination was used in the initial period, but the formation of harmful carcinogenic by-products and various other disadvantages were examined. Therefore, the photocatalytic method is adapted for the disinfection process, and it has the ability to inactivate a certainly wide range of detrimental microorganisms [78]. It also finds its wide range of applications in the medical field [79], pharmaceutical and food industries etc. [80]. The well-studied disinfectant NMs such as titanium dioxide (TiO_2), CuO, Ag NPs are often used for such disinfectants. This process is innocuous, chemically stable and it can be recyclable without losing its catalytic properties. Recently, the visible light photocatalysis has gained much attention because of its solar energy conversion by using visible light [81]. Hence this technique is quite adaptable and efficient, it is not only a technology that is substituting the traditional method but also a fine approach for solving the disinfection problems or other related problems. DDS are NMs based biomedical devices, the toxicity of the nanostructured materials have been one of the growing concerns because of their

Materials Research Forum LLC
https://doi.org/10.21741/9781644901571-1

use in the biomedical field [82,83]. Therefore bioinspired NMs have gained huge attraction in the biomedical field as a safe drug delivery system [84,85]. The nanocarriers have several advantages in the drug delivery because they are potent in tumor treatment [86]. A few bioinspired procedures to prepare DDS are namely the conventional encapsulation methods (evaporation/extraction, interphase polymerization, concertation) and layer by layer approach method (LBL) [86,87]. Once the DDS arrives at the targeted site they release therapeutic agents and its release can be regulated using an external media.

Next one is fluorescent sensors, which is widely used in monitoring and clinical diagnostics etc. [87,88]. Day by day, the demand for these sensors has been increasing because of their ability of imaging within the living cells and to detect different targets [89]. DNA and protein based NMs have good sensing ability to detect various metal ions. More precisely, the highly toxic Hg (II) ions detection was effectively carried out in the presence of proteins and DNA-assemblies. Also, in the detection of H_2O_2 as a main product obtained in most of the enzymatic reactions carried *via* protein developed noble NMs [90].

4.2 Energy and environmental applications

Catalysis plays an important role in both academic and industrial applications *viz.*, for the preparation of medicinally important organic compounds, drug molecules, agrochemicals, dyes etc. [91]. Traditionally catalysis falls into two major regions such as homogeneous and heterogeneous catalysts depending upon its nature in solvent medium and reaction. However, the homogeneous catalysts have vast applications because of their uniform nature with solvent medium and also it increases the reaction rate efficiency [83,92]. However, it has few disadvantages with the inefficiency in recyclable nature which is important in medicinal applications. Also, it should obey all the 12 principles of green chemistry [92]. In this regard heterogeneous catalysts, especially metal NPs have been prepared for a longer time. Earlier it was prepared by using organic stabilizers and environmental hazardous solvents which creates serious negative impact on the eco-system. So, as an alternative to hazardous solvents, environmentally comfortable green stabilizers such as surfactants, plant extracts, and biomolecules such as carbohydrates, proteins, nucleic acids, green algae, and lipids have started to be utilize in the NMs synthesis. The great advantage with this bio system that it is biodegradable [93]. Also water has been used as green solvent for the preparation of metal NPs which is environmental friendly [94]. In few cases, solvents such as alcohols were also studied as a green medium for stabilizing bio based NMs [95].

The electrocatalysis is being developed and enhanced with the help of bioinspired NMs. In addition, researchers are now seeking opportunities from bio-templates such as bacteria, viruses, DNA and proteins as an alternative for traditionals [17]. These methods can introduce favorable complexity into the products at nanoscale or atomic scale, which can remarkably improve the electrocatalytic performances of the catalysis *via* increasing the charge transfer process [21]. For example, the DNA mediated synthesis of NMs was playing multiple roles such as binder, stabilizer and accelerator in the electrocatalytic water splitting [96,97]. The enzyme horseradish peroxidase (HRP) was simply mixed with DNA, chitosan and Fe_3O_4 nanoparticles, and then applied on the electrode surface to form an enzyme-incorporated polyion complex film. HRP was immobilized on the electrode surface by DNA/chitosan/Fe_3O_4 bio-polyion complex membrane [98]. The HRP on the electrode exhibited fast electron transfer rate results high affinity to H_2O_2 and good bioactivity toward H_2O_2 reduction.

Supercapacitors have evoked extensive research because of their high power density, fast recharge capability, and long cycle life [99]. To improve energy densities for practical applications, carbon based nanomaterials, particularly carbon nanotubes (CNTs), carbon nanofibers and graphene, have been widely used as electrodes [100,101]. The study with bioinspired deposited peptide nanotubes (PNT) arrays deposited by PVD technology on carbon electrodes demonstrates a pronounced enlargement of electrical double-layer capacitance by 50 times. A hybrid carbon NMs with hierarchical structures improves the electrochemical performance of carbon-based supercapacitors [102,103]. To avoid restacking, the intercalation of hard species such as biopolymers, and carbon nanotubes between the graphene sheets drastically enhance the super capacitors efficiency [104]. The insertion of carbon nanotubes into the chemically converted graphene sheets can significantly improve their electrochemical potential and allow the device operation at high rates. Modified smooth carbon electrodes with deposited PNT resulted in enlargement of double-layer capacitance density, compared to background carbon electrode.

Bioinspired NMs can also use in the electrochemical sensor applications. When the NPs were assembled with enzyme, DNA or protein based materials enhancing the properties such as selectivity and sensitivity [105]. One such case, the enzyme developed NPs will increase the immobilization of enzymes and increase the effect of sensitivity to progress high performance biosensor. In the real sample analysis, biosensor to detect glucose with NPs of different shape, size and composition are used as electrode materials for effective sensing [106,107]. By providing large surface area the graphene NPs stacks creating an enzyme biosensor to detect glucose. Also in the electrochemical genosensors, a single stranded DNA is assembled over the electrode and helps in capturing the target DNA through hybridization by signals being generated and detected electrochemically [108]. This DNA developed electrochemical bio sensor provides excellent sensitivity towards the product analysis.

Another application with this bioinspired NMs is fluorescent sensing and imaging. Quantum dots (QD) are semiconducting NPs which are having unique physical properties and great potential in fabrication of next generation optoelectronic devices [109]. In this line, the bioconjugation benefits specificity and versatility of QD based in-vitro sensing. For the detection of proteins, viruses, bacteria and other molecules, fluoroimmuno assay based QD-antibody conjugates are used [110]. The high specificity between QD-DNA conjugates helps for multiplexing detection of samples [111]. QD with bioconjugates have great potential in ultra-sensitive genetic target analysis through multiplexed DNA detection scheme. Fluorescence resonance energy transfer (FRET) sensing strategies based on fluorescent protein enables detection to monitor molecular level interactions [112]. Now, when we look at cells, multimodal and vivo imaging, bio functionalized QDs found applications in these types of imaging system. When QDs conjugates with nano moieties, they become highly specific and label the receptors enable to monitor the dynamics of proteins and lipids. Functionalized QDs found application in vivo tracking, vasculature imaging, tumor imaging and targeted therapy [113].

Another important application with bioinspired NMs is Surface Enhanced Raman Spectroscopy (SERS) studies [114]. It is a surface sensitive technique when the analyte molecules adsorb over the rough metal surface such as Au, Ag and Pt etc. [115,116]. The LASER beam hit on the metal surface causes plasmonic resonance which interacts with the analyte molecules and gets detected even in very low concentration (even in ppm levels) [117]. Various biomolecules such as proteins, enzymes, amino acids and peptides were used in the biomediated NMs synthesis. The biomaterial controls the nucleation and growth of the NMs over the surface [118,119]. In particular, for preparing dimers NPs, DNA having high control over the interparticle distance of below 5 nm. Also, biomolecules increases the enhancement factor (EF) in a few order with the increased number of hot spots in SERS [120]. The biomolecule DNA based NMs were effective in the SERS, due to the cluster like formation on the double helix-structure of DNA results in a perfect chain like assembly for detecting the probe molecules. Kundu *et al.* were pioneer for applying DNA based NMs such as Au, Ag, Pt, Rh, Ru, Ir, Pd and Os for SERS application. Following are the challenges associated with the bioinspired NMs.

5.   Challenges and opportunities in bioinspired nanomaterials

5.1   Challenges

The fusion of bio molecules and nanoparticles has given rise to a new class of interesting materials called the bio inspired NMs which have abundant applications in diverse fields. Even after intense research, the biomediated nanoparticles are just complementing the known conventional methods and have not replaced them yet. Some of the major challenges hindering the extensive application of bio nanoparticles are discussed below.

Specificity

Specificity is the ability of a species to target only the desired sites for analysis. Biological molecules like proteins, enzymes and DNA are highly specific, whereas the metal NPs are not very specific. So combining the bio molecules with metal nanoparticles might increase their activity but it reduces their specificity. This is one of the major challenges encountered in the synthesis of bio mediated NMs. One such example, Glucose oxidase (enzyme) converts only glucose but when it's combined with Au nanoparticles, the combination reduces several reducing sugars as well [121]. This lack of specificity leads to undesired side reactions which decreases the efficiency of the NMs. In the medical field it could even be fatal.

Orientation

One of the most common difficulties observed in the preparation of bio inspired NMs is the positioning of the biological molecules on the surface or in the interfaces, three dimensionally. The bio molecules can either be single or a clusters, depending on the requirement [121]. Positioning of single nanoparticles on DNA self-assemblies is a way of achieving the required orientation but in this method the kinetics is poor and cannot be used for large dimensional structures. Simply, superstructures with longer chains cannot be synthesized by this method [122]. It is still obscure, whether the optical properties of the single nanoparticles will be retained in their biologically modified superstructures [122].

Bio conjugation

Bio conjugation is the technique used to bind the biological molecules to the NPs. Till now there is no established method for bio conjugation. Bio conjugation plays a major role in bio sensing and targeted drug delivery system, thus developing clear synthesis routes in bio conjugation is important. Very accurate concentration of reactants must be used. Even a slight variation leads to formation of aggregated structures [123]. Bio conjugation can be beneficial as well as detrimental. For example, the oxidase activity of ceria nanoparticles increases on conjugation with polysaccharides like dextran whereas the Au NPs lose their glucose oxidation capacity when conjugated to polymers [121]. Cleavable and non-cleavable bio conjugated NPs have profound application in targeted drug delivery but their detailed mechanisms are not yet known [123]. If one of the reacting species is hydrophobic and the other one is hydrophilic, it is very difficult to conjugate and the stability of the thus formed bio nanoparticles will be very low.

Protein Corona

Protein corona is the layer that forms on the nanoparticle surface when it interacts with the biological molecules especially proteins. The layer thus formed masks the surface properties of the base NPs. So it will be difficult to analyze the interaction of the NPs with the biological species. This causes big trouble in the synthesis of bio mediated nanoparticles [124]. The protein corona of *in vitro* conditions greatly defer from the protein corona of *in vivo* condition making it difficult for the use of bio NMs in therapeutic applications. Targeting molecules on the surface of the NPs are masked by the protein corona in complex biological fluids hereby reducing their effectiveness in targeted drug delivery [124]. It is difficult to determine the conformational changes in protein after their interaction with the NPs. Finding that the conformational changes are reversible or irreversible also possess a big challenge. But it is essential to have an insight on these attributes since the protein function is explicitly linked with the conformational changes [125].

5.2   Opportunities

Though we face a lot of challenges in the field of bio NMs, there is a myriad of opportunities to look for in the future. The following are some of the new developments in this field of biobased NMs. Especially, in the medicinal application neural stem cells; it is a great boon to treat them with nanotechnology. Very recently, studies were carried out on NMs for cancer vaccines with high specificity towards increasing immune system. The large surface area and particle size are the great advantage to interact with the cellular system and finds applications in different pharmaceutical areas including bio-sensing, therapies and diagnostics.

Nanozymes

Nanozymes are novel artificial NMs, which mimic the functions of enzymes. Nanozymes can easily detect the substrate/analytes and hence are widely used in bio sensor and targeted drug delivery applications [126]. For example, F^- increases the oxidase activity of CeO_2 nanozyme and hence F^- is detected using Ceria nanozyme [121]. The cost effectiveness of nanozymes when compared to natural enzymes is an added advantage, because NPs are way cheaper [126]. Also, iron oxide NPs are used instead of peroxidase enzymes in ELISA (Enzyme Linked Immunosorbent Assay) [121]. Peroxidase mimic nanozymes are even used in tumor imaging and several nanozymes have therapeutic values. One more interesting application is in the field of electronics where nanozymes are used in fabricating logic gates [126].

In Medicine

Bio NMs have surplus applications in the field of medicine. The ability of researchers to position a particular bio molecule in three dimensions assists for engineering the protein molecules on virus structures. These are helpful in vaccine development [127]. The idea of detection and diagnosis using a single bio nano assembly is of profound interest to researchers. Multiplexed bio sensors having the ability to detect, quantify and treat disease, in a miniature assembly is looked upon as the technology for future diagnosis [128]. Bio NMs can also be used in wound cares, as they have the potential to give both mechanical support and therapeutic effect. Their high surface to volume ratio and similar pore size to that of human tissue are the added benefits and aid in healing [128]. Bio Quantum dots have applications in cell labeling, tracking cell migration and as whole animal contrast agent.

Multifunctional bionano superstructures

With the varied new discoveries in the field of bio NMs, the focus is now shifting towards the integration of benefits from different bio nano assemblies to form single superstructures with multiple functions. For example, in biomedical applications, controlled drug delivery NPs are incorporated into specific targeting nanoparticles forming super structures [129].

In nature we find plants and animals having beneficial properties like anti fogging, super hydrophobicity, self-healing, self-cleaning etc. Research ascertains that these are due to the nanostructure found in them [130]. The adoption and integration of these provide abundance of opportunity for bio inspired NMs in future.

6.   Summary and outlook

Inspired from nature, engineering of bioinspired NPs for various applications has been acquiring recent research interest. The new-fangled bioinspired strategies having many advantages compared to traditional methods those are the use of mild conditions, greener biomolecules, stabilizer, etc. In this book chapter, we have discussed the emerging trends in the preparation of bioinspired NPs by using different strategies, their properties and their promising applications in different fields such as in energy sector, medicinal field, agrochemicals and pharmaceuticals and so on. From the outline of this chapter, it is concluded that green strategies in the preparation of NPs need lesser energy and environmental friendly conditions. The detailed explanation in the bioinspired NMs preparations, open a new view to control the size and shape of metal NPs which provides great hope to the medicinal sector, particularly in the drug delivery systems. Though, bioinspired NPs having many applications, it has few disadvantages such as the restriction to produce in bulk scale for industrial applications. However, in current strategy, the development in the preparation of bioinspired NMs for various applications is ongoing research interest. Overall this chapter concludes the following:

- The bio inspired NMs have great advantage with the greener approach towards the biological system and increase the selectivity towards the drug delivery.

- Bio-molecules such as microorganisms, virus, fungi, proteins and amino acids were largely studied as a template for the fabrication of NMs in a controlled morphology.

- The nano bio-confined size and shape of NMs with their surface charges meets the biological system which have the same factors such as immunoglobulins are presented.

- The NMs entered in the bio system selectively absorbs the biomolecules to treat the defected cellular organisms effectively.

- For energy applications, the structural confinement simultaneously increases the charge transfer efficiency that reduces the resistance of the systems with bio-NMs.

- In future, the bio-inspired NMs could show better opportunity to deliver highly selective drugs and favorable sensitivity towards bio system.

Thus, it is clear that the bio-templated NMs are highly active in the biological system to treat various problems. In the energy sectors, the bio-based materials show high catalytic efficiency in many organic conversions. On the other hand, bio-templated NMs can also be used in supercapacitor and water splitting applications where it increases the surface area of the NMs significantly with minimum catalytic loading. In future, these above discussed designing processes of bio-inspired NMs could be used as an attractive and sustainable process for the alternative of many hazardous conventional methods.

References

[1]   M.F.H. Jr, D.W. Mogk, J. Ranville, I.C. Allen, G.W. Luther, L.C. Marr, B.P. Mcgrail, M. Murayama, N.P. Qafoku, K.M. Rosso, N. Sahai, P.A. Schroeder, P. Vikesland, P. Westerhoff, Y. Yang, Natural, incidental, and engineered nanomaterials

and their impacts on the Earth system, Science. 363 (2019) 6434.
https://doi.org/10.1126/science.aau8299

[2]   J.R. Heath, Nanoscale Materials, Acc. Chem. Res. 32 (1999) 388

[3]   C. Aur, T. Nesma, P. Juanes-velasco, A. Landeira-viñuela, H. Fidalgo-gomez, V.
Acebes-fernandez, R. Gongora, A. Parra, R. Manzano-roman, M. Fuentes, Interactions
of Nanoparticles and Biosystems: Microenvironment of Nanoparticles and
Biomolecules in Nanomedicine, Nanomaterials. 9 (2019) 1365

[4]   C.M. Niemeyer, Nanoparticles, Proteins, and Nucleic Acids: Biotechnology Meets
Materials Science, Angew. Chem. Int. Ed. 40 (2001) 4128 - 4158

[5]   H. Hosein, D.R. Strongin, T. Douglas, K. Rosso, A Bioengineering Approach to
the Production of Metal and Metal Oxide Nanoparticles, 2005

[6]   A.R. Tao, S. Habas, P. Yang, Shape Control of Colloidal Metal Nanocrystals,
Small. 4 (2008) 310–325. https://doi.org/10.1002/smll.200701295

[7]   C. Contado, Nanomaterials in consumer products: a challenging analytical
problem, Front. Chem. 3 (2015) 48. https://doi.org/10.3389/fchem.2015.00048

[8]   X. Xu, Y. Jian, Y. Li, X. Zhang, Z. Tu, Z. Gu, Bio-Inspired Supramolecular
Hybrid Dendrimers Self-Assembled from Low-Generation Peptide Dendrons for
Highly E ffi cient Gene Delivery and Biological Tracking, ACS Nano. 8 (2014) 9255–
9264

[9]   M.C. Roco, C.A. Mirkin, M.C. Hersam, Nanotechnology research directions for
societal needs in 2020 : summary of international study, J Nanopart Res. (2011) 897–
919. https://doi.org/10.1007/s11051-011-0275-5

[10]   T.L. Theis, S. Ghosh, Toward Sustainable Nanoproducts, J. Ind. Ecol. 12 (2008)
329–359. https://doi.org/10.1111/j.1530-9290.2008.00046.x

[11]   R.J. White, R. Luque, V.L. Budarin, J.H. Clark, D.J. Macquarrie, Supported metal
nanoparticles on porous materials. Methods and applications, Chem.Soc.Rev. 38
(2009) 481–494. https://doi.org/10.1039/b802654h

[12]   M.J. Gallagher, a J.T.B. Caley Allen, T.A. Qiu, P.L. Clement, O.P. Krause, L.M.
Gilbertson, Research highlights: applications of life-cycle assessment as a tool for
characterizing environmental impacts of engineered nanomaterials, Environ. Sci.
Nano. 4 (2017) 276–281. https://doi.org/10.1039/C7EN90005H

[13]   M.P. Tsang, E. Kikuchi-uehara, G.W. Sonnemann, C. Aymonier, M. Hirao,
Evaluating nanotechnology opportunities and risks through integration of life-cycle
and risk assessment, Nat. Nanotechnol. 12 (2017) 734.
https://doi.org/10.1038/nnano.2017.132

[14]   C.J. Murphy, Sustainability as an emerging design criterion in nanoparticle synthesis and applications, J. Mater. Chem. 18 (2008) 2173–2176. https://doi.org/10.1039/b717456j

[15]   M.J. Eckelman, J.B. Zimmerman, P.T. Anastas, Toward Green Nano Syntheses E-factor Analysis of Several Nanomaterial Syntheses, J. Ind. Ecol. 12 (2008) 316–328. https://doi.org/10.1111/j.1530-9290.2008.00043.x

[16]   M. Razavi, Bio-based nanostructured materials, Elsevier Ltd., 2018. https://doi.org/10.1016/B978-0-08-100716-7.00002-7

[17]   J. Huang, L. Lin, D. Sun, H. Chen, D. Yang, Q. Li, Bio-inspired synthesis of metal nanomaterials and applications, Chem.Soc.Rev. 44 (2015) 6330–6374. https://doi.org/10.1039/c5cs00133a

[18]   J.E. Hutchison, Greener Nanoscience: A Proactive Approach to Advancing Applications and Reducing Implications of Nanotechnology, ACS Nano. 2 (2008) 395–402

[19]   M. Romero-franco, H.A. Godwin, M. Bilal, Y. Cohen, Needs and challenges for assessing the environmental impacts of engineered nanomaterials (ENMs), Beilstein J. Nanotechnol. 8 (2017) 989–1014. https://doi.org/10.3762/bjnano.8.101

[20]   J.-L. Wang, J.-W. Liu, S.-H. Yu, Recycling Valuable Elements from Chemical Synthesis Process of Nanomaterials: A Sustainable View, ACS Mater. Lett. 5 (2019) 541–548. https://doi.org/10.1021/acsmaterialslett.9b00283

[21]   L.M. Gilbertson, J.B. Zimmerman, D.L. Plata, J.E. Hutchison, P.T. Anastas, Designing nanomaterials to maximize performance and minimize undesirable implications guided by the Principles of Green Chemistry, Chem. Soc. Rev. 16 (2015) 5758–5777. https://doi.org/10.1039/C4CS00445K

[22]   K.D. Gilroy, A. Ruditskiy, H. Peng, D. Qin, Y. Xia, Bimetallic Nanocrystals: Syntheses, Properties, and Applications, Chem. Rev. 116 (2016) 10414−10472. https://doi.org/10.1021/acs.chemrev.6b00211

[23]   P.C. Ke, R. Qiao, Carbon nanomaterials in biological systems, J. Phys. Condens. Matter. 19 (2007) 373101. https://doi.org/10.1088/0953-8984/19/37/373101

[24]   A.E. Nel, L. Mädler, D. Velegol, T. Xia, E.M. V Hoek, P. Somasundaran, F. Klaessig, V. Castranova, M. Thompson, Understanding biophysicochemical interactions at the nano–bio interface, Nat. Mater. 8 (2009) 543–557. https://doi.org/10.1038/nmat2442

[25]   C. Sanchez, H. Arribart, M.M.G. Guille, G. Guille, Biomimetism and bioinspiration as tools for the design of innovative materials and systems, Nat. Mater. 4 (2005) 277–288

[26]   M. Sarikaya, C. Tamerler, A.K. Jen, K. Schulten, F. Baneyx, Molecular biomimetics: nanotechnology through biology, Nat. Mater. 2 (2003) 577–585

[27]   R. Peter, Continuous Biopahrma Manufacturing, BioPharm Int. 28 (2015) 1–56

[28]   M. Ding, G. Chen, W. Xu, C. Jia, H. Luo, Nano Materials Science Bio-inspired synthesis of nanomaterials and smart structures for electrochemical energy storage and conversion, Nano Mater. Sci. 11 (2019) 1–17. https://doi.org/10.1016/j.nanoms.2019.09.011

[29]   P. Mohanpuria, K.N.K. Rana, S.K. Yadav, Biosynthesis of nanoparticles: technological concepts and future applications, J Nanopart Res. 10 (2008) 507–517. https://doi.org/10.1007/s11051-007-9275-x

[30]   X. Li, H. Xu, Z. Chen, G. Chen, Biosynthesis of Nanoparticles by Microorganisms and Their Applications, J. Nanomater. 2011 (2011) 270974. https://doi.org/10.1155/2011/270974

[31]   A. Fariq, T. Khan, A. Yasmin, Microbial synthesis of nanoparticles and their potential applications in biomedicine, J. Appl. Biomed. 15 (2017) 241–248. https://doi.org/10.1016/j.jab.2017.03.004

[32]   X. Zhang, S. Yan, R.D. Tyagi, R.Y. Surampalli, Chemosphere Synthesis of nanoparticles by microorganisms and their application in enhancing microbiological reaction rates, Chemosphere. 82 (2011) 489–494. https://doi.org/10.1016/j.chemosphere.2010.10.023

[33]   I.W. Lin, C. Lok, C. Che, Biosynthesis of silver nanoparticles from silver(I) reduction by the periplasmic nitrate reductase c- type cytochrome subunit NapC in a silver-resistant E. coli, Chem. Sci. 5 (2014) 3144–3150. https://doi.org/10.1039/c4sc00138a

[34]   S.K. Das, J. Liang, M. Schmidt, F. Laffir, E. Marsili, Biomineralization Mechanism of Gold by Zygomycete Fungi Rhizopous oryzae, ACS Nano. 7 (2012) 6165–6173

[35]   R. Ramanathan, A.P.O. Mullane, R.Y. Parikh, P.M. Smooker, S.K. Bhargava, V. Bansal, Bacterial Kinetics-Controlled Shape-Directed Biosynthesis of Silver Nanoplates Using Morganella psychrotolerans, Langmuir 2011,. 27 (2011) 714–719. https://doi.org/10.1021/la1036162

Bioinspired Nanomaterials
Materials Research Foundations **111** (2021) 1-35

Materials Research Forum LLC
https://doi.org/10 21741/9781644901571-1

[36]   B. Nair, T. Pradeep, Coalescence of Nanoclusters and Formation of Submicron Crystallites Assisted by Lactobacillus Strains, Cryst. Growth Des. 4 (2002) 293–298

[37]   M.E. Bayer, M.H. Bayer, Lanthanide Accumulation in the Periplasmic Space of Escherichia coli B, J. Bacteriol. 173 (1991) 141–149

[38]   M.I. Husseiny, M.A. El-aziz, Y. Badr, M.A. Mahmoud, Biosynthesis of gold nanoparticles using Pseudomonas aeruginosa, Spectrochim. Acta Part A Mol. Biomol. Spectrosc. 67 (2007) 1003–1006. https://doi.org/10.1016/j.saa.2006.09.028

[39]   P. Mukherjee, A. Ahmad, D. Mandal, S. Senapati, S.R. Sainkar, M.I. Khan, R. Parishcha, P. V Ajaykumar, M. Alam, R. Kumar, M. Sastry, Fungus-Mediated Synthesis of Silver Nanoparticles and Their Immobilization in the Mycelial Matrix: A Novel Biological Approach to Nanoparticle Synthesis, Nano Lett. 1 (2001) 515–519. https://doi.org/10.1021/nl0155274

[40]   A. Ahmad, P. Mukherjee, S. Senapati, D. Mandal, M.I. Khan, M. Sastry, Extracellular biosynthesis of sil v er nanoparticles using the fungus Fusarium oxysporum, Colloids Surfaces B Biointerfaces. 28 (2003) 313–318

[41]   K. Sangeetha, S.S. Sankar, K. Karthick, S. Anantharaj, S.R. Ede, S. Wilson T., S. Kundu, Synthesis of ultra-small Rh nanoparticles congregated over DNA for catalysis and SERS applications, Colloids Surfaces B Biointerfaces. 173 (2019) 249–257. https://doi.org/10.1016/j.colsurfb.2018.09.052

[42]   M. Kowshik, S. Ashtaputre, S. Kharrazi, W Vogel, J. Urban, S.K. Kulkarni, K. Paknikar, Extracellular synthesis of silver nanoparticles by a silver-tolerant yeast strain MKY3, Nanotechnology. 14 (2003) 95–100

[43]   H. Chen, J. Huang, D. Huang, D. Sun, M. Shao, Q. Li, Novel AuPd nanostructures for hydrogenation of 1,3-butadiene, J. Mater. Chem. A. 3 (2015) 4846–4854. https://doi.org/10.1039/C4TA06226D

[44]   E. Dujardin, C. Peet, G. Stubbs, J.N. Culver, S. Mann, Organization of Metallic Nanoparticles Using Tobacco Mosaic Virus Templates, Nano Lett. 3 (2003) 413–417

[45]   A.K. Manocchi, N.E. Horelik, B. Lee, H. Yi, Simple, Readily Controllable Palladium Nanoparticle Formation on Surface- Assembled Viral Nanotemplates, Langmuir. 26 (2010) 3670–3677. https://doi.org/10.1021/la9031514

[46]   A. Merzlyak, S. Indrakanti, S. Lee, Genetically Engineered Nanofiber-Like Viruses For Tissue Regenerating Materials, Nano Lett. 2 (2009) 846–852

[47]   M. Knez, A.M. Bittner, F. Boes, C. Wege, H. Jeske, E. Mai, K Kern, Biotemplate Synthesis of 3-nm Nickel and Cobalt Nanowires, Nano Lett. 3 (2003) 1079–1082

[48] S. Balci, K. Hahn, P. Kopold, A. Kadri, C. Wege, K. Kern1, A.M. Bittner, Electroless synthesis of 3 nm wide alloy nanowires inside Tobacco mosaic virus, Nanotechnology. 23 (2012) 45603. https://doi.org/10.1088/0957-4484/23/4/045603

[49] A.A.A. Aljabali, J.E. Barclay, O. Cespedes, A. Rashid, S.S. Staniland, G.P. Lomonossoff, D.J. Evans, Charge Modifi ed Cowpea Mosaic Virus Particles for Templated Mineralization, Adv. Funct. Mater. 2011,. 21 (2011) 4137–4142. https://doi.org/10.1002/adfm.201101048

[50] A.K. Manocchi, S. Seifert, B. Lee, H. Yi, In Situ Small-Angle X-ray Scattering Analysis of Palladium Nanoparticle Growth on Tobacco Mosaic Virus Nanotemplates, Langmuir 2011,. 27 (2011) 7052–7058

[51] C. Lin, Y. Liu, S. Rinker, H. Yan, DNA Tile Based Self-Assembly: Building Complex Nanoarchitectures, ChemPhysChem. 7 (2006) 1641–1647. https://doi.org/10.1002/cphc.200600260

[52] E. Braun, Y. Eichen, U. Sivan, G. Ben-Yoseph, DNA-templated assembly and electrode attachment of a conducting silver wire, Nature. 391 (1998) 775-778

[53] Z. Kang, X. Yan, Y. Zhang, J. Pan, J. Shi, X. Zhang, Y. Liu, J.H. Choi, D.M. Porterfiel, Single-Stranded DNA Functionalized Single-Walled Carbon Nanotubes for Microbiosensors via Layer-by-Layer Electrostatic Self- Assembly, ACS Appl. Mater. Interfaces. 6 (2014) 3784–3789

[54] Z. Chen, C. Liu, F. Cao, J. Ren, X. Qu, DNA metallization: principles, methods, structures, and applications, Chem. Soc. Rev. 47 (2018) 4017–4072. https://doi.org/10.1039/C8CS00011E

[55] O.I. Wilner, I. Willner, Functionalized DNA Nanostructures, Chem. Rev. 112 (2012) 2528–2556

[56] N.C. Seeman, Nucleic Acid Junctions and Lattices, J. Theor. Biol. 99 (1982) 237–247

[57] J.D. Moroz, P. Nelson, Torsional directed walks, entropic elasticity, and DNA twist stiffness, Proc. Natl. Acad. Sci. 94 (1997) 14418–14422

[58] S. Kumaravel, P. Thiruvengetam, S.R. Ede, K. Karthick, S. Anantharaj, S.S. Sankar, S. Kundu, Cobalt tungsten oxide hydroxide hydrate (CTOHH) on DNA scaffold: an excellent bi-functional catalyst for oxygen evolution reaction (OER) and aromatic alcohol oxidation, Dalt. Trans. 48 (2019) 17117–17131. https://doi.org/10.1039/c9dt03941d

[59] K. Sangeetha, P. Thiruvengetam. K. Karthick, S.S. Sankar, S. Kundu, Detection of Lignin Motifs with RuO2-DNA as Active Catalyst via Surface Enhanced Raman

Scattering Studies, ACS Sustain. Chem. Eng. (2019).
https://doi.org/10.1021/acssuschemeng.9b04414

[60]   S. Anantharaj, U. Nithiyanantham, S.R. Ede, S. Kundu, Osmium organosol on
DNA: Application in catalytic hydrogenation reaction and in SERS studies, Ind. Eng.
Chem. Res. 53 (2014) 19228–19238. https://doi.org/10.1021/ie503667y

[61]   K. Sakthikumar, S. Anantharaj, S.R. Ede, K. Karthick, S. Kundu, Highly stable
Rhenium Organosol on DNA Scaffold for Catalytic and SERS Applications, J. Mater.
Chem. C. 4 (2016) 6309–6320. https://doi.org/10.1039/C6TC01250G

[62]   D. Majumdar, A. Singha, P.K. Mondal, S. Kundu, DNA-Mediated Wirelike
Clusters of Silver Nanoparticles: An Ultrasensitive SERS Substrate, ACS Appl. Mater.
Interfaces. 5 (2013) 7798–7807

[63]   K. Sakthikumar, S. Anantharaj, S.R. Ede, K. Karthick, S. Kundu, A highly stable
rhenium organosol on a DNA scaffold for catalytic and SERS applications †, J. Mater.
Chem. C. 4 (2016) 6309–6320. https://doi.org/10.1039/c6tc01250g

[64]   S. Anantharaj, K. Sakthikumar, A. Elangovan, G. Ravi, T. Karthik, S. Kundu,
Ultra-small rhenium nanoparticles immobilized on DNA scaffolds: An excellent
material for surface enhanced Raman scattering and catalysis studies, J. Colloid
Interface Sci. 483 (2016) 360–373. https://doi.org/10.1016/j.jcis.2016.08.046

[65]   N. Sai, R. Satyavolu, L.H. Tan, Y. Lu, DNA-Mediated Morphological Control of
Pd-Au Bimetallic Nanoparticles, J. Am. Chem. Soc. 138 (2016) 16542–16548.
https://doi.org/10.1021/jacs.6b10983

[66]   J. Xie, Y. Zheng, J.Y. Ying, Protein-Directed Synthesis of Highly Fluorescent
Gold Nanoclusters, J. Am. Chem. Soc. 131 (2009) 888–889

[67]   J. Sheng, L. Wang, Y. Han, W. Chen, H. Liu, M. Zhang, L. Deng, Y.-N. Liu, Dual
Roles of Protein as a Template and a Sulfur Provider: A General Approach to Metal
Sulfides for Efficient Photothermal Therapy of Cancer, Small. 14 (2017) 1702529.
https://doi.org/10.1002/smll.201702529

[68]   T. Yang, L. Liu, X. Lv, Q. Wang, H. Ke, Y. Deng, H. Yang, X Yang, G. Liu, Y.
Zhao, H. Chen, Size-Dependent Ag2S Nanodots for Second Near-Infrared
Fluorescence/Photoacoustics Imaging and Simultaneous Photothermal Therapy, ACS
Nano. 11 (2017) 1848–1857. https://doi.org/10.1021/acsnano.6b07866

[69]   K.M. Hindi, A.J. Ditto, M.J. Panzner, D.A. Medvetz, D.S. Han, C.E. Hovis, J.K.
Hilliard, J.B. Taylor, Y.H. Yun, C.L. Cannon, W.J. Youngs, Biomaterials The
antimicrobial efficacy of sustained release silver – carbene complex-loaded L -tyrosine
polyphosphate nanoparticles : Characterization , in vitro and in vivo studies,

Biomaterials. 30 (2009) 3771–3779.
https://doi.org/10.1016/j.biomaterials.2009.03.044

[70]   K. Blecher, A. Nasir, A. Friedman, The growing role of nanotechnology in in combating infectious disease, Virulence. 5 (2011) 395–401

[71]   J. Singh, A.S. Dhaliwal, Novel Green Synthesis and Characterization of the Antioxidant Activity of Silver Nanoparticles Prepared from Nepeta leucophylla Root Extract, Anal. Lett. 52 (2019) 213–230.
https://doi.org/10.1080/00032719.2018.1454936

[72]   D.F. Ollis, E. Pelizzetti, N. Serpone, Destructuction of Water contaminants, Environ. Si. Technol. 25 (1991) 1532

[73]   A. Mills, R.H. Davies, D. Worsley, Water Purification by Semiconductor Photocatalysis, Chem. Soc. Rev. 22 (1982) 417–425

[74]   D.S. Bhatkhande, V.G. Pangarkar, A.A.C.M. Beenackers, Photocatalytic degradation for environmental applications – a review, Chem Technol Biotechnol. 116 (2002) 102–116. https://doi.org/10.1002/jctb.532

[75]   I.K. Konstantinou, T.A. Albanis, TiO2 -assisted photocatalytic degradation of azo dyes in aqueous solution: kinetic and mechanistic investigations A review, Appl. Catal. B Environ. 49 (2004) 1–14. https://doi.org/10.1016/j.apcatb.2003.11.010

[76]   M.A. Rauf, M.A. Meetani, S. Hisaindee, An overview on the photocatalytic degradation of azo dyes in the presence of TiO2 doped with selective transition metals, Desalination. 276 (2011) 13–27. https://doi.org/10.1016/j.desal.2011.03.071

[77]   A.M. Braun, M. Gilson, M. Krieg, M. Maurette, P. Murasecco, E. Oliveros, Electron and Energy Transfer from Phenothiazine Triplets, in: ACS Symp. Ser. Am. Chem. Soc., 1985: pp. 79–97

[78]   L.I.U. Wei, W. Xiangfei, C.A.O. Lixin, S.U. Ge, Z. Lan, W. Yonggang, Microemulsion synthesis and photocatalytic activity of visible light-active BiVO4 nanoparticles, Sci. China. 54 (2011) 724–729. https://doi.org/10.1007/s11426-010-4156-z

[79]   K.P.K. A, I.F. Chaberny, K. Massholder, M. Stickler, V.W. Benz, H.-G. Sonntag, L. Erdinger, Disinfection of surfaces by photocatalytic oxidation with titanium dioxide and UVA light, Chemosphere. 53 (2003) 71–77. https://doi.org/10.1016/S0045-6535(03)00362-X

[80]   S.C. Roy, O.K. Varghese, M. Paulose, C.A. Grimes, Toward Solar Fuels: Photocatalytic Conversion of Hydrocarbons, ACS Nano. 4 (2010) 1259–1278

[81]   F. Dong, Y. Sun, L. Wu, Z. Wu, Facile transformation of low cost thiourea into nitrogen-rich graphitic carbon nitride nanocatalyst with high visible light photocatalytic performance, Catal. Sci. Technol. 2 (2012) 1332–1335. https://doi.org/10.1039/c2cy20049j

[82]   C. Alvarez-lorenzo, A. Concheiro, Bioinspired drug delivery systems, Curr. Opin. Biotechnol. 24 (2013) 1167–1173. https://doi.org/10.1016/j.copbio.2013.02.013

[83]   S. V Patwardhan, Biomimetic and bioinspired silica: recent developments and applications, Chem. Commun. 47 (2011) 7567–7582. https://doi.org/10.1039/c0cc05648k

[84]   S.C. Jang, O.Y. Kim, C.M. Yoon, D. Choi, T. Roh, J. Park, J. Nilsson, J. Lo, Y.S.G. M, Bioinspired Exosome-Mimetic Nanovesicles for Targeted Delivery of Chemotherapeutics to Malignant Tumors, ACS Nano. 7 (2013) 7698-7710

[85]   K. Ariga, K. Kawakami, M. Ebara, Y. Kotsuchibashi, Q. Ji, J.P. Hill, Bioinspired nanoarchitectonics as emerging drug delivery systems, NewJ. Chem. 38 (2014) 5149–5163. https://doi.org/10.1039/C4NJ00864B

[86]   C.R. Steven, G.A. Busby, C. Mather, B. Tariq, M.L. Briuglia, D.A. Lamprou, A.J. Urquhart, M. Helen, S. V Patwardhan, Bioinspired silica as drug delivery systems and their biocompatibility, J. Mater. Chem. B. 2 (2014) 5028–5042. https://doi.org/10.1039/c4tb00510d

[87]   X. Zhou, P. Ma, A. Wang, C. Yu, T. Qian, S. Wu, J. Shen, Biosensors and Bioelectronics Dopamine fl uorescent sensors based on polypyrrole/graphene quantum dots core/shell hybrids, Biosens. Bioelectron. 64 (2015) 404–410. https://doi.org/10.1016/j.bios.2014.09.038

[88]   M. Gao, B.Z. Tang, Fluorescent Sensors Based on Aggregation-Induced Emission: Recent Advances and Perspectives, ACS Sens. 10 (2017) 1382–1399. https://doi.org/10.1021/acssensors.7b00551

[89]   E. Tomat, S.J. Lippard, Imaging mobile zinc in biology, Curr. Opin. Chem. Biol. 14 (2010) 225–230. https://doi.org/10.1016/j.cbpa.2009.12.010

[90]   N.I. Georgiev, V.B. Bojinov, P.S. Nikolov, Dyes and Pigments The design, synthesis and photophysical properties of two novel 1 , 8-naphthalimide fl uorescent pH sensors based on PET and ICT, Dye. Pigment. 88 (2011) 350–357. https://doi.org/10.1016/j.dyepig.2010.08.004

[91]   S.K. Das, M.M.R. Khan, A.K. Guha, N. Naskar, Bio-inspired fabrication of silver nanoparticles on nanostructured silica: characterization and application as a highly

efficient hydrogenation catalyst, Green Chem. 15 (2013) 2548–2557.
https://doi.org/10.1039/c3gc40310f

[92]   J. Sun, J. Zhang, M. Zhang, M. Antonietti, X. Fu, X. Wang, Bioinspired hollow
semiconductor nanospheres as photosynthetic nanoparticles, Nat. Commun. 3 (2012)
1239. https://doi.org/10.1038/ncomms2152

[93]   A. Latorre-sanchez, J.A. Pomposo, Recent bioinspired applications of single-chain
nanoparticles, Polym Int. 65 (2016) 855–860. https://doi.org/10.1002/pi.5078

[94]   C. Burcu, D. Ekiz, E. Piskin, G. Demirel, Green catalysts based on bio-inspired
polymer coatings and electroless plating of silver nanoparticles, J. Mol. Catal. A
Chem. 350 (2011) 97–102. https://doi.org/10.1016/j.molcata.2011.09.017

[95]   D.P. Debecker, C. Faure, M. Meyre, A. Derre, E.M. Gaigneaux, A New Bio-
Inspired Route to Metal-Nanoparticle-Based Heterogeneous Catalysts, Small. 4 (2008)
1806–1812. https://doi.org/10.1002/smll.200800304

[96]   S. Anantharaj, P.E. Karthik, B. Subramanian, S. Kundu, Pt Nanoparticle Anchored
Molecular Self-Assemblies of DNA: An Extremely Stable and Efficient HER
Electrocatalyst with Ultralow Pt Content, ACS Catal. 6 (2016) 4660–4672.
https://doi.org/10.1021/acscatal.6b00965

[97]   S. Anantharaj, M. Jayachandran, S. Kundu, Unprotected and interconnected Ru0
nano-chain networks: Advantages of unprotected surfaces in catalysis and
electrocatalysis, Chem. Sci. 7 (2016) 3188–3205. https://doi.org/10.1039/c5sc04714e

[98]   L. Lai, J.R. Potts, D. Zhan, L. Wang, K. Poh, C. Tang, H. Gong, Z. Shen, J. Lin,
R.S. Ruoff, Exploration of the active center structure of nitrogen-doped graphene-
based catalysts for oxygen reduction reaction, Energy Environ. Sci. 5 (2012) 7936–
7942. https://doi.org/10.1039/c2ee21802j

[99]   C. Gong, S. Sun, Y. Zhang, L. Sun, Z. Su, A. Wu, G. Wei, Hierarchical
nanomaterials via biomolecular self-assembly and bioinspiration for energy and
environmental applications, Nano Scale. 11 (2019) 4147–4182.
https://doi.org/10.1039/C9NR00218A

[100] S.R. Ede, S. Kundu, Microwave Synthesis of SnWO4 Nanoassemblies on DNA
Scaffold: A Novel Material for High Performance Supercapacitor and as Catalyst for
Butanol Oxidation, ACS Sustain. Chem. Eng. 3 (2015) 2321–2336.
https://doi.org/10.1021/acssuschemeng.5b00627

[101] S.R. Ede, A. Ramadoss, U. Nithiyanantham, S. Ananthara, S. Kundu, Bio-
molecule Assisted Aggregation of ZnWO4 Nanoparticles (NPs) into Chain-like
Assemblies: Material for High Performance Supercapacitor and as Catalyst for Benzyl

Alcohol Oxidation, Inorg. Chem. 54 (2015) 3851–3863.
https://doi.org/10.1021/acs.inorgchem.5b00018

[102] L. Wang, C. Gong, X. Yuan, G. Wei, Controlling the Self-Assembly of Biomolecules into Functional Nanomaterials through Internal Interactions and External Stimulations: A Review, Nanomaterials. 9 (2019) 285. https://doi.org/10.3390/nano9020285

[103] U. Nithiyanantham, A. Ramadoss, R. Ede, S. Kundu, DNA mediated wire-like clusters of self-assembled TiO2 nanomaterials: supercapacitor and dye sensitized solar cell applications, Nanoscale. 6 (2014) 8010–8023. https://doi.org/10.1039/c4nr01836b

[104] A. Samanta, I.L. Medintz, Nanoparticles and DNA – a powerful and growing functional combination in bionanotechnology, Nanoscale. 8 (2016) 9037–9095. https://doi.org/10.1039/C5NR08465B

[105] P. Simon, Y. Gogotsi, Materials for electrochemical capacitors, Nat. Mater. 7 (2008) 845–854

[106] M.A. Zhen-zhen, Y.U. Hui-cheng, W.U. Zhao-yang, W.U. Yan, X. Fu-bing, A Highly Sensitive Amperometric Glucose Biosensor Based on a Nano-cube Cu2O Modified Glassy Carbon Electrode, Chinese J. Anal. Chem. 44 (2016) 822–827. https://doi.org/10.1016/S1872-2040(16)60934-9

[107] Y. Li, R. Zhao, L. Shi, G. Han, Y. Xiao, Acetylcholinesterase biosensor based on electrochemically inducing 3D graphene oxide network/multi-walled carbon nanotube composites for detection of pesticides, RSC Adv. 7 (2017) 53570–53577. https://doi.org/10.1039/C7RA08226F

[108] W. Zhao, J. Xu, H. Chen, Photoelectrochemical DNA Biosensors, Chem. Rev. 114 (2014) 7421–7441. https://doi.org/10.1021/cr500100j

[109] V.I. Klimov, S.A. Ivanov, J. Nanda, M. Achermann, I. Bezel, J.A. Mcguire, A. Piryatinski, Single-exciton optical gain in semiconductor nanocrystals, Nature. 447 (2007) 441–446. https://doi.org/10.1038/nature05839

[110] F. Zhang, H. Jiang, X. Li, X. Wu, H. Li, Amine-Functionalized GO as an Active and Reusable Acid–Base Bifunctional Catalyst for One-Pot Cascade Reactions, ACS Catal. 4 (2014) 394–401

[111] V. Pardo-yissar, E. Katz, J. Wasserman, I. Willner, Acetylcholine Esterase-Labeled CdS Nanoparticles on Electrodes: Photoelectrochemical Sensing of the Enzyme Inhibitors, J. Am. Chem. Soc.. 125 (2003) 622–623

[112] S.W. Hell, Toward fluorescence nanoscopy, Nat. Biotechnol. 21 (2003) 1347–1355. https://doi.org/10.1038/nbt895

[113] N.N. Mamedova, N.A. Kotov, Albumin − CdTe Nanoparticle Bioconjugates: Preparation , Structure, and Interunit Energy Transfer with Antenna Effect, Nano Lett. 1 (2001) 281–286

[114] Y. Yan, A.I. Radu, W. Rao, H. Wang, G. Chen, K. Weber, D. Wang, D. Cialla-May, J. Popp, P. Schaaf, Mesoscopically Bi-continuous Ag−Au Hybrid Nanosponges with Tunable Plasmon Resonances as Bottom-Up Substrates for Surface- Enhanced Raman Spectroscopy, Chem. Mater. 28 (2016) 7673−7682. https://doi.org/10.1021/acs.chemmater.6b02637

[115] S. Chen, P. Xu, Y. Li, J. Xue, S. Han, W. Ou, L. Li, W. Ni, Rapid Seedless Synthesis of Gold Nanoplates with Microscaled Edge Length in a High Yield and Their Application in SERS, Nano-Micro Lett. 8 (2016) 328–335. https://doi.org/10.1007/s40820-016-0092-6

[116] F. Pu, Y. Huang, Z. Yang, H. Qiu, J. Ren, Nucleotide-Based Assemblies for Green Synthesis of Silver Nanoparticles with Controlled Localized Surface Plasmon Resonances and Their Applications, ACS Appl. Mater. Interfaces. 10 (2018) 9929–9937. https://doi.org/10.1021/acsami.7b18915

[117] L. Sun, D. Zhao, Z. Zhang, B. Li, D. Shen, DNA-based fabrication of density-controlled vertically aligned ZnO nanorod arrays and their SERS applications, J. Mater. Chem. 21 (2011) 9674–9681. https://doi.org/10.1039/c1jm10830a

[118] Z.Y. Jiang, X.X. Jiang, S. Su, X.P. Wei, S.T. Lee, Z.Y. Jiang, X.X. Jiang, S. Su, Y. He, Silicon-based reproducible and active surface-enhanced Raman scattering substrates for sensitive , specific , and multiplex DNA detection Silicon-based reproducible and active surface-enhanced Raman scattering substrates for sensitive , specific , and mul, Appl. Phys. Lett. 100 (2012) 203104. https://doi.org/10.1063/1.3701731

[119] J. Shen, J. Su, J. Yan, B. Zhao, D. Wang, S. Wang, S. Mathur, K. Li, M. Liu, C. Fan, Y. He, S. Song, Bimetallic nano-mushrooms with DNA-mediated interior nanogaps for high-efficiency SERS signal amplification, Nano Res. 8 (2014) 731–742. https://doi.org/10.1007/s12274-014-0556-2

[120] L. Zhang, H. Ma, L. Yang, Design and fabrication of surface-enhanced Raman scattering substrate from DNA − gold nanoparticles assembly with 2-3 nm interparticle gap, RSC Adv. 4 (2014) 45207–45213. https://doi.org/10.1039/C4RA06947A

[121] Y. Zhou, B. Liu, R. Yang, J. Liu, Filling in the Gaps between Nanozymes and Enzymes: Challenges and Opportunities, Bioconjugate Chem. 28 (2017) 2903−2909. https://doi.org/10.1021/acs.bioconjchem.7b00673

[122] Y. Gao, Z. Tang, Design and Application of Inorganic Nanoparticle Superstructures: Current Status and Future challenges, Small. 15 (2011) 2133–2146. https://doi.org/10.1002/smll.201100474

[123] E.T. Issue, D. Conjugates, Challenges in antibody – drug conjugate discovery: a bioconjugation and analytical perspective, Bioanalysis. 7 (2015) 1561–1564

[124] S. Zanganeh, R. Spitler, M. Erfanzadeh, M. Mahmoudi, Protein Corona: Opportunities and Challenges, Int. J. Biochem. Cell Biol. 75 (2016) 143–147. https://doi.org/10.1016/j.biocel.2016.01.005

[125] M. Mahmoudi, I. Lynch, M.R. Ejtehadi, M.P. Monopoli, F.B. Bombelli, S. Laurent, Protein-Nanoparticle Interactions: Opportunities and Challenges, Chem. Rev. 111 (2011) 5610–5637. https://doi.org/10.1021/cr100440g

[126] X. Wang, Y. Hu, H. Wei, Nanozymes in bionanotechnology: from sensing to therapeutics and abd beyond, Inorg. Chem. Front. 3 (2015) 41–60. https://doi.org/10.1039/C5QI00240K

[127] Q. Zhao, S. Li, H. Yu, N. Xia, Y. Modis, Virus-like particle-based human vaccines: quality assessment based on structural and functional properties, Trends Biotechnol. 31 (2013) 654–663. https://doi.org/10.1016/j.tibtech.2013.09.002

[128] R. Feiner, T. Dvir, Soft and fibrous multiplexed biosensors, Nat. Biomed. Eng. 4 (2020) 135-136. https://doi.org/10.1038/s41551-020-0522-0

[129] A.M. El-toni, M.A. Habila, P.A.Alo. Labis, M. Alhoshan, F. Zhang, Design, synthesis and applications of core–shell, hollow core, and nanorattle multifunctional nanostructures, Nanoscale. 8 (2015) 2510-2531. https://doi.org/10.1039/c5nr07004j

[130] X. Mo, Y. Wu, J. Zhang, T. Hang, M. Li, Bioinspired Multifunctional Au Nanostructures with Switchable Adhesion, Langmuir. 31 (2015) 10850–10858. https://doi.org/10.1021/acs.langmuir.5b02472

Materials Research Forum LLC
https://doi.org/10.21741/9781644901571-2

Chapter 2

Bioinspired Metal Nanoparticles for Microbicidal Activity

Rajalakshmi Subramaniyam*[1], Shairam Manickaraj[1], Prasaanth Ravi Anusuyadevi*[2]

[1]Chemical Biology and Nanobiotechnology Laboratory, AU-KBC Research centre, Anna University, MIT Campus, Chromepet, Chennai-600 044, India

[2]KTH Royal Institute of Technology, Department of Fibre and Polymer Technology, Teknikringen 56, SE-100 44, Stockholm, Sweden

Rajalakshmi Subramaniyam (rajimaniyam@gmail.com); Prasaanth Ravi Anusuyadevi (ranu.prasaanth@gmail.com)

Abstract

The broad reception for nanotechnology is due to their appreciable size and versatile applications in the interdisciplinary areas. In this modern era one of the major problems is microorganisms possessing antibiotic resistance, nanoparticles (NPs) are a lucrative option to solve this. In materials science, "green synthesis" has gained extensive attention as a reliable, sustainable, and eco-friendly protocol for synthesizing a wide range of materials, especially metals, and metal oxides nanomaterials, hybrid materials and bioinspired materials. As such, green synthesis is regarded as an important tool to reduce the destructive effects associated with the traditional methods for synthesis of nanoparticles commonly utilized in laboratory and industry. Bio-inspired NPs held edges over conventionally synthesized nanoparticles due to their low cost, easy synthesis and low toxicity. This chapter elaborates the developments on the biosynthesis of NPs using natural extracts with particular emphasis on their application as microbiocidal agents. This chapter has very specifically dealt with coinage metals such as Cu, Ag, Au due to their significance of antimicrobial activities. Succeeding, reported the developments in the synthetic methodologies of metal-oxide (Titanium dioxide, TiO_2) nanoparticles using novel plant extracts with high medicinal value and their corresponding ability to degrade bacterial pathogens through advanced oxidation process (AOPs) based on heterogeneous photocatalysis.

Bioinspired Nanomaterials
Materials Research Foundations 111 (2021) 36-62
Materials Research Forum LLC
https://doi.org/10.21741/9781644901571-2

Keywords

Biosynthesis, Metal/Metal Oxide Nanoparticle, Microbicidal Activities, Copper, Silver, Gold, Titanium Dioxide, Photocatalysis

Contents

1. Introduction

The world is witnessing a systematic regular pattern of pandemics occurring at a constant phase, be it the Spanish flu (1918), Asian flu (1957) or the current COVID-19 [1]. The concurrent development of drugs (organic compounds or metal nanoparticle formulations) to cure the existing disease and technologies to prevent the future pandemics to happen in this world have been the top priority of the scientific community. One of the major potential threat to future pandemic could be caused by "Superbugs" which refers to the Anti-biotic Resistant Bacteria (ARB). Conventionally, ARB causes severe illness and it is held accountable for more than 2 million infections to human [2]. One of the major channels, these bacteria comes in contact with human is through unrefined wastewater streams. Hence development of technologies or process for the purification of water streams contaminated with these sorts of bacteria has become the need of the hour. Amongst the water refinement-based environmental remediation strategies, the Advanced Oxidation Processes (AOPs) based on semiconductor mediated heterogeneous photocatalysis involving the generation of highly active reactive oxidation species (like hydroxyl radicals HO·) is the most beneficial and cheap method. As it has the propensity to utilize the abundant solar energy for the degradation of such pathogenic bacterial species.

This book chapter specifically elucidates the biogenic synthesis of metal oxide nanoparticles (TiO_2) and other key metal nanoparticles (Cu, Ag and Au) which on exposure to photonic irradiations generates Reactive Oxidative Species (ROS) which effectively destroys the bacteria.

2. Introduction to Biogenic Synthesis of Metal Nanoparticles using Plant Extracts

The plants or substances derived from the plants (plant extracts) possesses myriad range of redox potential and therefore exhibits its proclivity as a reducing agent. Thus, resulting in the biogenic production of metallic or metal-based nanoparticles through the reduction of the corresponding metal ions. Metal ions are vital for many biological processes in the human system and it has been used to cure diseases since the good old times. Conventionally, the synthesis of these nanoparticles within plants or using the plant extracts involves the following main subsequent steps. They are:

1. Nucleation of metal atoms, which obtained through reduction of metal ions and this step is termed as activation step.

2. Unconstrained integration/coalescence of generated nuclei (small size), resulting in the generation of particles with large size.

This eventually results in the augmentation of thermodynamic stability of the as-obtained nanoparticles, a process termed as Ostwald Ripening. This entire step of generation of large-size nanoparticles starting from the amalgamation of small size particles up to the Ostwald ripening process is termed as growth step. The final step of the production of nanoparticles is termed as termination step, which involves the termination of further growth of the thermodynamically stable nanoparticles and the final size of the synthesized particles has been reached [3]. For an example, Figure 1 illustrates the schematic process of gold nanoparticles formation where the fast growth is the final step of the stable nanoparticle formation.

While the current bio-based scientific and engineering fields have frequently involved in the synthesis of metal nanoparticles for the application in modern medicine (allopathy). As the metal nanoparticles from the bio-based sources have surface functionalized with non-toxic substances, contrast to the metal nanoparticles produced from chemical methods [4]. Astonishingly, ayurvedic system of medicine, which has originated from Indian tradition long back of about 5000 years B.C. This novel system has reported drugs for diverse diseases through the formulations composed of substances from different

Bioinspired Nanomaterials

Materials Research Forum LLC

Materials Research Foundations **111** (2021) 36-62

https://doi.org/10 21741/9781644901571-2

origins: 1. Metal, 2. Animal and 3. Herbal. Rasahastra an integral part of the ayurvedic system have intensively reported on the usage of heavy metals and drugs from mineral origins for medical applications [5]. Through these ayurvedic works developed by ancient seers of Indian origin, metals such as Zinc (Zn), Copper (Cu), Gold (Au) and Silver (Ag) have known to exhibit antimicrobial activity for many centuries [6]. Additionally, metal-oxide based semiconductor nanoparticles: Titanium dioxide (TiO_2), Silver oxide (Ag_2O), Cupric oxide (CuO), Zinc oxide (ZnO), Calcium oxide (CaO), Magnesium oxide (MgO) and even Silicon (Si) have found to demonstrate antimicrobial activity against a wide range of microorganisms [7-9].

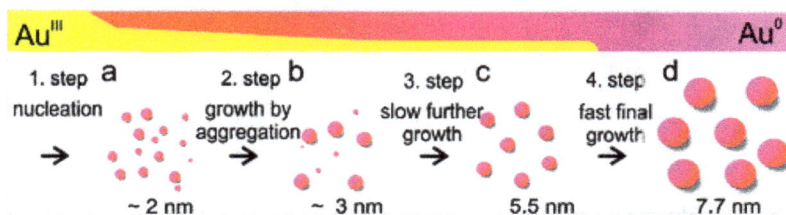

Figure 1. *Schematic illustration for the deduced process of gold nanoparticle formation. [Reprinted with permission. Copyright Chemical Reviews, 2014, 114(15), 7610-30.]*

In this decade, biogenic synthesis of metal nanoparticles, especially metals like silver, gold and copper using nano-factories like plant extracts has become predominant subject of study for bio-nanotechnologist. Even though the above said metal nanoparticles has been intensively reported in this book chapter. We have also elucidated the key findings of metal-oxide based nanoparticles synthesized using plant extracts in this work. This is of prime importance not only for microbicidal application but also for crucial implementations involving the paradigm shift of targeting the usage of renewable energy, like heterogeneous photocatalysis [10]. Currently, the green synthesized NPs have appealed researchers for its environmentally safe nature, simple preparation methods and being inexpensive. (Figure 2) [11, 12]. During the synthesis, extracts from algae, fungi, actinomycetes, bacteria and plants are employed as capping agents and also for the purpose of bio-reduction [13]. These biological extracts are non-toxic and non-volatile in nature which makes them suitable for synthesis. The synthesized NPs obtain their high stability and enhanced dispersity from the biological extracts which contains a blend of carbohydrates, proteins, vitamins, polymers and natural surfactants [14]. The compounds found in the extract to synthesis the NPs influences the size as well as the size distribution of nanoparticles. The existence of a potent reductant found in the extract stimulates the rate of the reaction and facilitates the synthesis of nanoparticles with smaller size. A stable new nuclei or secondary nuclei formation and a precise size distribution can be observed if the biomolecules reduce the salt rapidly. On the other hand, in a slow reaction the secondary nucleation is suppressed over the primary nucleation. In order to prevent the agglomeration, the bio molecules distribute itself as a monolayer on the surface of nanoparticles. Recent studies reported the surface functionalization of nanoparticles synthesized via., antibiotic mediated method which demonstrates encouraging biocidal activity [15,16]. Thus, the biomolecules available on the surface of nanoparticles complement the antimicrobial activity of NPs synthesized using extract.

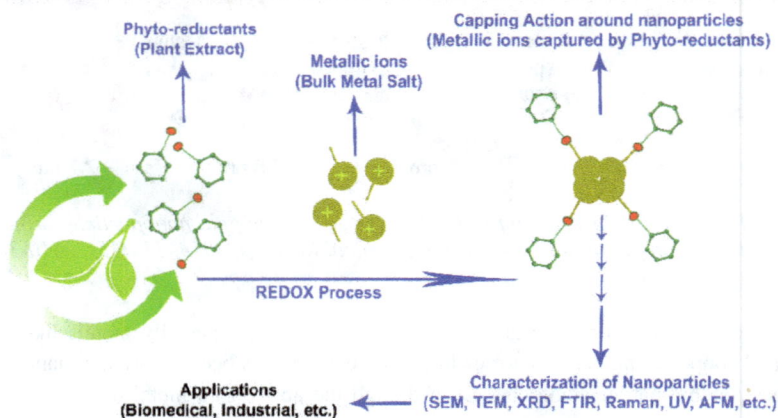

Figure 2. A scheme describing the mechanism of in vitro green synthesis of nanoparticles. [Reprinted with permission. Copyright Materials, 2018, 11(6), 940.]

The pathways focused in the antibacterial activity of nanoparticles are still not deciphered. However, biogenic nanoparticles pursue two different methods to establish their antibacterial capacity: either by generating ions (in case of metallic NPs) or by releasing reactive oxygen species (in case of metal oxide NPs). The released ions and ROS from the NPs damage the cell wall, genetic material or membrane lipid through oxidation (Figure 3) [17]. The nanoparticles get reduced during synthesis by bioactive compounds like vitamins, proteins, polyphenols, carbohydrates, polymers etc., that are present in the plant and microbe extracts [18]. Apart from providing stability to the

Bioinspired Nanomaterials   Materials Research Forum LLC
Materials Research Foundations **111** (2021) 36-62   https://doi.org/10.21741/9781644901571-2

synthesized NPs these compounds also add different functional groups to the nanoparticle surface. These functional groups help with the formation of chemical bonds and act as an active site for nanoparticles and bacterial cell communication, thereby it aids inhibition. [19]. The NPs get attached with the bacterial cell through the electrostatic and lipophilic interactions. Subsequently, the biogenic nanoparticles showed improved antimicrobial properties than chemically synthesized NPs. In a particular study, it was reported that biogenic NPs showed impressive inhibitory effect against all the bacterial strains studied like. *P. vulgaris, K. pneumonia, P. aeruginosa, S. aureus* and *V. cholera* while chemically synthesized nanoparticles were found to be ineffective against *P. vulgaris, P. aeruginosa* and *K. pneumonia* [20].

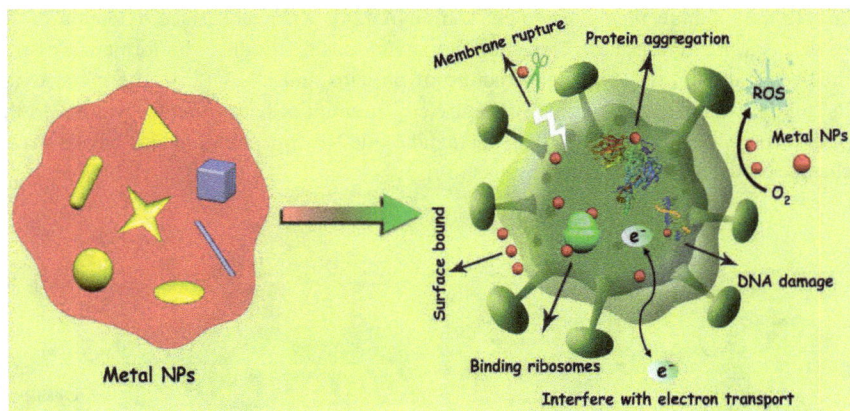

Figure 3. Schematic diagram for antibacterial mechanism of metal nanoparticles. [Reprinted with permission. Copyright Journal of Materials Chemistry B, 2020, 8(22), 4764-4777].

The mechanism of nanoparticle activity against microbes makes them unique. The cell wall of the bacteria is assembled in such a way as to afford strength, rigidity, and shape and to defend the cell from mechanical damage or osmotic rupture. Based on the chemical structure and functions, the bacterial cell wall can be classified as either Gram-positive or Gram-negative. Refer Figure 4 for the chemical composition of Gram-positive and Gram-negative bacteria. The cell wall structure plays a trivial role in the tolerance or vulnerability of bacteria or microbes in the presence of various external chemical components, including NPs. Hence, the antibacterial activities of NPs could depend on two main factors that are: physicochemical properties of NPs and the type of bacteria [21]. The exact mechanism of NP toxicity against various microbes remains poorly understood. The strong interaction of NPs with the membrane of the bacteria will lead to attachment of NPs to the wall that disturbs the rigidity of the membrane [22]. The attachment of NPs induced oxidative stress by free radical formation (reactive oxygen species) which activates the breakdown of the cell wall which is shown in the Figure 4. Several factors, such as the solution pH, temperature, air, and concentration, affect the antibacterial activity of Cu NPs. It was often observed that the de-aggregation leads to enhancement of toxicity since the surface area of non-aggregated NPs is higher than the aggregated particles that might have better interaction with the bacterial surface [23]. The mechanistic studies suggest that the metal ions of Cu generate hydroxyl radicals which damages the biomolecules like protein and DNA [24], whereas surface oxidation of Au NPs and Ag NPs releases the corresponding metal ions that induce the formation of free radicals and disrupt the essential structure of proteins and DNA [25]. The metal oxide NPs' toxicity is mainly governed by the toxicity strength of heavy metals. Moreover, the size and number of colony as well as the metal oxide NP concentration also have a strong influence on the antimicrobial properties [26].

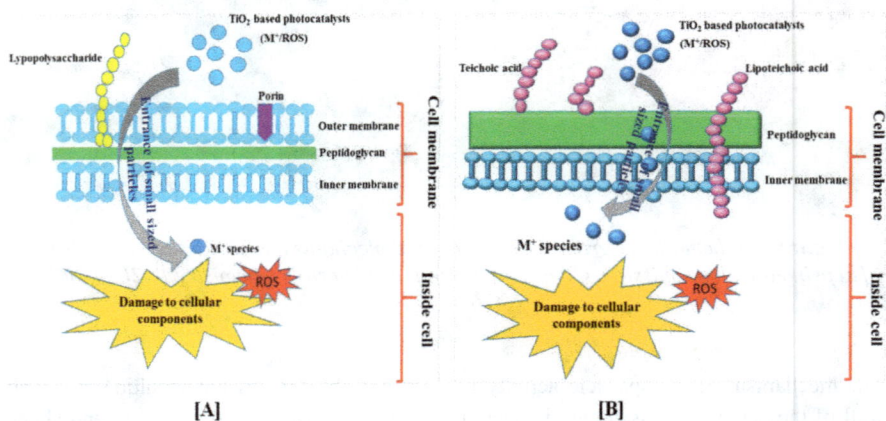

Figure 4. Membrane structural composition and proposed mechanism in the Gram-negative bacteria [A] and Gram-positive bacteria [B]. [Reprinted with permission. Copyright Environmental Research, 2017, 154, 296-303.]

Bioinspired Nanomaterials   Materials Research Forum LLC
Materials Research Foundations 111 (2021) 36-62   https://doi.org/10.21741/9781644901571-2

2.1   Silver nanoparticle

Microbicidal activity of silver is very famous due to the redox potential of silver ions and this silver nanoparticles (Ag NPs) possesses almost all significant antimicrobial activities [27]. Green synthesis using glucan as a biopolymer produces a spherical Ag NPs capped with glucan with 2.445 ± 1.08 nm. And it exhibited only 0.68% hemolysis to human RBCs at its LD_{50} dosage [28]. Thus, Ag NPs–glucan conjugates established to be compatible with human RBCs. In another study, polysaccharide capped Ag NPs of 2.78 ± 1.47 nm in size were synthesized from a heteropolysaccharide. These nanoparticles were also compatible with human RBCs at its LD50 dosage [30].

The interaction between Ag NPs and microorganisms starts with adhesion between Ag NPs and the microbial cell membrane, this act is based on the electrostatic attraction between the charged microbial cell membrane and charged Ag NPs [31]. Figure 2 explains the redox process and the mechanistic flow of metal nanoparticle formation. The communication between Ag NPs and the bacterial cells also presented by a significant decrease in zeta potential in both Gram-positive and Gram-negative bacteria [32, 33]. The degree of membrane damage helps the Ag NPs to enter the cell and disturb its vital functions by interacting with DNA and proteins [34]. One of the suggested mechanisms for antimicrobial activity of Ag NPs relies on silver ion release from the nanoparticles, which has detrimental effects on DNA and proteins [34, 35]. Feng et al. studies showed that silver ions led to the transformation of the bacterial DNA to a condensed state from the naturally relaxed state by which the DNA molecule quits from replication [36].

The intracellular effect of Ag NPs is not only limited to DNA damage. Proteomic studies have demonstrated Ag NPs role on proteins and protein synthesis. As previously quoted, Ag NPs and Ag^+ ions have an inclining to react with thiol groups of proteins. Thiol or thiolate groups are found in the amino acid cysteine as the functional group. Cysteine is one amongst the least abundant amino acids but it routinely acts as a highly conserved residue within functional sites in proteins. Accordingly, its importance in biological reactions is because of its high-affinity metal-binding nature, its nucleophilic role, and ability to form disulfide bonds which is essential for the three-dimensional structure of proteins [37, 38]. Yan et al. showed that the expression of the many membrane proteins was significantly up-regulated or down-regulated in P. aeruginosa as a result of treatment with Ag NP [39]. These proteins are necessarily responsible for ion binding, transport, flagellum assembly, pore formation, antibiotic resistance, and membrane stabilization. The most markedly up-regulated proteins functioned in phospholipid synthesis, ATP synthesis, and transmembrane transport. Additionally, to assist the transport of Ag NPs into the cell, a series of metal transporters including copper, iron, zinc, and magnesium transporters were repressed. Wigginton et al. conducted a proteomic study to find the interaction between *E. coli* proteins and Ag NPs [40]. Das *et al.* studied and tested the Ag NPs against multidrug-resistant *P. aeruginosa* cells, and the antimicrobial activity of Ag NPs with respect to ROS formation. For this study, they utilized fluorescence assay which is based on oxidation of non-fluorescent 2′,7′-dichlorofluorescein diacetate (H2-DCFDA) to highly fluorescent 2′,7′-dichlorofluorescein (DCF). Ag NP-treated *P. aeruginosa* cells turns into DCF positive, pointing out the intracellular ROS formation [41].

The ROS generation arbitrated by Ag^+ ions released from Ag NPs can bring about flaw with the bacterial electron transport chain and proton motive force by hindering the enzymes involved in the reactions [42, 43]. Additionally, to disrupt the membrane functions, Kim *et al.* showed that Ag NP-mediated ROS generation convinced protein leakage by unsettling the membrane permeability [44]. Treatment with Ag NPs resulted in cell death due to protein leakage. Ag NPs with bio-coating and in sequence with its size demonstrates antifungal activity. Ishida *et al.* studies showed the antifungal activity of green Ag NPs synthesized from *Fusarium oxysporum*. Green synthesized Ag NPs demonstrated noteworthy level of antifungal activity against Candida and Cryptococcus, and it had a minimum inhibitory concentration (MIC) values of ≤1.68 µg/ml [45]. Investigation using the disc diffusion method the colloidal silver showed potential antifungal activity against three species of *Candida* namely *C. albicans*, *C. glabrata* and *C. tropicalis* [46]. Ag NPs synthesized from *Pilimelia columellifera* displayed antifungal activity against fungi causing superficial mycoses. The smallest MIC of Ag NPs was noticed against *M. furfur* (16 µg/ml) [46].

2.2 Copper nanoparticle

Copper nanoparticles (Cu NPs) have recently interested researchers, because of the availability of copper has made it a suitable material to work with, and it shares properties similar to those of other expensive noble metals, including silver and gold. In addition, copper nanoparticles are noted to have antimicrobial activity against a number of species of bacteria and fungi [47].

As the Cu NPs showed potent activity against a wide range of microbes it can also be used as an alternative antimicrobial especially against crop pathogens [48]. Diverse formulations of Cu NPs as nano-fungicides, nano-antimicrobials, or nano-fertilizers helps with dual purpose to control the infections as well as nutrient for plants. Kanhed *et al.* reported the *in-vitro* antifungal activity of chemically synthesized Cu NPs with a commercially available antifungal agent against four different plant pathogenic fungi, viz., F. oxysporum, C. lunata, A. alternata and P. destructiva. The synthesized Cu NPs showed activity against all the plant pathogenic fungi used in the experiment. The results exhibited that *C. lunata* and *A. alternata* were comparatively resistant to the commercial antifungal agent (bavistin) but showed sensitivity towards Cu NPs as well. In this study, Cu NPs demonstrated the high antifungal activity against *C. lunata* followed by *A. alternata*, and the minimum activity against *P. destructiva*. It was also found that the antifungal activity of bavistin increases in fusion with Cu NPs in case of *F. oxysporum*. The present study also proved the antifungal nature of Cu NPs against plant pathogenic fungi tested [49]. In another study, Usman *et al.* reported the synthesis of Cu NPs in the presence of a chitosan stabilizer using chemical method with the evaluation of the antimicrobial activity of the nanoparticles by using several test microorganisms like methicillin-resistant *S. aureus*, *B. subtilis*, *P. aeruginosa*, *Salmonella choleraesuis* and *Candida albicans* [50]. Sampath *et al.* used polyvinylpyrrolidone (PVP) as a capping agent and synthesized a jasmine bud-shaped Cu NPs by a green chemical reduction method, L-ascorbic acid (AA) was used as a reducing agent and antioxidant agent, isonicotinic acid hydrazide (INH) as a reducing agent. The antibacterial activity of the Copper nano buds was checked by testing against *E. coli* and *S. aureus* [51]. Fujimori *et al.* investigated the antiviral activity of nanosized copper (I) iodide particles having an average size of 160 nm against an influenza A virus of swine-origin using plaque titration assay. They announced the dose-dependent activity on virus titer and the 50 % best concentration was found to be 17 g/ml for 60 min of exposure time. The SDS-PAGE analysis confirmed the inactivation of the virus due to the degradation of viral proteins such as hemagglutinin and neuraminidase by nanosized copper (I) iodide particles [52].

It is widely believed and also proved from the present study, that Cu NPs have compelling antimicrobial activity, but its mechanism of action still less understood and needs further research efforts. According to Das et al. Cu NPs enters into the cell due to their small size and it inactivates their proteins/ enzymes, generating hydrogen peroxide, which indeed causes bacterial cell death [53]. In another study, it was reported that the inactivation of proteins was due to the interaction of Cu NPs with its –SH (sulfhydryl) group [54]. Similarly, Cu NPs also interacted with DNA molecules leading to its degradation by upsetting its helical structure [55]. The cell membrane integrity depends on its electrochemical potential, but Deryabin *et al.* reported that Cu NPs decrease the electrochemical potential of the cell membrane, which basically affects cell membrane integrity [56]. It was also believed that metal nanoparticles releasing their respective ions and such heavy metal ions resulted an adverse effect on bacterial cells [57]. Accumulation of Cu NPs and Cu ions on cell surface form pits in the membrane, which largely leads to the leakage of the cellular component from the cell and finally causes cell death. Another vital reason suggested for cell death due to the action of Cu NPs was due to oxidative stress [56].

Copper is one of the favorable nutrients needed for the normal functioning of the human body so, it is maintained in homeostasis. If the intake of copper outpaces the range of human tolerance, it may exert toxic effects like hemolysis, jaundice, and eventually death. Similarly, if the intake of copper nanoparticles exceeds in the human body by ingestion or inhalation, it causes toxic effects in the respiratory tract, gastrointestinal tract, and in other tissues as well [58]. The acute toxicity of micro (17 μm) and nano (23.5 nm) copper particles in mice *in-vivo* was determined and found that copper nanoparticles were found to be more toxic than microparticles of copper because the nanoparticles can easily get inside the body through skin contact, inhalation, and ingestion [59]. It is important to assess the biocompatibility and toxicity of any nanoparticle before being used for any biological application [60]. The biocompatibility of green synthesized *C. gigantea* extract CuO NP was checked with HEK293 human embryonic kidney cell line by comparing them with commercial CuO NP effects. The LC_{50} as calculated by the results was 410 μg/ml which was significantly higher than the 112 μg/ml LC_{50} of commercial CuO NP. LC_{50} of CuO NP was also higher than the LC_{50} of CuO NP synthesized by conventional methods as previously reported [61]. Xu et al. reported a decrease in PC12 cell viability with increasing concentrations of copper nanoparticles and the decrease of cell viability is directly proportional to concentration and treatment time. These results demonstrated that copper nanoparticles are toxic in a size and concentration dependent manner to Dorsal root ganglion neurons in rat and PC12 cells in mice [62].

2.3   Gold nanoparticle

Gold nanoparticles (Au NPs) tend to have unique surface morphologies, stable nature, and geometry [63]. Au NPs are also used in the detection, diagnosis, and treatments of several diseases [64]. Gold has been used for several centuries in the treatment of variety of disorders. Gold compounds were initially explored by Robert Koch for its biocidal potential [65].

A particular study indicated that the concentrations of Au NPs tested (1.25–200 g/ml), inhibited the growth of the tested gram-negative bacterial strains of *E. coli, P. aeroginosa* and *K. pneumonia*. Cultures neglecting nanoparticles did not show any growth inhibition. But complete inhibition was reported at a higher concentration of 200 g/ml when compared to the low concentration of 1.25 g/ml, when tested. It was found that at 200 g/ml 88% of *E. coli*, 86% of *P. aeroginosa*, and 94% of *K. pneumonia* growth was inhibited [66]. The plant *E. hirta* has shown to have antibacterial activity both in its aqueous and methanolic extracts [67]. Hence, it helps to conclude that the plant material has attached to the Au NPs and its antibacterial activity is due to the synergistic effect of

Materials Research Forum LLC
https://doi.org/10.21741/9781644901571-2

the two combined [68]. The gold nanoparticles synthesized by using *Solanum nigrum* leaf extract were considered as free radical scavengers and antibacterial static agents. These nanoparticles naturally inhibited the growth of pathogenic *S. saprophyticus* and *B. subtilis* (Gram-positive bacteria) and *E. coli* and *P. aeruginosa* (gram-negative bacteria) [69]. The antimicrobial activity of *Hibiscus cannabinus* stem extract induced gold nanoparticles were performed against *P. aeruginosa* and *S. aureus*. The effect of antibacterial activity was more effective in the case of *P. aeruginosa* which may be due to electrostatic attraction of positively charged nanoparticles and negatively charged cell surface of microorganisms [70]. The gold nanoparticles synthesized by *Abelmoschus esculentus* extract showed antifungal activity against *P. graminis*, *A. flavus*, *A. niger*, and *C. albicans* [71]. The radiation-induced synthesized Au NPs were proved to have 56.42%, 59.5%, and 64.19% inhibition for *S. aureus* MRSA (Gram+ve), *P. aeruginosa* (Gram -ve) and *A. baumaninii/haemolyticus* (Gram -ve), respectively [70]. It can be suggested that Gram-negative *P. aeruginosa* and *A. baumaninii/haemolyticus* with thin cell walls are more prone to cell wall damage compared to Gram-positive *S. aureus* with the thick cell wall and therefore more inhibition was reported for Gram-negative bacteria [72] (Refer Figure 4).

The mechanisms of action of Au NPs mainly react with sulfur or phosphorus-holding bases, which are the widely known region for the Au NPs attack. When NPs attach to thiol functional groups of enzymes (nicotinamide adenine dinucleotide (NADH) dehydrogenases), they suspend the respiratory chains by the generation of a high number of free radicles, leading to cell death [73]. A hypothesis for cellular death is that these NPs decrease the ATPase activities; Au NPs may also inhibit the binding of tRNA to ribosomal subunit [74]. An elevated number of electrons are produced by Au NPs while killing *Leishmania*, which yields ROS (O-2 and OH). These radicals destroy DNA and other cellular components of the micobes [75]. Another possible mechanism is that these Au NPs hinders the transmembrane H+ efflux [76]. The smaller size of the nanopaticles could also be credited to its antimicrobial potential, mostly 250 times less than the size of the bacterial cell, which makes it easier to adhere to the cell wall and disrupt the cellular process causing cellular death [77]. Herdt *et al.* reported that the gold surface can degrade DNA after interaction with its surface [78].

2.4   TiO$_2$ NPs

Titanium dioxide (TiO$_2$) nanoparticles doesn't possess any antiquity for its usage as novel material for antimicrobial activities. However, in the recent times there has been an augmenting interest in usage of TiO$_2$ nanoparticles in the field of microbicidal studies [79]. TiO$_2$ is a n-type, interesting transition metal-oxide semiconducting material and exist in three different crystallite structures: anatase, rutile and brookite. It is an inert material and exhibit robust stability in both alkaline and acidic conditions. Due to its interesting optical, electrical and magnetic properties, TiO_2 has found intensified implementation in solar cells, heterogeneous photocatalysts for environmental remediation, cosmetics and skin care products for the protection against UV rays, paints, food colorants, toothpastes and even in chemical sensors [80,81]. The TiO_2 nanoparticles possess a large surface area, excellent surface morphology, non-toxicity and excellent bio-related activity with a myriad range of bacterial species [82]. The so far explored synthetic protocols of attaining TiO_2 nanoparticles have successfully displayed the facile control over the crystalline structure of the as-synthesized TiO_2 which ultimately ends up in tuning the properties of TiO_2 nanoparticles by varying the crystalline structure and size.

The recent investigations reported in the field of green synthesis of TiO_2 using the plant extracts have produced TiO_2 nanoparticles with well defined morphology, size, monodispersed size distribution and simultaneous in a cost-effective mode compared to chemical synthetic methods [82]. As extracts derived from various parts of the plant like leaves, barks, flowers, root, seeds, fruits and flowers exhibit their capacity as a potential reducing and capping agents [83,84]. The above extracts contain multitude of bioorganic compounds: flavonoids, terpenoids, alkaloids, steroids, polyphenols and proteins, which are observed to play the role of either reducing or capping agents during the green nanoparticle synthesis [85,86]. Very importantly within the bio-based green synthetic regime of metallic nanoparticles, plant-extract based synthesis is more effective than the microbial based synthesis [4]. Plant extracts mediated green synthesis of TiO_2 has been recently reported on extracts obtained from different plants and also from diverse parts of the plant. In this book chapter, we summarize the key TiO_2 nanoparticles synthesized from diverse plant extracts which have exhibited exalted antimicrobial activity, see Table 1. The plant extracts are primarily opted on the information obtained from the phytochemical screening of the extracts derived from plant confirming the presence of reducing or capping agents. Subsequently, mostly the plants with medicinal value are chosen from which extracts are derived. Having mentioned the basis on which plant extract is opted for the green synthesis of metal nanoparticles. The most interesting and beneficial plant extracts used for the synthesis of TiO_2 nanoparticles are mentioned in this section.

Table 1. Biological synthesized TiO2 nanoparticles for antibacterial activity

S. No	Plant Name	Part of Plant	Size of the NPs	Shape of NPs	Anti-microbial activity	Ref
1.	*Morinda citrifolia*	Leaf	10 nm (avg crystal line size)	Quasi-Spherical	Evaluated by agar well-diffusion method Exhibited superior anti-microbial activity against Gram-positive Bacteria than Gram-negative bacteria and Fungal pathogens	[82]
2.	*Trigonell a foenum graecum*	Leaf	20 to 90 nm	Spherical	Evaluated by standard disc diffusion method TiO2 NPs exhibit good anti-microbial activity against 1. *Yersinia enterocolitica, 2. Escherichia coli, 3. Staphylococcus aureus, 4. Enterococcus faecalis and 5. Streptococcus faecalis*	[81]
3	*Psidium guajava*	Leaf	32.58 nm	Spherical	Evaluated by disc-diffusion method TiO2 exhibit enhanced microbicals activity against these two human pathogens: Staphylococcus aureus and Escherichia coli	[80]
4	*Cynodon dactylon*	Leaf	16 nm	Hexagon al	Agar well-diffusion method TiO2 NPs exhibited anti-microbial activity against the growth of E.coli	[98]

Morinda citrifolia (Indian Mulberry or Noni Tree) has a rich medicinal history in India. The fruit of this plant has been profoundly found in many parts of ancient medicinal texts of Ayurveda and Siddha. This multi-purpose plant possesses more than 150 nutraceuticals like americanol, glycosides, anthraquinones, bio-ligands, sterol derivatives, etc., [87]. The various part of this plant possesses anti-oxidant, anti-microbial, anti-cancer and anti-inflammatory properties. The extracts obtained from this plant are used for reducing anxiety (anxiolytic effect), treating wound infections, arteriosclerosis and diabetes. Having realized its intensified medicinal potential of Morinda citrifolia for the human race. In various parts of south India, especially in Hindu temples of Tamil Nadu, this tree is grown along with other trees like *Ocimum tenuiflorum* (Tulsi) and *Aegle marmelos* (Bilva or Indian bael) [88].

In this decade the extracts various part of *Morinda citrifolia* plant as profoundly used for the synthesis of silver nanoparticles for the antifungal and medical treatment applications. For the first time in 2015, Thodhal *et al.* used the aqueous root extract derived from

Morinda citrifolia as low-cost reducing agent for the synthesis of TiO_2 nanoparticles. This TiO_2 nanoparticles displayed anatase crystalline structure and were produced in diverse morphologies: spherical, triangle and oval. This TiO_2 nanoparticles exhibited larvicidal activity against the larvae of Anopheles stephensi, Aedes aegypti and Culex quinquefasciatus insects [89].

Succeeding to this, Sundrarajan *et al.* developed modified hydrothermal technique, in which for the first time, they used the leaf extracts of *Morinda citrifolia* for the synthesis of TiO_2 nanoparticles. Here the leaf extracts contained phenolic compounds like flavonoids, vitamin C and anthraquinones played an effective role of stabilizing/capping agent and this resulted in generation of small size (see Table 1), well-defined quasi-spherical shaped particle and with rutile crystalline phase. This TiO_2 nanoparticles exhibited good antimicrobial activity against diverse microorganism like Bacillus subtilis, Staphylococcus aureus, Candida albicans, Escherichia coli, Aspergillus niger and Pseudomonas aeruginosa [82].

Like Noni tree, *Trigonella foenum-graecum leguminosse* (Fenugreek) possess an intense rich history in Ayurveda, Siddha and Unani, the Indian system of medicine. It is been used as aphrodisiac agent for revitalization of humans from problems associated with digestive & respiratory (bronchial) systems, gout and arthritis. The fenugreek seeds are used to treat liver cancer, blood pressure and diabetes, simultaneously have found its place in myriad of medical formulations all across the world [90,91]. Subhapriya *et al.* used the leaf extracts of fenugreek plants and a high temperature sintering method for the synthesis of TiO_2 nanoparticles. The as-obtained bio-synthesized TiO_2 nanoparticles with aqueous leaf extracts possess spherical shape (for size, see Table 1) with very high phase purity exhibiting the rutile crystalline structure. This nanoparticle was studied for its anti-microbial activity with various microorganisms Enterococcus faecalis, Staphylococcus aureus, Streptococcus faecalis, Bacillus subtilis, Yersinia enterocolitica, Proteus vulgaris, Escherichia coli, Pseudomonas aeruginosa, Klebsiella pneumoniae and Candida albicans. The results indicated high anti-microbial activity for Gram-positive bacteria than Gram-negative bacteria while they were subjected to TiO_2 nanoparticles from fenugreek leaf extracts. The weak anti-bacterial activity observed here was attributed to the presence of very thin cell wall constituting several layers of lipopolysaccharide and few layers of peptidoglycan for gram-negative bacteria and on the contrast a very thick cell wall is observed for gram-positive bacteria [81] which can be seen from the Figure 4.

Compared to the phase purity TiO_2 nanoparticles obtained from the plant extracts mentioned in the above two cases. Velayutham *et al.* synthesized TiO_2 nanoparticles using aqueous leaf extracts from *Catharanthus roseus* which possessed a mixture of both anatase and rutile crystalline forms and exhibited anti-parasitic activity [92]. The study of microbiocidal activity through the usage of TiO_2 nanoparticles has been heavily undertaken by the research community as the surface of nanosized TiO_2 generate strong oxidizing agents upon exposure to photons with appropriate energy, which exhibit a non-selective oxidation of chemicals that comes in contact with TiO_2 nanoparticles surface [93, 94]. Figure 5 clearly demonstrates the detailed mechanism of TiO_2 against bacterial cell wall along with their photocatalytic mechanism. In this prospective it is utmost importance to have adequate knowledge of plant extracts in the bio-based synthesis of TiO_2 as the optical-response based anti-microbial properties of TiO_2 are dependent on its band gap which in turn is dependent on the crystalline structure and their corresponding mixture of crystalline structures.

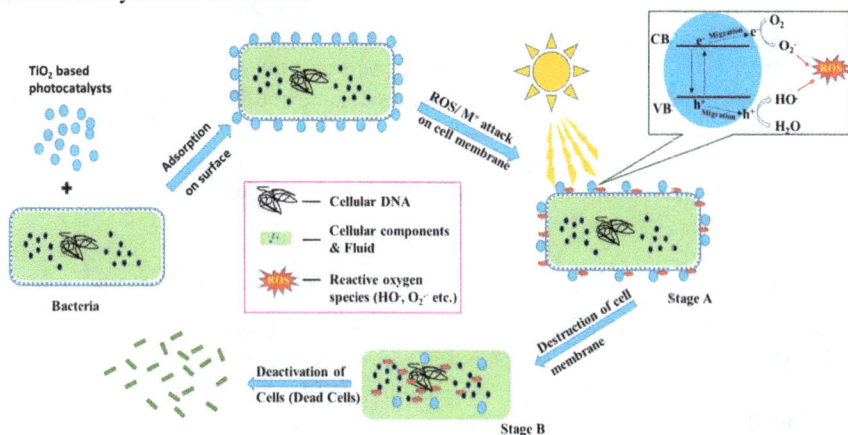

Figure 5. *The possible mechanism of the photocatalytic disinfection of bacteria. [Reprinted with permission. Copyright Environmental Research, 2017, 154, 296-303.]*

Aligning to this discussion, Thirunavukkarasu *et al.* synthesized TiO_2 nanoparticles through plant extract mediated synthesis. Where the resultant TiO_2 is composed of both rutile and anatase crystal structures with spherical morphology using aqueous leaf extracts from *Psidium guajava* (Guava). The *Psidium guajava* based extracts was opted here because it possesses immense medicinal value. The leaves of *Psidium guajava* was commonly used as common medicine for diarrhoea and for treating ulcer and rheumatic pain [80, 95]. The green synthesis of silver nanoparticles using the leaf extracts of *Psidium guajava* exhibited enhanced antibacterial properties than their chemical counterparts. The leaf extracts of *Psidium guajava* contains flavonoids, eugenol and tannins, which act as reducing agents for the synthesis of metal nanoparticles [96, 97].

Bioinspired Nanomaterials   Materials Research Forum LLC
Materials Research Foundations **111** (2021) 36-62   https://doi.org/10.21741/9781644901571-2

The TiO_2 nanoparticles produced using the aqueous leaf extracts of *Psidium guajava* was tested for microbicidal activity against human pathogens: P. aeruginosa, E. coli, P. mirabilis, S. aureus and A. hydrophila using the disc diffusion method. The enhanced microbicidal activity exhibited by the TiO_2 nanoparticles (for size and morphology, see Table 1) here was against Staphylococcus aureus and Escherichia coli, compared to other pathogens studied here [80].

3. Summary

To conclude, the biogenic synthesis of metal nanoparticles (Cu, Ag and Au) and semiconductor based heterogeneous photocatalysts (especially TiO_2), possessing microbicidal activity have gathered an intriguing attention among the nano-biotechnologist and chemical engineers due to their non-toxic properties compared to the traditionally synthesized nanoparticles using organic and inorganic chemicals. The importance of the use of nanoparticles is due to its size and morphology whereas metal helps in the redox activity after the material enters inside cell. Furthermore, nanoparticles possess better advantage due to the low dosage level also because of their biological synthesis lacks any harmful effects.

Future Perspective

Copper and Silver nanoparticles due to their huge availability and cost effectiveness can be easily scaled up for industrial purposes employing biogenic methods. With the advent of 'Floating Photocatalysts' in heterogeneous photocatalytic field targeting the water purification [99], eliminating the bottle necks associated with ROS employed photocatalysis, especially with the usage of plant's structural components like cellulose nanofibers for attaining the buoyant system [93]. Along with the recent report on the utilization of lignin (pivotal structural component of plant's cell wall) for the fabrication of nanocomposite photocatalysts with enhanced photocatalytic activity compared pristine semiconductor photocatalysts [100]. The future perspective of such novel systems in photocatalysis can attain commercial implementation of water purification in real time aqueous system with employment of biogenically synthesized photocatalysts.

References

[1]   Live Science, 2020. 20 of the worst epidemics and pandemics in history. Retrieved from: https://www.livescience.com/worst-epidemics-and-pandemics-in-history.html

[2]   S. Varnagiris, M. Urbonavicius, S. Sakalauskaite, R. Daugelavicius, L.
        Pranevicius, M. Lelis, D. Milcius, Floating TiO2 photocatalyst for efficient
        inactivation of E. coli and decomposition of methylene blue solution, Sci. Total
        Environ. 720 (2020) 137600. https://doi.org/10.1016/j.scitotenv.2020.137600

[3]   V. V Makarov, A.J. Love, O. V Sinitsyna, S.S. Makarova, I. V Yaminsky, M.E.
        Taliansky, N.O. Kalinina, "Green" nanotechnologies: synthesis of metal
        nanoparticles using plants, Acta Naturae (Англоязычная Версия). 6 (2014).
        https://doi.org/10.32607/20758251-2014-6-1-35-44

[4]   M. Rai, C.A. dos Santos, Nanotechnology applied to pharmaceutical technology,
        Springer, 2017. https://doi.org/10.1007/978-3-319-70299-5

[5]   M.B. Galib, M. Mashru, C. Jagtap, B.J. Patgiri, P.K. Prajapati, Therapeutic
        potentials of metals in ancient India: A review through Charaka Samhita, J.
        Ayurveda Integr. Med. 2 (2011) 55. https://doi.org/10.4103/0975-9476.82523

[6]   S.S.N. Fernando, T. Gunasekara, J. Holton, Antimicrobial Nanoparticles:
        applications and mechanisms of action, (2018).
        https://doi.org/10.4038/sljid.v8i1.8167

[7]   J.T. Seil, T.J. Webster, Antimicrobial applications of nanotechnology: methods
        and literature, Int. J. Nanomedicine. 7 (2012) 2767.
        https://doi.org/10.2147/IJN.S24805

[8]   A. Khezerlou, M. Alizadeh-Sani, M. Azizi-Lalabadi, A. Ehsani, Nanoparticles and
        their antimicrobial properties against pathogens including bacteria, fungi, parasites
        and viruses, Microb. Pathog. 123 (2018) 505–526.
        https://doi.org/10.1016/j.micpath.2018.08.008

[9]   P. Panchal, D.R. Paul, A. Sharma, D. Hooda, R. Yadav, P. Meena, S.P. Nehra,
        Phytoextract mediated ZnO/MgO nanocomposites for photocatalytic and
        antibacterial activities, J. Photochem. Photobiol. A Chem. 385 (2019) 112049.
        https://doi.org/10.1016/j.jphotochem.2019.112049

[10]   C. Xu, P. R. Anusuyadevi, C. Aymonier, R. Luque, S. Marre, Nanostructured
        materials for photocatalysis. *Chemical Society Reviews* 2019, *48* (14), 3868-3902.
        https://doi.org/10.1039/C9CS00102F

[11]   H. Duan, D. Wang, Y. Li, Green chemistry for nanoparticle synthesis, Chem. Soc.
        Rev. 44 (2015) 5778–5792. https://doi.org/10.1039/C4CS00363B

[12]   S. Ahmed, S.A. Chaudhry, S. Ikram, others, A review on biogenic synthesis of
        ZnO nanoparticles using plant extracts and microbes: a prospect towards green
        chemistry, J. Photochem. Photobiol. B Biol. 166 (2017) 272–284.
        https://doi.org/10.1016/j.jphotobiol.2016.12.011

[13]   M.S. Akhtar, J. Panwar, Y.-S. Yun, Biogenic synthesis of metallic nanoparticles by plant extracts, ACS Sustain. Chem. Eng. 1 (2013) 591–602. https://doi.org/10.1021/sc300118u

[14]   D. Sharma, S. Kanchi, K. Bisetty, Biogenic synthesis of nanoparticles: A review, Arab. J. Chem. 12 (2019) 3576–3600. https://doi.org/10.1016/j.arabjc.2015.11.002

[15]   A. Rai, A. Prabhune, C.C. Perry, Antibiotic mediated synthesis of gold nanoparticles with potent antimicrobial activity and their application in antimicrobial coatings, J. Mater. Chem. 20 (2010) 6789–6798. https://doi.org/10.1039/c0jm00817f

[16]   M. Demurtas, C.C. Perry, Facile one-pot synthesis of amoxicillin-coated gold nanoparticles and their antimicrobial activity, Gold Bull. 47 (2014) 103–107. https://doi.org/10.1007/s13404-013-0129-2

[17]   L. Sintubin, B. De Gusseme, P. der Meeren, B.F.G. Pycke, W. Verstraete, N. Boon, The antibacterial activity of biogenic silver and its mode of action, Appl. Microbiol. Biotechnol. 91 (2011) 153–162. https://doi.org/10.1007/s00253-011-3225-3

[18]   P.K. Gautam, A. Singh, K. Misra, A.K. Sahoo, S.K. Samanta, Synthesis and applications of biogenic nanomaterials in drinking and wastewater treatment, J. Environ. Manage. 231 (2019) 734–748. https://doi.org/10.1016/j.jenvman.2018.10.104

[19]   S. Baker, A. Pasha, S. Satish, Biogenic nanoparticles bearing antibacterial activity and their synergistic effect with broad spectrum antibiotics: Emerging strategy to combat drug resistant pathogens, Saudi Pharm. J. 25 (2017) 44–51. https://doi.org/10.1016/j.jsps.2015.06.011

[20]   S. Sudhasree, A. Shakila Banu, P. Brindha, G.A. Kurian, Synthesis of nickel nanoparticles by chemical and green route and their comparison in respect to biological effect and toxicity, Toxicol. Environ. Chem. 96 (2014) 743–754. https://doi.org/10.1080/02772248.2014.923148

[21]   M.J. Hajipour, K.M. Fromm, A.A. Ashkarran, D.J. de Aberasturi, I.R. de Larramendi, T. Rojo, V. Serpooshan, W.J. Parak, M. Mahmoudi, Antibacterial properties of nanoparticles, Trends Biotechnol. 30 (2012) 499–511. https://doi.org/10.1016/j.tibtech.2012.06.004

[22]   A. Thill, O. Zeyons, O. Spalla, F. Chauvat, J. Rose, M. Auffan, A.M. Flank, Cytotoxicity of CeO2 nanoparticles for Escherichia coli. Physico-chemical insight of the cytotoxicity mechanism, Environ. Sci. Technol. 40 (2006) 6151–6156. https://doi.org/10.1021/es060999b

[23]   A. Pramanik, D. Laha, D. Bhattacharya, P. Pramanik, P. Karmakar, A novel study of antibacterial activity of copper iodide nanoparticle mediated by DNA and membrane damage, Colloids Surfaces B Biointerfaces. 96 (2012) 50–55. https://doi.org/10.1016/j.colsurfb.2012.03.021

[24]   S. Wang, R. Lawson, P.C. Ray, H. Yu, Toxic effects of gold nanoparticles on Salmonella typhimurium bacteria, Toxicol. Ind. Health. 27 (2011) 547–554. https://doi.org/10.1177/0748233710393395

[25]   C.E. Santo, N. Taudte, D.H. Nies, G. Grass, Contribution of copper ion resistance to survival of Escherichia coli on metallic copper surfaces, Appl. Environ. Microbiol. 74 (2008) 977–986. https://doi.org/10.1128/AEM.01938-07

[26]   Y.W. Baek, Y.-J. An, Microbial toxicity of metal oxide nanoparticles (CuO, NiO, ZnO, and Sb2O3) to Escherichia coli, Bacillus subtilis, and Streptococcus aureus, Sci. Total Environ. 409 (2011) 1603–1608. https://doi.org/10.1016/j.scitotenv.2011.01.014

[27]   T.C. Dakal, A. Kumar, R.S. Majumdar, V. Yadav, Mechanistic basis of antimicrobial actions of silver nanoparticles, Front. Microbiol. 7 (2016) 1831. https://doi.org/10.3389/fmicb.2016.01831

[28]   I.K. Sen, A.K. Mandal, S. Chakraborti, B. Dey, R. Chakraborty, S.S. Islam, Green synthesis of silver nanoparticles using glucan from mushroom and study of antibacterial activity, Int. J. Biol. Macromol. 62 (2013) 439–449. https://doi.org/10.1016/j.ijbiomac.2013.09.019

[29]   D.K. Manna, A.K. Mandal, I.K. Sen, P.K. Maji, S. Chakraborti, R. Chakraborty, S.S. Islam, Antibacterial and DNA degradation potential of silver nanoparticles synthesized via green route, Int. J. Biol. Macromol. 80 (2015) 455–459. https://doi.org/10.1016/j.ijbiomac.2015.07.028

[30]   A. Abbaszadegan, Y. Ghahramani, A. Gholami, B. Hemmateenejad, S. Dorostkar, M. Nabavizadeh, H. Sharghi, The effect of charge at the surface of silver nanoparticles on antimicrobial activity against gram-positive and gram-negative bacteria: a preliminary study, J. Nanomater. 2015 (2015). https://doi.org/10.1155/2015/720654

[31]   A. Ahmad, Y. Wei, F. Syed, K. Tahir, A.U. Rehman, A. Khan, S. Ullah, Q. Yuan, The effects of bacteria-nanoparticles interface on the antibacterial activity of green synthesized silver nanoparticles, Microb. Pathog. 102 (2017) 133–142. https://doi.org/10.1016/j.micpath.2016.11.030

[32]   B. Ramalingam, T. Parandhaman, S.K. Das, Antibacterial effects of biosynthesized silver nanoparticles on surface ultrastructure and nanomechanical properties of gram-negative bacteria viz. Escherichia coli and Pseudomonas

aeruginosa, ACS Appl. Mater. Interfaces. 8 (2016) 4963–4976.
https://doi.org/10.1021/acsami.6b00161

[33]   S.K. Gogoi, P. Gopinath, A. Paul, A. Ramesh, S.S. Ghosh, A. Chattopadhyay, Green fluorescent protein-expressing escherichia c oli as a model system for investigating the antimicrobial activities of silver nanoparticles, Langmuir. 22 (2006) 9322–9328. https://doi.org/10.1021/la060661v

[34]   Y.-H. Hsueh, K.-S. Lin, W.-J. Ke, C.-T. Hsieh, C.-L. Chiang, D.-Y. Tzou, S.-T. Liu, The antimicrobial properties of silver nanoparticles in Bacillus subtilis are mediated by released Ag+ ions, PLoS One. 10 (2015) e0144306. https://doi.org/10.1371/journal.pone.0144306

[35]   O. Bondarenko, A. Ivask, A. Käkinen, I. Kurvet, A. Kahru, Particle-cell contact enhances antibacterial activity of silver nanoparticles, PLoS One. 8 (2013). https://doi.org/10.1371/journal.pone.0064060

[36]   Q.L. Feng, J. Wu, G.Q. Chen, F.Z. Cui, T.N. Kim, J.O. Kim, A mechanistic study of the antibacterial effect of silver ions on Escherichia coli and Staphylococcus aureus, J. Biomed. Mater. Res. 52 (2000) 662–668. https://doi.org/10.1002/1097-4636(20001215)52:4<662::AID-JBM10>3.0.CO;2-3

[37]   K. Siriwardana, A. Wang, M. Gadogbe, W.E. Collier, N.C. Fitzkee, D. Zhang, Studying the effects of cysteine residues on protein interactions with silver nanoparticles, J. Phys. Chem. C. 119 (2015) 2910–2916. https://doi.org/10.1021/jp512440z

[38]   K. Poole, Efflux-mediated antimicrobial resistance, J. Antimicrob. Chemother. 56 (2005) 20–51. https://doi.org/10.1093/jac/dki171

[39]   X. Yan, B. He, L. Liu, G. Qu, J. Shi, L. Hu, G. Jiang, Antibacterial mechanism of silver nanoparticles in Pseudomonas aeruginosa: proteomics approach, Metallomics. 10 (2018) 557–564. https://doi.org/10.1039/C7MT00328E

[40]   N.S. Wigginton, A. de Titta, F. Piccapietra, J.A.N. Dobias, V.J. Nesatyy, M.J.F. Suter, R. Bernier-Latmani, Binding of silver nanoparticles to bacterial proteins depends on surface modifications and inhibits enzymatic activity, Environ. Sci. Technol. 44 (2010) 2163–2168. https://doi.org/10.1021/es903187s

[41]   B. Das, S.K. Dash, D. Mandal, T. Ghosh, S. Chattopadhyay, S. Tripathy, S. Das, S.K. Dey, D. Das, S. Roy, Green synthesized silver nanoparticles destroy multidrug resistant bacteria via reactive oxygen species mediated membrane damage, Arab. J. Chem. 10 (2017) 862–876. https://doi.org/10.1016/j.arabjc.2015.08.008

[42] M.N. Gallucci, J.C. Fraire, A.P.V.F. Maillard, P.L. Páez, I.M.A. Mart\'\inez, E.V.P. Miner, E.A. Coronado, P.R. Dalmasso, Silver nanoparticles from leafy green extract of Belgian endive (Cichorium intybus L. var. sativus): Biosynthesis, characterization, and antibacterial activity, Mater. Lett. 197 (2017) 98–101. https://doi.org/10.1016/j.matlet.2017.03.141

[43] S. Belluco, C. Losasso, I. Patuzzi, L. Rigo, D. Conficoni, F. Gallocchio, V. Cibin, P. Catellani, S. Segato, A. Ricci, Silver as antibacterial toward Listeria monocytogenes, Front. Microbiol. 7 (2016) 307. https://doi.org/10.3389/fmicb.2016.00307

[44] S.-H. Kim, H.-S. Lee, D.-S. Ryu, S.-J. Choi, D.-S. Lee, others, Antibacterial activity of silver-nanoparticles against Staphylococcus aureus and Escherichia coli, Korean J. Microbiol. Biotechnol. 39 (2011) 77–85.

[45] K. Ishida, T.F. Cipriano, G.M. Rocha, G. Weissmüller, F. Gomes, K. Miranda, S. Rozental, Silver nanoparticle production by the fungus Fusarium oxysporum: nanoparticle characterisation and analysis of antifungal activity against pathogenic yeasts, Mem. Inst. Oswaldo Cruz. 109 (2014) 220–228. https://doi.org/10.1590/0074-0276130269

[46] N.T. Khan, M. Mushtaq, Determination of antifungal activity of silver nanoparticles produced from Aspergillus Niger, Biol. Med. 9 (2017) 1. https://doi.org/10.4172/0974-8369.1000363

[47] M. Wypij, J. Czarnecka, H. Dahm, M. Rai, P. Golinska, Silver nanoparticles from Pilimelia columellifera subsp. pallida SL19 strain demonstrated antifungal activity against fungi causing superficial mycoses, J. Basic Microbiol. 57 (2017) 793–800. https://doi.org/10.1002/jobm.201700121

[48] M.S. Usman, M.E. El Zowalaty, K. Shameli, N. Zainuddin, M. Salama, N.A. Ibrahim, Synthesis, characterization, and antimicrobial properties of copper nanoparticles, Int. J. Nanomedicine. 8 (2013) 4467. https://doi.org/10.2147/IJN.S50837

[49] H.-J. Lee, G. Lee, N.R. Jang, J.H. Yun, J.Y. Song, B.S. Kim, Biological synthesis of copper nanoparticles using plant extract, Nanotechnology. 1 (2011) 371–374.

[50] P. Kanhed, S. Birla, S. Gaikwad, A. Gade, A.B. Seabra, O. Rubilar, N. Duran, M. Rai, In vitro antifungal efficacy of copper nanoparticles against selected crop pathogenic fungi, Mater. Lett. 115 (2014) 13–17. https://doi.org/10.1016/j.matlet.2013.10.011

[51] M. Sampath, R. Vijayan, E. Tamilarasu, A. Tamilselvan, B. Sengottuvelan, Green synthesis of novel jasmine bud-shaped copper nanoparticles, J Nanotechnol. 2014 (2014). https://doi.org/10.1155/2014/626523

Bioinspired Nanomaterials   Materials Research Forum LLC
Materials Research Foundations **111** (2021) 36-62   https://doi.org/10.21741/9781644901571-2

[52]   Y. Fujimori, T. Sato, T. Hayata, T. Nagao, M. Nakayama, T. Nakayama, R.
       Sugamata, K. Suzuki, Novel antiviral characteristics of nanosized copper (I)
       iodide particles showing inactivation activity against 2009 pandemic H1N1
       influenza virus, Appl. Environ. Microbiol. 78 (2012) 951–955.
       https://doi.org/10.1128/AEM.06284-11

[53]   R. Das, S. Gang, S.S. Nath, R. Bhattacharjee, Linoleic acid capped copper
       nanoparticles for antibacterial activity, J. Bionanoscience. 4 (2010) 82–86.
       https://doi.org/10.1166/jbns.2010.1035

[54]   A.M. Schrand, M.F. Rahman, S.M. Hussain, J.J. Schlager, D.A. Smith, A.F. Syed,
       Metal-based nanoparticles and their toxicity assessment, Wiley Interdiscip. Rev.
       Nanomedicine Nanobiotechnology. 2 (2010) 544–568.
       https://doi.org/10.1002/wnan.103

[55]   J.-H. Kim, H. Cho, S.-E. Ryu, M.-U. Choi, Effects of metal ions on the activity of
       protein tyrosine phosphatase VHR: highly potent and reversible oxidative
       inactivation by Cu2+ ion, Arch. Biochem. Biophys. 382 (2000) 72–80.
       https://doi.org/10.1006/abbi.2000.1996

[56]   D.G. Deryabin, E.S. Aleshina, A.S. Vasilchenko, T.D. Deryabina, L. V Efremova,
       I.F. Karimov, L.B. Korolevskaya, Investigation of copper nanoparticles
       antibacterial mechanisms tested by luminescent Escherichia coli strains,
       Nanotechnologies Russ. 8 (2013) 402–408.
       https://doi.org/10.1134/S1995078013030063

[57]   N. Cioffi, L. Torsi, N. Ditaranto, G. Tantillo, L. Ghibelli, L. Sabbatini, T. Bleve-
       Zacheo, M. D'Alessio, P.G. Zambonin, E. Traversa, Copper nanoparticle/polymer
       composites with antifungal and bacteriostatic properties, Chem. Mater. 17 (2005)
       5255–5262. https://doi.org/10.1021/cm0505244

[58]   Z. Chen, H. Meng, G. Xing, C. Chen, Y. Zhao, G. Jia, T. Wang, H. Yuan, C. Ye,
       F. Zhao, others, Acute toxicological effects of copper nanoparticles in vivo,
       Toxicol. Lett. 163 (2006) 109–120. https://doi.org/10.1016/j.toxlet.2005.10.003

[59]   B.M. Prabhu, S.F. Ali, R.C. Murdock, S.M. Hussain, M. Srivatsan, Copper
       nanoparticles exert size and concentration dependent toxicity on somatosensory
       neurons of rat, Nanotoxicology. 4 (2010) 150–160.
       https://doi.org/10.3109/17435390903337693

[60]   B. Sarkar, S.K. Verma, J. Akhtar, S.P. Netam, S.K. Gupta, P.K. Panda, K.
       Mukherjee, Molecular aspect of silver nanoparticles regulated embryonic
       development in Zebrafish (Danio rerio) by Oct-4 expression, Chemosphere. 206
       (2018) 560–567. https://doi.org/10.1016/j.chemosphere.2018.05.018

[61] J. Sun, S. Wang, D. Zhao, F.H. Hun, L. Weng, H. Liu, Cytotoxicity, permeability, and inflammation of metal oxide nanoparticles in human cardiac microvascular endothelial cells, Cell Biol. Toxicol. 27 (2011) 333–342. https://doi.org/10.1007/s10565-011-9191-9

[62] P. Xu, J. Xu, S. Liu, G. Ren, Z. Yang, In vitro toxicity of nanosized copper particles in PC12 cells induced by oxidative stress, J. Nanoparticle Res. 14 (2012) 906. https://doi.org/10.1007/s11051-012-0906-5

[63] A.N. Grace, K. Pandian, Antibacterial efficacy of aminoglycosidic antibiotics protected gold nanoparticles—A brief study, Colloids Surfaces A Physicochem. Eng. Asp. 297 (2007) 63–70. https://doi.org/10.1016/j.colsurfa.2006.10.024

[64] A.K. Khan, R. Rashid, G. Murtaza, A. Zahra, Gold nanoparticles: synthesis and applications in drug delivery, Trop. J. Pharm. Res. 13 (2014) 1169–1177. https://doi.org/10.4314/tjpr.v13i7.23

[65] B. DJ Glišić, M.I. Djuran, Gold complexes as antimicrobial agents: an overview of different biological activities in relation to the oxidation state of the gold ion and the ligand structure, Dalt. Trans. 43 (2014) 5950–5969. https://doi.org/10.1039/C4DT00022F

[66] M. Nadeem, B.H. Abbasi, M. Younas, W. Ahmad, T. Khan, A review of the green syntheses and anti-microbial applications of gold nanoparticles, Green Chem. Lett. Rev. 10 (2017) 216–227. https://doi.org/10.1080/17518253.2017.1349192

[67] M.R. Bindhu, M. Umadevi, Antibacterial activities of green synthesized gold nanoparticles, Mater. Lett. 120 (2014) 122–125. https://doi.org/10.1016/j.matlet.2014.01.108

[68] M. Venkatachalam, K. Govindaraju, A.M. Sadiq, S. Tamilselvan, V.G. Kumar, G. Singaravelu, Functionalization of gold nanoparticles as antidiabetic nanomaterial, Spectrochim. Acta Part A Mol. Biomol. Spectrosc. 116 (2013) 331–338. https://doi.org/10.1016/j.saa.2013.07.038

[69] A. Muthuvel, K. Adavallan, K. Balamurugan, N. Krishnakumar, Biosynthesis of gold nanoparticles using Solanum nigrum leaf extract and screening their free radical scavenging and antibacterial properties, Biomed. Prev. Nutr. 4 (2014) 325–332. https://doi.org/10.1016/j.bionut.2014.03.004

[70] M.R. Bindhu, P.V. Rekha, T. Umamaheswari, M. Umadevi, Antibacterial activities of Hibiscus cannabinus stem-assisted silver and gold nanoparticles, Mater. Lett. 131 (2014) 194–197. https://doi.org/10.1016/j.matlet.2014.05.172

[71] C. Jayaseelan, R. Ramkumar, A.A. Rahuman, P. Perumal, Green synthesis of gold nanoparticles using seed aqueous extract of Abelmoschus esculentus and its

antifungal activity, Ind. Crops Prod. 45 (2013) 423–429.
https://doi.org/10.1016/j.indcrop.2012.12.019

[72] A.I. El-Batal, A.-A.M. Hashem, N.M. Abdelbaky, Gamma radiation mediated green synthesis of gold nanoparticles using fermented soybean-garlic aqueous extract and their antimicrobial activity, Springerplus. 2 (2013) 1–10. https://doi.org/10.1186/2193-1801-2-129

[73] K.M. Kumar, B.K. Mandal, M. Sinha, V. Krishnakumar, Terminalia chebula mediated green and rapid synthesis of gold nanoparticles, Spectrochim. Acta Part A Mol. Biomol. Spectrosc. 86 (2012) 490–494. https://doi.org/10.1016/j.saa.2011.11.001

[74] Y. Cui, Y. Zhao, Y. Tian, W. Zhang, X. Lü, X. Jiang, The molecular mechanism of action of bactericidal gold nanoparticles on Escherichia coli, Biomaterials. 33 (2012) 2327–2333. https://doi.org/10.1016/j.biomaterials.2011.11.057

[75] A. Ahmad, F. Syed, M. Imran, A.U. Khan, K. Tahir, Z.U.H. Khan, Q. Yuan, Phytosynthesis and antileishmanial activity of gold nanoparticles by M aytenus Royleanus, J. Food Biochem. 40 (2016) 420–427. https://doi.org/10.1111/jfbc.12232

[76] T. Ahmad, I.A. Wani, I.H. Lone, A. Ganguly, N. Manzoor, A. Ahmad, J. Ahmed, A.S. Al-Shihri, Antifungal activity of gold nanoparticles prepared by solvothermal method, Mater. Res. Bull. 48 (2013) 12–20. https://doi.org/10.1016/j.materresbull.2012.09.069

[77] R. Geethalakshmi, D.V.L. Sarada, Characterization and antimicrobial activity of gold and silver nanoparticles synthesized using saponin isolated from Trianthema decandra L., Ind. Crops Prod. 51 (2013) 107–115. https://doi.org/10.1016/j.indcrop.2013.08.055

[78] A.R. Herdt, S.M. Drawz, Y. Kang, T.A. Taton, DNA dissociation and degradation at gold nanoparticle surfaces, Colloids Surfaces B Biointerfaces. 51 (2006) 130–139. https://doi.org/10.1016/j.colsurfb.2006.06.006

[79] W. Ahmad, K.K. Jaiswal, S. Soni, Green synthesis of titanium dioxide (TiO2) nanoparticles by using Mentha arvensis leaves extract and its antimicrobial properties, Inorg. Nano-Metal Chem. (2020) 1–7. https://doi.org/10.1080/24701556.2020.1732419

[80] T. Santhoshkumar, A.A. Rahuman, C. Jayaseelan, G. Rajakumar, S. Marimuthu, A.V. Kirthi, K. Velayutham, J. Thomas, J. Venkatesan, S.-K. Kim, Green synthesis of titanium dioxide nanoparticles using Psidium guajava extract and its antibacterial and antioxidant properties, Asian Pac. J. Trop. Med. 7 (2014) 968–976. https://doi.org/10.1016/S1995-7645(14)60171-1

Bioinspired Nanomaterials
Materials Research Foundations **111** (2021) 36-62

Materials Research Forum LLC
https://doi.org/10.21741/9781644901571-2

[81]   S. Subhapriya, P. Gomathipriya, Green synthesis of titanium dioxide (TiO2) nanoparticles by Trigonella foenum-graecum extract and its antimicrobial properties, Microb. Pathog. 116 (2018) 215–220. https://doi.org/10.1016/j.micpath.2018.01.027

[82]   M. Sundrarajan, K. Bama, M. Bhavani, S. Jegatheeswaran, S. Ambika, A. Sangili, P. Nithya, R. Sumathi, Obtaining titanium dioxide nanoparticles with spherical shape and antimicrobial properties using M. citrifolia leaves extract by hydrothermal method, J. Photochem. Photobiol. B Biol. 171 (2017) 117–124. https://doi.org/10.1016/j.jphotobiol.2017.05.003

[83]   G. Sathishkumar, C. Gobinath, K. Karpagam, V. Hemamalini, K. Premkumar, S. Sivaramakrishnan, Phyto-synthesis of silver nanoscale particles using Morinda citrifolia L. and its inhibitory activity against human pathogens, Colloids Surfaces B Biointerfaces. 95 (2012) 235–240. https://doi.org/10.1016/j.colsurfb.2012.03.001

[84]   B.-N. Su, A.D. Pawlus, H.-A. Jung, W.J. Keller, J.L. McLaughlin, A.D. Kinghorn, Chemical Constituents of the Fruits of Morinda c itrifolia (Noni) and Their Antioxidant Activity, J. Nat. Prod. 68 (2005) 592–595. https://doi.org/10.1021/np0495985

[85]   Y. Zhou, W. Lin, J. Huang, W. Wang, Y. Gao, L. Lin, Q. Li, L. Lin, M. Du, Biosynthesis of gold nanoparticles by foliar broths: roles of biocompounds and other attributes of the extracts, Nanoscale Res. Lett. 5 (2010) 1351. https://doi.org/10.1007/s11671-010-9652-8

[86]   N. Durán, P.D. Marcato, M. Durán, A. Yadav, A. Gade, M. Rai, Mechanistic aspects in the biogenic synthesis of extracellular metal nanoparticles by peptides, bacteria, fungi, and plants, Appl. Microbiol. Biotechnol. 90 (2011) 1609–1624. https://doi.org/10.1007/s00253-011-3249-8

[87]   E.M. Yahia, Postharvest biology and technology of tropical and subtropical fruits: fundamental issues, Elsevier, 2011. https://doi.org/10.1533/9780857092762

[88]   S. Pratima, P. Shrikanth, Organoleptic and Preliminary Phytochemical Study of Achchhuka (Morinda Citrifolia), Int J Ayurvedic Herb Med. 5 (2015) 2029–2032.E.M. Yahia, Postharvest biology and technology of tropical and subtropical fruits: fundamental issues, Elsevier, 2011.

[89]   J.R. Peralta-Videa, Y. Huang, J.G. Parsons, L. Zhao, L. Lopez-Moreno, J.A. Hernandez-Viezcas, J.L. Gardea-Torresdey, Plant-based green synthesis of metallic nanoparticles: scientific curiosity or a realistic alternative to chemical synthesis?, Nanotechnol. Environ. Eng. 1 (2016) 4. https://doi.org/10.1007/s41204-016-0004-5

[90]   M.S. Baliga, P.L. Palatty, M. Adnan, T.S. Naik, P.S. Kamble, Anti-Diabetic Effects of Leaves of Trigonella foenum-graecum L.(Fenugreek): Leads from Preclinical Studies, J Food Chem Nanotechnol. 3 (2017) 67–71. https://doi.org/10.17756/jfcn.2017-039

[91]   U.C.S. Yadav, N.Z. Baquer, Pharmacological effects of Trigonella foenum-graecum L. in health and disease, Pharm. Biol. 52 (2014) 243–254. https://doi.org/10.3109/13880209.2013.826247

[92]   K. Velayutham, A.A. Rahuman, G. Rajakumar, T. Santhoshkumar, S. Marimuthu, C. Jayaseelan, A. Bagavan, A.V. Kirthi, C. Kamaraj, A.A. Zahir, Evaluation of Catharanthus roseus leaf extract-mediated biosynthesis of titanium dioxide nanoparticles against Hippobosca maculata and Bovicola ovis, Parasitol. Res. 111 (2012) 2329–2337. https://doi.org/10.1007/s00436-011-2676-x

[93]   P. R. Anusuyadevi, A. V. Riazanova, M. S. Hedenqvist, A. J. Svagan, Floating Photocatalysts for Effluent Refinement Based on Stable Pickering Cellulose Foams and Graphitic Carbon Nitride (g-C3N4). ACS Omega. 5 (35) (2020) 22411-22419. https://doi.org/10.1021/acsomega.0c02872

[94]   P.R. Anusuyadevi, Synthesis of Novel Nanophotocatalyst in Micro/Millifludic Supercritical Reactor, (2018).

[95]   R. de O. Teixeira, M.L. Camparoto, M.S. Mantovani, V.E.P. Vicentini, Assessment of two medicinal plants, Psidium guajava L. and Achillea millefolium L., in in vitro and in vivo assays, Genet. Mol. Biol. 26 (2003) 551–555. https://doi.org/10.1590/S1415-47572003000400021

[96]   U.K. Parashar, V. Kumar, T. Bera, P.S. Saxena, G. Nath, S.K. Srivastava, R. Giri, A. Srivastava, Study of mechanism of enhanced antibacterial activity by green synthesis of silver nanoparticles, Nanotechnology. 22 (2011) 415104. https://doi.org/10.1088/0957-4484/22/41/415104

[97]   S.K. Basha, K. Govindaraju, R. Manikandan, J.S. Ahn, E.Y. Bae, G. Singaravelu, Phytochemical mediated gold nanoparticles and their PTP 1B inhibitory activity, Colloids Surfaces B Biointerfaces. 75 (2010) 405–409. https://doi.org/10.1016/j.colsurfb.2009.09.008

[98]   D. Hariharan, K. Srinivasan, L.C. Nehru, Synthesis and characterization of Tio2 nanoparticles using cynodon dactylon leaf extract for antibacterial and anticancer (A549 cell lines) activity, J. Nanomedicine Res. 5 (2017) 138–142. https://doi.org/10.15406/jnmr.2017.05.00138

Bioinspired Nanomaterials
Materials Research Foundations **111** (2021) 63-95

Materials Research Forum LLC
https://doi.org/10.21741/9781644901571-3

Chapter 3

Bioinspired Nanomaterials for Drug Delivery

Balaji Maddiboyina[1]*, Jeyabalan Shanmugapriya[2], Swetha Rasala[3], Gandhi Sivaraman[4,*]

[1]NRK & KSR Gupta College of Pharmacy, Tenali, Guntur, Andhra Pradesh-522201, India

[2]Department of chemistry, Madura College, Madurai-62500⁻, India

[3]CÚRAM, SFI Research Centres for Medical Devices, Biomedical Sciences, National University of Ireland, Galway, H92 W2TY Ireland

[4]Department of chemistry, Gandhigram rural Institute-Deemed to be university, Gandhigram - 624302

raman474@gmail.com

Abstract

Over the preceding few decades therapeutic/drug delivery systems were explored and investigated as a tactic to advance the efficiency and safety of therapeutic agents for various biomedical applications. Nano-engineering on the various biomaterials are reported and are under investigation to enhance the pharmacokinetics and pharmacodynamics of many drugs, with proven enhancements in terms of objective facility, therapeutic efficacy, reduction in dosing frequency and associated drug side effects. Bioinspired materials from various sources (biomass, plants, animals, cells, biotechnology interventions) are of great interest with additive advantages over synthetic materials in terms of biocompatibility, biodegradation, nontoxicity, non-immunogenic and are cost effective systems. Bioinspired nano platforms are proceeding round the world to contrive novel drug delivery carriers using different strategies. This chapter encompasses encroachments in the diverse types of bioinspired polymers and their nano delivery systems. Comprehensive evidence is also concise on delivery systems morphological, biological functionalities from respectively material and their potentialities as persuasive carriers for drug delivery systems.

Keywords

Nanomaterials, Nanodelivery Systems, Nanoparticles, Nanotherapies

Contents

Bioinspired Nanomaterials   Materials Research Forum LLC
Materials Research Foundations **111** (2021) 63-95   https://doi.org/10.21741/9781644901571-3

1.   Introduction

1.1   Bioinspired biomimetic materials

Although many therapeutics show high pharmacological activity toward specific pathologies, their use in the native form is often limited by practical reasons. Major drawbacks associated to free drug administration are related to the poor stability, limited biodistribution, low barrier penetration capabilities and lacking targeting properties of active molecules. Cited restrictions have pushed the introduction of nanoscience in medicine, with the development of nano-sized controlled drug delivery systems.

Nanoparticles for drug delivery effectively reduce hurdles of free drug administration, promoting the use of existing drugs, already developed and tested, but unused in free form for their induced side effects, with a significant reduction of costs related to drug discovery. Nanoparticles are homing systems that hide and protect poorly stable active molecules from physiological environment, preserving the pharmacological activity. Additional benefits associated to nanoparticles in drug delivery are related to their customizability and functionalizability, which provide specific properties to address poor bioavailability, issues related to solubility and the off target deposition in healthy tissues. In recent years, the exploitation of nanoparticles, from synthetic or biological derivation, or biohybrid, is significantly widespread, thanks to significant benefits that vectors can provide.

At first, materials for natural derivation were processed to obtain carriers for drug delivery. Natural materials, from plants (e.g., alginates) or animals (e.g., chitosan), were extensively used for their degradability and biocompatibility. Also liposomes were considered valuable carriers to entrap and transport drugs. Liposomes can be obtained from amphiphilic natural molecules by means of well consolidate procedures. Preparation procedures for carriers from natural derivation were optimized for a wide range of starting materials, architectures and final dimensions. Although their functional behaviour was acceptable, natural carriers were successfully improved by introducing chemical modifications and then obtaining hybrid materials.

The development of hybrid materials pushed the discovery of novel and more performant synthetic polymers that were, in consequence, evaluated as starting materials for carrier preparation. Artificial drug carriers, composed of degradable polymers or inorganic materials, are constantly under investigation and achieved good results pushed a number of systems into the market. Natural and artificial particles represent the first generation of drug carriers. The basic concept for the design of first-generation nanoparticles was to provide a homing system for active molecules, to protect therapeutics from the physiological environment and modulate the delivery kinetics.

However, synthetic nanoparticle can be modified and functionalized to improve stealth and targeting properties. Such novel characteristics led scientists to develop more complex systems, with engineered surfaces, decorated with natural molecules or synthetic bioinspired units. Beside synthetic nanoparticles, natural systems derived from pathogens or cells possess suitable properties to improve drug therapies. Biological vectors have intrinsic characteristics enabling long circulation and targeting of specific districts. Natural vectors evolved over years to accomplish their tasks, refining their peculiar characteristics. Scientists have now developed specific protocols to obtain safe carriers from living cells, as well as improved techniques to maximize drug loading without compromise, native carrier properties ghost or living cells are currently under investigation with interesting results. While functional properties of natural and artificial nanoparticles are well known and already optimized with consolidate strategies [1], functional performances of biological systems are not yet well recognized [2]. Moreover, the increasing advances in cell biology allow today to merge beneficial aspects of all classes, to obtain combined biohybrid structures with improved characteristics.

1.2   Nano engineering of biomaterials and their biomedical applications

Nano engineering of nanomaterials with diverse sources are used to accomplish nanocomposite. Carbon-based nanomaterials (carbon nanotubes or CNTs, graphene, nanodiamonds), polymeric nanoparticles (dendrimers and hyper branched polymers), inorganic/ceramic nanoparticles (hydroxyapatite, silicates and calcium phosphate) and metal/metaloxide nanoparticles (gold, silver and ironoxides) are specific instances of these nanomaterials (Figure 1). This integration adds unique properties into hydrogels including the following [3]:

- The biomaterials are principally used for implantation solicitations such as dental implant, mechanical heart valve, intraocular lens, hip, knee and shoulder joints, etc.

Bioinspired Nanomaterials   Materials Research Forum LLC
Materials Research Foundations **111** (2021) 63-95   https://doi.org/10.21741/9781644901571-3

- They remain similarly used to nurture cells in culture, to assay for blood proteins in the clinical laboratory, in equipment for dealing out biomolecules for biotechnological presentations, for implants to normalise fertility in cattle, in diagnostic gene arrays, in the aquaculture of oysters and for investigational cell-silicon "biochips".

- The bioactive ceramic Nano-sized Hydroxyapatite (HA) is available commercially for use in bone replacement applications. HA [Ca10(PO4)6(OH)2] is a synthetic calcium phosphate material that has similarities to the mineral component of human bone.

- The nanosized biomaterials can also diminish the drug-release rate, revealing sustained releasing actions. Decisively, the rational particle size and spherical morphology bequeath with intravenous transport capability. Nano-biomaterials are considerably further seemly for the encapsulation and delivery of hydrophobic therapeutic agents because the carbonaceous framework and also can form specific π-π supramolecular interactions with aromatic drug molecules.

- Photothermal therapy (PTT) has remained broadly pragmatic to indulgence cancer by using NIR-resonant nano-agents, which might absorb the NIR light and transform it into cytotoxic heat to kill the cancerous cells and cut the invasive impairment to normal cells.

- Hyaluronic acid (HA) functionalized undeviating mesoporous carbon spheres (UMCS) were fashioned for targeted enzyme amenable drug delivery disbursing a facile electrostatic attraction approach.

- Gene therapy is assumed to be an operative and benign mode to overwhelm oncogenes and confine the proliferation of intractable tumors through emerging exogenous nucleic acids as therapeutic agents. Polyethylenimine (PEI) revised oxidized mesoporous carbon nanosphere (OP) for combined photothermal and gene therapy.

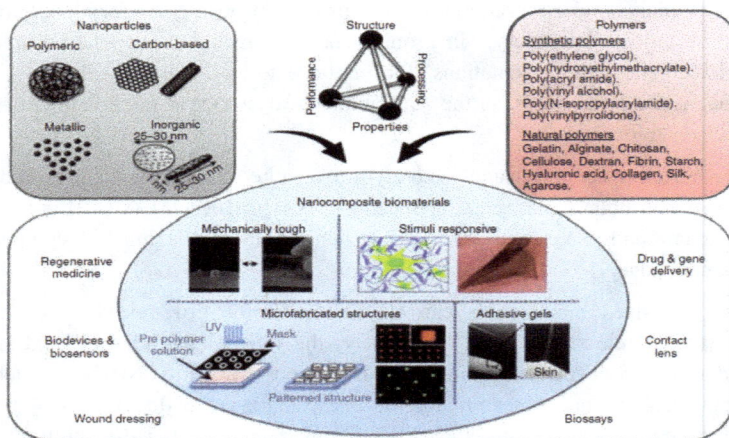

Figure 1 Nanocomposite hydrogels for biomedical applications.

1.3 Advantages of bioinspired materials

Advantages of nanoscience and nanotechnology have overwhelmingly headed to the improvement of functional materials in current years, which obligate institute solicitations extending from biomedical to ecological engineering and high-energy storage as well as gathered benefits in fundamental science. Amongst these efficient nanomaterials are carbon nanotubes (CNTs), graphene, fullerenes, soft, polymeric nanoparticles, metal organic nanomaterials, self-assembled and supramolecular nanostructures, and their results to label a little. Their exceptional physico-chemical properties such as catalytic, dielectric, optical and mechanical give rise to their distinct solicitations in sensors, drug delivery, proteomics and biomolecular electronics. In exact, their biological solicitations have expanded fundamental indulgent of biomolecular systems such as vesicles, viruses and cells as well as enthused the intention of nanomaterials with biological functions. The former ones obligate been frequently called bioinspired nanomaterials [4].

1.4 Advanced bioinspired nanodelivery systems

Inspired by nature's astonishing delicacy, scholars obligate remained scheming nanomaterials with variability of solicitations in biomedicine. The principal intention for the astonishing utilities of these bioinspired nanomaterials stems from the point that the human biological structure is made up of nanoscale self-assembly of biological molecules. There has remained notable evolvement in the earlier decades in the area of biomimics and bioinspired materials such as organs-on-chips, smart robotic devices, a new class of materials that mimic the homeostatic skills of living organisms to acclimatise and self-regulate, and nanomaterials for tissue engineering and orthopaedic implants. Biological models are architectures that execute as vessels such as viral capsids. Explicitly, these biological containers can utility as carriers for DNA assays and immunoassays, drugs, catalysts, and be used in novel material synthesis [5].

2.   Advanced bioinspired nano delivery systems

2.1   Albumin based nano drug delivery systems

Albumin is the furthermost profuse form of serum protein frequently institute in the blood; albumin is diverse from further plasma proteins since it is not glycosylated. Crucial benefits of albumin-based targeting carriers embrace biodegradation, non-immunogenicity, and significant stability over a wide range of pH (4–9) and temperature (10–60 °C). Albumin retains a cryoprotectant outcome, which is practicable for lyophilization of formulations; it also has a long half-life, which is required for prolonged renal clearance. Furthermore, albumin can conveyance a variability of molecules and plays a crucial role in sustaining homeostasis [6]. As such, albumin is deliberated an ultimate carrier component to enrich pharmacokinetic profiles of various drugs. Approaches have remained advanced to synthesize albumin-based drug carriers by binding or conjugating drug cargos to endogenous or exogenous albumin. A paclitaxel bound albumin nanoparticle is a model example for instituting the probable of albumin-based delivery. It binds to the gp60 receptor extant at the cell surface and activates caveolin-1 mediated transcytosis, which also transports some of the unbound plasma constituents. This system can be subjugated for transporting the cargo to the brain via adsorptive mediated transcytosis. Moreover, the albumin-bound paclitaxel had a ~4-fold increase in the cellular uptake of endothelial cells as paralleled to clinical formulation [7].

2.2   Examples of bio-inspired delivery systems in clinical trials

Albumin has turn into one of the utmost imperative drug delivery carriers in cancer therapy [6]. Numerous albumin-based products have prepared it to clinical trials or level commercial claims. For example, INNO-206, an albumin bound prodrug, is enduring Phase I clinical trials for treating sarcoma and gastric cancer. This is an acid-sensitive hydrazone imitative of DOX, retentive high plasma stability in its albumin bound form

Bioinspired Nanomaterials
Materials Research Foundations **111** (2021) 63-95

Materials Research Forum LLC
https://doi.org/10.21741/9781644901571-3

and dropping cardiotoxicity of DOX. Abraxane produced by Celgene is another well-established albumin-based paclitaxel nanoparticle system used for treating solid tumors. This product has confirmed lower toxicity and higher antitumor efficacy than unbound paclitaxel [8].

2.3 Polysaccharide based nano drug delivery systems

Carbohydrates, sometimes called "sugars," are an imperative class of biomolecules found profusely in nature. Monosaccharides, the simplest sugars, are the basic structural units of carbohydrates. These units have three to nine carbons and a distinguishing carbonyl group, which can be moreover an aldehyde, in aldoses, or a ketone, in ketoses. Monosaccharides occur predominantly in cyclic form, and can be linked together via "α" or "β" glycosidic bonds, forming linear or branched chains of oligosaccharides (2–20 units), with the general formula $(CH2O)n$.

Chains with more than 20 monosaccharide residues are stated to as polysaccharides. The generic term "glycan" is frequently used to denote to any oligo- and polysaccharide, either free or covalently linked to other molecules, such as proteins or lipids, in the form of glycoconjugates. The field of glycobiology has been emerging at a prodigious step over the past decades. Though carbohydrates were principally deliberated principally as storage and structural materials, it is currently clear that they reveal a plurality of biological activities [9]. This is in large part accompanying with their excessive diversity, for which quite a lot of factors subsidize. There is an inclusive array of monosaccharides. The ones frequently institute in animal glycans embrace: (i) neutral sugars pentoses and hexoses; (ii) hexosamines hexoses with a free or N-acetylated amino group; (iii) deoxyhexoses hexoses without a hydroxyl group at position (iv) uronic acids hexoses with a negatively charged carboxyl group; and (v) sialic acids family of 9-carbon acidic sugars [10]. Apart from the previously stated adaptations, hydroxyl groups of monosaccharides can also be chemically revised by methylation and esterification (phosphate, acyl, and sulfate esters). Furthermore, the existence of asymmetrical (chiral) carbons in monosaccharides gives rise to diverse isomeric forms, with diverse biochemical properties. To practice higher-order structures, mono-saccharides can be linked organised in many diverse ways, because of the several potential isomers. Bioinspired Materials for Medical Applications that can be designed between two units. The glycosidic linkage can encompass substitute stereoisomers (α or β) at the anomeric carbon of the earlier unit, and the several hydroxyl groups at the ensuing unit consent several potentials of isomerization. The collective existence of diverging also subsidizes to the structural diversity of glycans. This way, provisional on the type of glycosidic linkage, sugar chains of alike alignment can adopt very diverse conformations and

Bioinspired Nanomaterials

Materials Research Foundations **111** (2021) 63-95

Materials Research Forum LLC

https://doi.org/10.21741/9781644901571-3

bioactivities. A classic example is that of starch and cellulose, which are both homopolymers of glucose found in plants, where they play storage and structural roles, respectively. While α1 4 linkages and branching in starch outcomes in helical chains and a added disordered three-dimensional (3D) structure, β1–4 linkages in cellulose result in a straight chain 3D structure, strengthen by interchain hydrogen bonds. These structural adaptations explanation for their quite dissimilar biochemical properties and biological function. The foremost classes of animal glycans include glycosaminoglycans (GAGs) that, with the allowance of hyaluronic acid, occur as proteoglycans and other conjugates such as glycoproteins and glycolipids. Glycans can mediate a wide range of biological developments by virtue of their physical properties, such as charge, molecular conformation, mass or gel-forming ability, and their biochemical function is indomitable by their nanoscale organization. On the other hand, several of the more detailed functions of glycans involve recognition by glycan-binding proteins (GBPs), such as lectins and GAG-binding proteins. In nature, all cells and various macromolecules carry a set of covalently linked glycans. The existence of glycans at the cell surface and in the ECM, place them in optimal station for facilitating a range of processes principal cell–cell, cell–matrix, and cell–molecule interactions, not only within an organism, but also between different organisms. Inspired by their biological roles, different biomimetic materials have been designed using native, modified, and synthetic glycans as building blocks [11].

3.   Design of glycan-based delivery systems

Glycan-based biomaterials, from more unassuming oligosaccharides to more complex polysaccharides, can be engineered with exclusive properties and distinct arrangements, being predominantly engaging for the scheme of innovative drug-delivery systems. In these presentations, glycans from animal, nonanimal, and synthetic origins have been used, often chemically altered to accomplish detailed and consistent physicochemical properties and bioactivity [12]. To meet different needs, glycan based biomaterials have been administered into innumerable shapes, comprising micelles, nano/microparticles, hydrogels, nano/micro-fibers, and porous 3D scaffolds.

The design of glycan-based drug-delivery systems exploits unlike properties of this extremely diverse family of natural compounds. Provisional on the type of carbohydrate, some key properties may embrace (i) gel-forming facility, frequently used in the advance of matrix-type drug carriers; (ii) hydrophilic nature, which can be reconnoitred to enrich the circulatory half-lives of diverse types of drugs; (iii) polyelectrolyte nature, to encourage bottom up nano-assembly and/or complexation between glycans and drugs of opposite charge via electrostatic interactions; (iv) bioadhesiveness, recurrently subjugated as a means to increase drug retaining at certain localities, namely at mucosal surfaces;

and (v) affinity for GBPs, frequently engaged in the strategy of targeted carriers, such as glycan-decorated particles. There are also sugar-based compounds with very precise properties [13]. This is the case of cyclodextrins, a family of cyclic oligosaccharides made up of glucose monomers bound organised in a ring. Given the exceptional nature divulged by their structure, where the interior is substantially less hydrophilic than the exterior, cyclodextrins are able to form host–guest complexes with hydrophobic molecules, improving their solubility, physical chemical stability, and bioavailability. As such, these compounds have newly found a large number of solicitations in the drug-delivery field. Hydrogel-based matrix-type drug carriers have been gaining accumulative acceptance as ECM mimics for regenerative medicine and tissue engineering. In the ECM, gel-forming glycans provide hydration, structural stability, and selective permeability [14]. They are also elaborate in explicit, noncovalent binding of several endogenous molecules,

4. Hyaluronic acid

Hyaluronic acid (HA) is a natural, linear, endogenous polysaccharide that plays imperative physiological and biological roles in the human body. Currently, among biopolymers, HA is emergent as an alluring starting material for hydrogels proposal due to its biocompatibility, native biofunctionality, biodegradability, non-immunogenicity, and versatility. Since HA is not able to form gels alone, chemical adaptations, covalent crosslinking, and gelling agents are constantly desired in order to acquire HA-based hydrogels. Therefore, in the last decade, unlike approaches for the design of physical and chemical HA hydrogels have been advanced, such as click chemistry reactions, enzymatic and disulfide crosslinking, supramolecular assembly via inclusion complexation, and so on. HA-based hydrogels turn out to be adaptable platforms, extending from static to smart and stimuli-responsive systems, and for these reasons, they are extensively probed for biomedical applications like drug delivery, tissue engineering, regenerative medicine, cell therapy, and diagnostics. Additionally, the overexpression of HA receptors on many tumor cells makes these platforms favourable drug delivery systems for targeted cancer therapy. The aim of the present chapter is to highlight and confer recent improvements made in the last years on the design of chemical and physical HA-based hydrogels and their solicitation for biomedical purposes, in precise, drug delivery. Prominent devotion is given to HA hydrogel-based drug delivery systems for targeted therapy of cancer and osteoarthritis [15].

Bioinspired Nanomaterials
Materials Research Foundations **111** (2021) 63-95

Materials Research Forum LLC
https://doi.org/10.21741/9781644901571-3

5. Keratin

Keratin derives from the Greek word "kera" which means a horn. In the 1950s, the word keratin first looked in the literature to define a substantial made up of hard tissues. In 1905, a United States patent was allotted labelling the process of keratin extraction from animal hooves with the help of lime. Since then, much research -based methods have been advanced with the aim of extracting keratin using primarily oxidative and reductive methodologies [16]. Primarily, these skills were pragmatic to extract keratin from animal-based sources such as horns, hooves, chicken feathers, and finally human hairs. These keratin-rich sources are challenging to vitiate as the polypeptide in their edifice is closely packed in α-helix (α-keratin) or β-sheet (β-keratin) into supercoiled chains which are toughly stabilized by numerous hydrogen bonds and hydrophobic interactions, in adding to the disulfide bonds, hooves, and horns [17].

In nature, the chicken feather is conceivably the greatest profuse and easily existing keratinous biopolymer. A keratin-based chicken feather from butchery accounts for more than 5 million tons per year global in the form of waste material [18]. Apart from its insignificant custom in low-grade products such as glue, corrugated paper, cardboard, animal feed, fertilizers, etc., the landfill disposal of poultry feather poses a substantial ecological and ecological threat. On the other hand, from the economic and ecological point of outlook, it is also anticipated to create an operative practice for the solicitation of such abandoned natural resource. The existence of multifunctional groups in keratin, such as disulfide, amino, thiol, phenolic, and carboxylic, makes it reactive under appropriate reaction situations. In a falling atmosphere, the amino and some of the further groups declared above in the keratin make its surface positive, and thus solubilization takes place. With its exceptional properties of biodegradability and nontoxic nature, keratin is amid the adaptable biopolymers that can be improved and advanced into several products of interests. Thus, a prominent amount of evidence on the features and hydrolysis of keratin has developed existing where intractable keratinous wastes are renewed into respected products [19]. The awareness on keratin-rich litters has been toughly increased and their use in cosmetics or in medicines to enrich drug delivery, and production of biodegradable films, are amid the principal and emergent biotechnological and biomedical applications [20].

In recent years, the emergent research comforts in the advance of keratin-based biomaterials are principally due to the exclusive properties of keratin that play a precarious role in the fabrication practice. Over the past few decades, exclusive physiochemical and biological characteristics of keratin have been emergent as factors for research based on biomaterials. To date, adequately of systematic work has been done and available by some experts on the advance and depiction of keratin and keratin-based novel products, for instance, keratin-based composites/blends, (hydro)-gels, thin films, nano- and microparticles, and 3-D scaffolds are of supreme prominence. In various cases, the aforesaid novel keratin-based constituents are publicised to retain multifunctional faces along with exceptional compatibility sorts. Newly, we have revealed novel enzyme-based approaches for modulating the physiochemical, thermomechanical, and biological features of keratin in order to advance keratin-based bio-composite materials that have applicable structures for their probable solicitation of interest [21]. Later on, lactase was engaged as a green catalyst to advance natural phenol embedded keratin-EC-based biocomposites via the surface dipping and fusion technique. This work verified that the range of various natural phenols that Stimuli Responsive Polymeric Nanocarriers for Drug Delivery uses embraces caffeic acid, gallic acid, p-4-hydroxybenzoic acid, and thymol could exclusively regulate the antibacterial probable of these newly established novel bio-composites. Figure 2, exemplifies a graphical depiction of advance and unique features of phenol-g-keratin-EC-based bio composites. Correspondingly, Verma and co-workers had reported construction, depiction, and biocompatibility of human hair-based keratin scaffolds for in vitro tissue engineering solicitations [22].

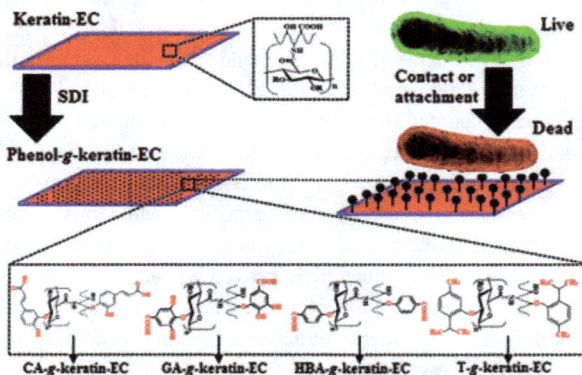

Figure 2 Development and novel characteristics of Phenol-g-Keratin-EC-based bio-composites.

6.   Cellulose

Cellulose being the first plentiful biopolymers in nature has many enthralling properties, counting low-cost, good biodegradability, and exceptional biocompatibility, which made

Bioinspired Nanomaterials   Materials Research Forum LLC
Materials Research Foundations **111** (2021) 63-95   https://doi.org/10.21741/9781644901571-3

cellulose a real probable substantial to generate nano-drug delivery systems (nano-DDS). In recent decades, cellulose has been expansively explored due to its approving properties, such as hydrophilicity, low-cost, biodegradability, biocompatibility, and non-toxicity, which marks it a good feedstock for the synthesis of biocompatible hydrogels. The plentiful hydrophilic functional groups (such as hydroxyl, carboxyl, and aldehyde groups) in the backbone of cellulose and its spinoffs can be used to concoct hydrogels easily with attractive structures and properties, prominent to burgeoning research interest in biomedical requests. By probing the research literatures over last decade, an assortment presented studies on cellulose based nano-DDS were brief and alienated into prodrugs, prodrug nanoparticles, solid or derived nanoparticles, amphiphilic copolymer nanoparticles, and polyelectrolyte complex nanoparticles [23].

Many types of cellulose-based nano-DDS can confirm proficient encapsulation of various drugs and then overawed the free drug molecule flaws. Among all the process designated, cellulose based amphiphilic nanoparticles are most recurrently used. These formulations have the higher drug loading capability, a simple and stretchy way to realise multifunctional. Apart from hydrophilic or hydrophobic reform, cellulose or its spinoffs can form nanoparticles with diverse small molecules and macromolecules, foremost to a large spectrum of cellulose-based nano-DDS and so long as some startling benefits. Thorough physicochemical portrayal and reflective indulgent of interactions of the cellulose-based nano-DDS with cells and tissues is vital. Furthermore, studies near technics constraint optimization and scale up from the laboratory to production level should be assumed. The advance of intravenous and orally pertinent cellulose-based nano-DDS will be an imperative research area, and these systems will have more viable status in the market [24].

Cellulose-based hydrogels are resultant from natural sources which are biodegradable and low-immunologic. These hydrogels are produced in four different ways: those acquired directly from native cellulose, those imitative from cellulose derivatives (methyl cellulose, carboxymethyl cellulose, hydroxy methyl cellulose, etc.), those acquired with other polymers as a fused, and lastly those gained from cellulose-inorganic hybrids. Cellulose hydrogels and its spinoffs have many desired properties such as high water retaining capacity, high crystallinity, fine fiber network, easy formability, and high tensile strength. In addition, some cellulose spinoffs display able performance against physiological variables such as pH and ionic strength. Cellulose-based hydrogels have gains such as better biocompatibility, less latent toxicity, and lower cost than the utmost synthetic polymer hydrogels. Because of these benefits, cellulose-based hydrogels are desired to be used in industrial pharmaceutics and biomedical fields [25].

7. Chitosan

Chitosan is attained from the deacetylation of chitin, and it is a copolymer of N-acetylglucosamine and d-glucosamine. Distinct chitin, chitosan is soluble, and its properties make it easy to handle and accomplish [26]. Both chitin and chitosan have been extensively used in biomedical solicitations with exact emphasis on drug delivery potentialities. Chitin has exposed highly hydrophobic properties caused by the N-acetylglucosamine polymeric structure which makes it a hard material, but it also has admirable electric properties which may be pragmatic to tissues requiring electrical conductance. Besides being a soluble polymer, chitosan presents novel features such as high biodegradability and biocompatibility, nonantigenicity, good adsorption properties, nontoxicity, and bio-functionality. The above stated characteristics of chitosan play a vital role in emergent smart therapeutic and health-related drug delivery systems. Likewise, it can be pragmatic to engineer novel carriers combining the polymer with carbon nanotubes, which increase the electrical conductivity of the scaffold [27].

8. Polyhydroxyalkanoates (PHAs)

Polymers that are formed by biological systems such as plants, animals, or microorganisms through metabolic-based engineering reactions are called natural polymers. Instances of biopolymers comprise carbohydrates, for example, cellulose and starch, proteins, for example, keratin and enzymes, and polyhydroxyalkanoates (PHAs), for example, poly-(3-hydroxybutyrate) [P(3HB)] [28]. A wider spectrum of such materials has been categorized consequently as natural or synthetic based on their nature of origin. The structure of a biopolymer distresses its functional characteristics where the functional capability is frequently reliant on the crystalline and amorphous nature of the materials. For example, cellulose or poly(β-d-glucose) is a structural polymer whose properties arise in part from its crystalline nature. Nevertheless, physiochemical and biological dealings can transform it into a beneficial structural material for many probable solicitations. Also, chemically revised celluloses, for example, ethyl cellulose and cellulose acetate, are found in a wide range of presentations. Modified cellulose is used in the manufacture of paint, plaster, adhesives, cosmetics, and pharmaceutical film coating and numerous other products. A family of compounds known as PHAs has acknowledged much courtesy as bio-sustainable materials because they are formed more certainly in extent by fermentation of carbon–rich substrates using microorganisms, particularly bacteria. Among the utmost likely and well characterized biopolymers, P(3HB) is of exact interest for the preparation of bio-based composite materials. There have been more than 150 monomers notorious as elements of PHAs [29]. The unhinged

Materials Research Forum LLC
https://doi.org/10.21741/9781644901571-3

nutritional supply origins the bacteria to collect PHAs in the form of granules as internal energy storage, as shown in figure 3.

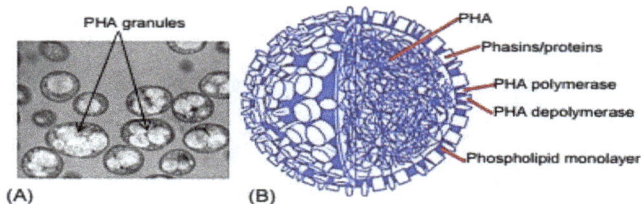

Figure 3: (A)-PHA granules, (B)- Schematic representation of PHA granules

9. Nucleic acid based nanodelivery systems

The enhanced appreciative of the genetic roots of plentiful diseases going hand in hand with the achievement of the human genome project opened the door for the discovery of novel therapeutics precisely curbing the expression of disease-relevant genes. Largely, this therapeutics can be considered into viral and non-viral formulations. The non-viral ones offering the overlook of eluding oncogenic risk and of dealing hypothetically larger payloads is either plasmid DNA (pDNA) encoding for therapeutic proteins such as GLP-1 or insulin and in case of DNA vaccines for antigens or RNA-based drugs with antisense oligonucleotides, short interfering RNA, microRNA, messenger RNA, and Aptamer. In disparity to all other drugs they can edit genes curing accordingly genetic defects. Additionally, they can be used to shut off assured gene expression. In the case of certain gastrointestinal diseases, such as inflammatory bowel disease (IBD) and colon cancer, where current drug cures are scant such gene-based therapeutics may require major clinical benefits [30].

Gene therapy with RNA and pDNA-based drugs is inadequate by poor enzymatic stability and poor cellular permeation. The provision of nucleic acids, in precise by the oral route, rests a major hurdle. These will emphasis on the barriers to the oral delivery of nucleic acids and the plans, in exact formulation strategies, which have been advanced to daze these barriers. Due to their very low oral bioavailability, the most noticeable and most explored biomedical claims for their oral delivery are related to the local treatment of inflammatory bowel diseases and colorectal cancers. Preclinical data but not yet clinical studies sustenance the potential use of the oral route for the local delivery of formulated nucleic acid-based drugs [31].

10. Lipid based nanodelivery systems

Lipids are a large class of materials that embraces fatty acids, glycerides, phospholipids, sphingolipids, waxes, and sterols. These compounds are frequently insoluble in water, or Design and preparation of biomimetic and bioinspired materials amphiphilic, and are recognised by their fatty-acid alignment, melting point, hydrophilic– lipophilic balance, and solubility in organic solvents. Lipid-based systems have expanded much interest in the modern years for drug-delivery resolves primarily due to their facility to expand the solubility and bioavailability of drugs with poor water solubility. Nevertheless, lipid-based systems have also verified superior ability for hydrophilic drugs, tailoring the release profile of the active contents in a biofunctional manner [32]. The most substantial dosage forms are liposomes, solid lipid nanoparticles (SLN), nanostructured lipid carriers (NLC), and self-emulsifying drug-delivery systems (SEDDS).

Liposomes are spherical vesicles poised of bilayers of phospholipids, cholesterol, and/or other lipids. Lecithin, phosphatidylglycerol, phosphatidylinositol, phosphatidylethanolamine, and phosphatidylserine are the chiefly used phospholipids. They can be categorized bestowing to their lamellarity as uni, oligo, and multilamellar, or by size as small, midway, and large. Due to its structure, they permit the assimilation of hydrophilic drugs in the aqueous core, and lipophilic drugs within the lipid bilayer as shown in figure 4. Retaining higher core, unilamellar liposomes are desired for encapsulation of hydrophilic drugs, while multilamellar liposomes are specifically used to encapsulate hydrophobic drugs due to the higher lipid content. Reliant on the number and composition of the bilayers and the incidence of coating, it is probable to acquire systems with modified release characteristics. Besides the marketed formulations, liposomes have been advised for administration of numerous drugs, comprising peptides and therapeutic proteins, as well as for gene therapy). Stealth liposomes like Doxil/Caelyx, Novantrone, or Lipoplatin are commercially existing examples of second-generation liposomes, surface-decorated with PEG moieties, ensuing in advance of blood circulation time and the therapeutic efficacy of many drugs through the evading of opsonisation, that is, exclusion by immune cells, and drip from reticuloendothelial system [33].

11. SLNs and NLC

Lipid nanoparticles mostly embrace two types of structures, SLN and NLC. They contain a solid lipid matrix, usually vastly refined triacylglycerols, complex acylglycerol mixtures, and even waxes, at room and body temperatures, dispersed in aqueous solution and stabilized with a layer of emulsifier agent, usually phospholipids. Lipid nanoparticles began as an substitute to liposomes because of the superior stability in biological fluids.

Bioinspired Nanomaterials
Materials Research Foundations **111** (2021) 63-95

Materials Research Forum LLC
https://doi.org/10.21741/9781644901571-3

They are colloidal carriers made of nontoxic [34], biodegradable and well-tolerated solid lipids dispersed either in water or in an aqueous surfactant solution. The lipid composition can delay degradation by impeding the quay of enzyme complexes, qualifying their probable for controlled drug delivery. Additional benefits of SLN are their particulate nature, facility to integrate both hydrophilic and hydrophobic drugs, the evading of organic solvents in the production procedures and the prospect to produce vastly concentrated lipid suspensions, lower cytotoxicity, and scale-up probability. Hydrophobic anticancer drugs such as camptothecin, anti-rheumatics such as methotrexate, or immune-suppressants as cyclosporine have been encapsulated into SLN for modulating there *in vivo* biodistribution and target drugs for their local of action. SLNs are also able to compress biopharmaceutical drugs with high aqueous solubility, retentive their structure after encapsulation and even freeze-drying. Therapeutic proteins or genetic material are presently formulated into solid lipid matrix, ensuing in biocompatible and environment-friendly conditions to alleviate those biological [35].

A second generation of lipid nanoparticles is the so-called NLC. These particles are primed not from a solid lipid but from a composite of solid and liquid (oils) lipids, which must be solid at least at 40°C. The foremost variance amid SLN and NLC is that the latter are articulated by nano-structuring the lipid matrix to increase drug loading and avoid drug expulsion. Using spatially diverse lipids leads to larger distances between the fatty acid chains of the acylglycerols and broad limitations in the crystal, on condition that more room to lodge drugs. The utmost drug loads could be accomplished by mixing solid lipids with small amounts of liquid lipids. Several drugs show a higher solubility in oils than in solid lipids, thus they can be dissolved in the oil and still be dwindling from degradation by the adjacent solid lipids [36].

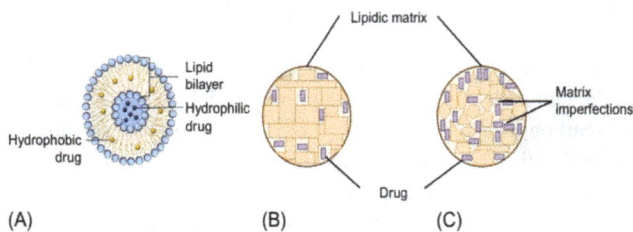

Figure 4 Schematic representation of (A)-liposome, (B)-SLN and (C)-NLC.

12. Peptide based nanodelivery systems

Peptides offer numerous benefits as building blocks for the intention of drug-delivery systems. They are endogenous molecules, which diminishes the risk of contrary effects; are poised of nonpolar, polar, or exciting amino acids, consenting a assured level of expectation of the self-assembly properties complete the meticulous medley of the peptide structure; may range from short to long and more stretchy chains assisting the assembly of structural diverse schedules, from solid crystals to soft disorderly materials; and are rather easy to synthesize and in some cases (short peptides) are commercially presented at practical prices. Moreover, the significant physicochemical properties of peptides enable them to initiate responsive materials to impetus such as temperature, pH, or the presence of specific molecules [37].

Short peptides have been vastly used in the tuition of both crystalline and of soft materials. Diphenylalanine occurred as probably the utmost adaptable, with presentation in nanoelectronics, tissue engineering, or as a model to inspect the molecular mechanisms of protein aggregation in amyloidogenisis. Numerous hydrophobic dipeptides, comprising diphenylalanine, are capable to create microporous crystals fashioned by hydrogen bond-induced head-to tail rally of dipeptides into helical arrangements. The crystal frameworks classically encompass 1D channel with a diameter of 3–10 Å. They were tested as adsorbents of numerous gases such as Xe, CO_2, CH_4, H_2, Ar, N_2, and O_2) and may find exciting biomedical presentation in the delivery of gasotransmitter molecules. Dipeptides were also fused in hybrid metal–organic materials performance exciting adsorption properties. Oligopeptides may self-assemble into numerous dissimilar architectures with perhaps even higher biomedical interest than microporous crystals, such as tubes, rods, fibrils, spheres, vesicles, and gels. Nanovesicular structures have been extremely probed for the delivery of hydrophobic drugs [38]. Also 3D cultures of nerve cells, endothelial cells, and chondrocytes were previously fruitfully shown in decidedly hydrated short-peptides-based scaffolds, like, for specimen, from the simple amphiphilic building blocks involving of dipeptides linked to fluorenylmethoxycarbonyl (Fmoc, roughly used as a protective group in peptide chemistry). The unification of small organic moieties or of unnatural amino acids has been recurrently used to normalise the physicochemical properties and to surge the proteolytic and thermal stability of these peptides. The properties of the final material are also overseen by the experimental disorders through the self-assembly. For request, a peptide amphiphile (PA) hydrogel produced by two diverse triggers, HCl and $CaCl_2$, ensuing in gels with parallel structure but bizarrely different viscoelastic properties. Whereas $CaCl_2$ produced a stronger gel with tighter inter- and intrafibril crosslinks, HCl encouraged a more flexible structure capable of hastily improving its shape after

Bioinspired Nanomaterials   Materials Research Forum LLC
Materials Research Foundations **111** (2021) 63-95   https://doi.org/10.21741/9781644901571-3

distortion. It is still a major encounter to design a novel peptide-based material with determined properties. Several design approaches are being advanced, mainly relating either the production of PAs or the presentation of the familiarity gained from protein secondary structural subjects, such as α-helix and β sheet. Though a greater number of revisions have concentrated on peptides folding into β-sheets, probably as a result of the high research motion on amyloid-like structures, α-helical folding has newly been unloading increasing attention [39]. This fact could be linked to the exact set of rules that have been customary for the assembly of α-helices that can lead to a rational molecular design.

13.   Bacteria/ Viral-based delivery systems

Bacterial cells enjoy exclusive characteristics that make them ideal contestants for cancer therapy. They can lance deep into remote tumor regions and colonize hypoxic and necrotic regions. They also expose evidence about the state of tumors and the ability of treatment because they are superficially noticeable. Some forms of bacteria such as Bacillus, Bifidobacterium sp., Listeria sp., Salmonella sp., Mycobacterium, and Clostridium are known to act as anticancer agents. Bacteria species can be added altered genetically to rally their properties and exploited as vehicles to deliver drugs, proteins, enzymes and genes for the management of cancer, with various cases reaching several phases of clinical trials, hence assembly them superb carriers for the production and targeted delivery of therapeutic cargos into cancer tissues. The exciting approach to genetically engineer bacteria with a drug release switch was used as a way to concurrently control the bacteria's population growth and simplify drug delivery [40].

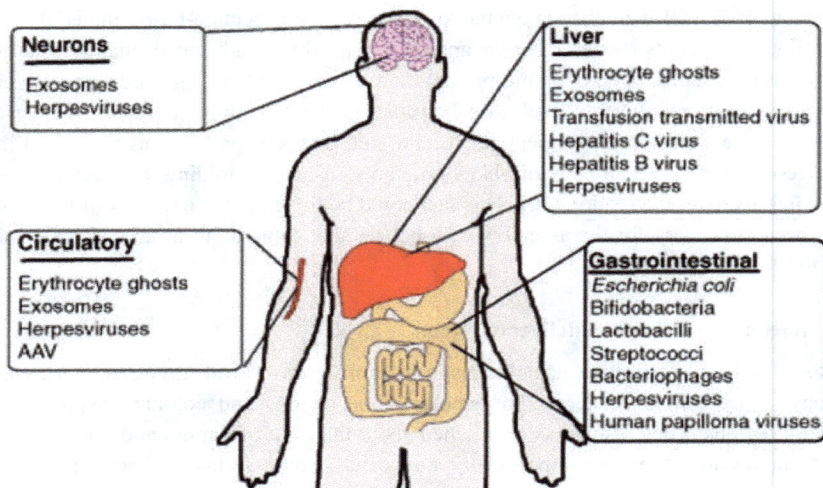

Figure 5 Examples of potential biological vehicles tolerated naturally in selected organs.

Bacterial ghosts are the utmost mutual form of cellular cloaks and are distinct as substrates imitative from Gram-negative bacteria lacking of genetic material. They are produced by controlled expression of the cloned lysis gene E from bacteriophage species. The use of bacterial ghosts in its place of live bacterial cells for drug delivery has some further benefits, so they have fascinated much consideration over the years. BGs do not inhabit the vibrant organs of the body, later reducing the risk of undesirable side effects; they are vastly stable and can realm their surface structures, thus retaining their immunomodulation capacities. Microbots Bacteria can also be used concurrently with nanoparticles to deliver therapeutic cargos into cells. Bacteria carry the drug on their surface conjugated to nanoparticles; hence the bacteria do not require genetic engineering for the delivery process. This practise takes lead of the hostile properties of bacteria, and this type of bacteria are known as microbots, which have the probable to selectively colonize the hypoxic areas of tumors that cannot be cured by predictable chemotherapeutic drugs. α-Helix is a key secondary structure of natural proteins that entails of a peptide chain coiled into a right-handed spiral conformation and stabilized by hydrogen bonds amid the N-H and the C=O groups in the backbone [41]. Methionine, alanine, leucine, glutamate, and lysine have distinct proclivity to be part of α-helix edifices while proline and glycine have poor helix-forming proclivities. A principally profuse α-helix-based structural motif is the coiled coil, in which the α-helix is regularly considered by a seven residue echoing unit of flashing hydrophobic and hydrophilic residues, often signified as (*abcdefg*)*n*.

Coiled coils have been used for drug delivery isolated or merged in liposomes and for the intention of supramolecular materials. Coiled coils display an inner hydrophobic core that can be discovered to convey hydrophobic drugs. The probable of loading cisplatin, a hydrophobic chemotherapeutic drug, into a right-handed coiled ccil (RHCC). RHCC encompassing the drug was able to bind and enter cells in vitro. Unsurprisingly stirring coiled coils, such as the leucine zipper, led to the gratitude of classification necessities for the congress of these structures [42].

Moreover, temperature receptive materials were previously designed Banwell et al., [41] by trading amino acids at this same fringe region; in one case they assimilated alanine to stimulate hydrophobic interactions between fibrils and in alternative by glutamine to adoptive hydrogen bonding. In both cases physical hydrogels were acquired, with the exactitude that glutamine-based gels were bent at low temperature although alanine-based gels were accomplished at high temperature. Thermo-responsive coiled-coil peptides were also inserted in liposome membranes to allow greater control over the release of fenced compounds in retort to temperature. In a recent work planned a supercharged coiled coil structure bearing numerous arginine residues that was magnificently complexed with plasmid DNA and encapsulated it in a liposome for gene therapy [43].

14. β-Sheet

β-Sheet is the further form of secondary edifice present in proteins and comprises of β-strands allied edgewise by backbone hydrogen bonds and arranged in a parallel or antiparallel fashion. Much like the α-helix, β-sheets can be made amphiphilic to simplify the design of design guidelines. Since in β-sheets, the side chains consecutively stick out of the plan in opposite directions, the HPHPHP outline forms β-sheets with a hydrophilic side and a hydrophobic side, which instinctively self-assemble. One of such rallies (RADA16, where R stands for arginine, A for alanine, and D for aspartic acid) is now promoted, mostly for exploration resolves, below the commercial name Pura Matrix. Numerous studies have engaged these hydrogels for cell culture, signifying it's impending for tissue engineering presentations [44]. It was also revealed that RADA16 is seemly for the slow delivery of proteins; the releasing kinetics is reliant on the size and charge of the macromolecules but it is also a function of the peptide density in the gel. By hosting a phenylalanine residue on the RADA16 structure, purposefully fashioned a motif for interaction with hydrophobic drugs. This group probed two peptide systems, RADAFI and RADAFII, the hydrogels were revealed to ensnare molecules comprising the phenyl group, ostensibly by π–π interaction, provided that another ground confirmation of the probable of these materials for drug delivery.

15. Peptide amphiphiles

Peptide amphiphiles (PAs) are double character molecules whose self-assembling mechanism is similar to that of phospholipids in cell membranes. To intention a PA, a hydrophobic important domain usually in the form of a polymer or alkyl chain, or less habitually, a structure of nonpolar amino acids-linked to hydrophilic peptides. When placed in aqueous environment such amphipathic character molecules tend to accumulate into supramolecular architectures such as spherical or cylindrical micelles. The facility to encapsulate hydrophilic molecules has previously been revealed by van Hell et al. [45] who testified the design of numerous PAs, which self-assemble into vesicles. In addition to on condition that a proficient carrier situation, these systems also extant the improvement of consenting a fine control of the properties of the muster surface by judicious selection of combining amino acids. The adaptability by combining lysines in the plan of a PA molecule to confer pH-responsiveness. The PAs self-assembled into micelle, ensnaring DOX, an antineoplastic drug that is released when placed in acidic environments due to electrostatic repulsions between the protonated lysine molecules. Other effective drug carriers based on PAs have been fashioned and a composing of carriers based on peptide self-assembly.

16. Virus-inspired drug delivery systems

Viruses are small infectious microorganisms entailing of nucleic acid molecules in a protein coat. They have been freshly used as vehicles for targeted delivery as they have the facility to transferal their genes into the host cells for imitation. Conjoining targeted viral vectors with drugs crafts an elevated anticancer effect since it synergizes the sympathetic aspects of both constituents. This type of combinatorial system epitomises an inventive slant for the design of tumor-targeted nanoparticles. Instances of viruses frequently used as viral vector systems comprise adenoviruses, adeno-associated viruses and retroviruses [46]. It was revealed in a revision that nanoparticles could be explicitly conjugated to diverse adenovirus capsid proteins and targeted to tumor cells. It was also perceived that these NP-labeled Ad vectors revealed the same level of targeting and infection proficiency to tumor cells as the unlabeled Ad vectors. For this purpose, it was resolved that Ad vectors can serve as a platform for the choosy self-assembly and targeted delivery of NPs to the target cells. By relating nanotechnology with gene concepts, it is probable to intention a multifunctional nanoscale device for cancer treatment. In alternative study, hyperthermia-inducing gold nanoparticles were involved

Bioinspired Nanomaterials   Materials Research Forum LLC

Materials Research Foundations **111** (2021) 63-95   https://doi.org/10.21741/9781644901571-3

to adenoviral vectors via m-covalent conjugation. These nanoparticles were engineered to target a tumor-associated carcino-embryonic antigen without mutable the infectivity of the viral vectors. There is a thorough discussion on the attractive virus-inspired delivery systems, which are reflected very innovative.

Virus-like particles (VLPs) are exceedingly orbicular self-assembled capsids consequent from viruses. VLPs are made of vacant shells lacking of genetic material and they are non-infectious. They are, nevertheless, adept of arriving target cells and can be used to deliver therapeutic cargoes such as peptides, antigens, and anticancer drugs. VLPs are anatomically stable and accepting to employment; their production development can be concluded at a low cost, thus well allocation as a carrier for drug molecules or building blocks for novel nanomaterials. In a study, paclitaxel was conjugated to VLPs that were derived from the bacteriophage MS2; paclitaxel did not conciliation viral capsid functionality [47].

Virosomes are a class of viruses that comprise a cohesive glycoprotein with a vacant inner compartment and they are also identified as re-formed viral particles. Virosomes are often created by solubilizing influenza virus with detergent and then reconstituting it with two influenza envelope glycoproteins: neuraminidase and haemagglutinin. These glycoproteins are liable for the structural homogeneity and stability, targeting, receptor-mediated endocytosis and endosomal escape after endocytosis of virosomes. One foremost benefit of virosomes as carrier systems is their talent to protect pharmaceutically active substances from proteolytic degradation and low pH within endosomes, submissive the satisfied to stay intact when attainment the cytoplasm. Freshly, an erythro-magneto-haemagglutinin virosome was premeditated for the delivery of the therapeutic formulation of decitabine. It was detected that this system explicitly elated the drug into tumor tissues and convinced tumor mass reduction in xeno-graft models of prostate cancer at a lower concentration than the therapeutic dose prerequisite by the free drug [48].

Virus-mimicking particles have over modest and conventional delivery carriers in terms of drug retention and targeting. Viruses are filamentous, and as such, filomicelles were synthesized to mimic the morphological features of viruses [49]. Filomicelles are serene of self-assembling amphiphilic block copolymers and can excellently evade the reticuloendothelial system, principal to a considerably longer blood circulation time in vivo. Earlier research reports perceived that paclitaxel loaded in filomicelles shrank tumors more excellently than the drug solution. The shape and flexibility of filomicelles may also impact their circulation and targeting competency as a capable as drug delivery carrier, while mechanisms for *in vivo* efficacy are being inspected. Virus-mimetic nanogels that are pH-sensitive have been synthesized. They mimic the structural and

Materials Research Forum LLC

https://doi.org/10.21741/9781644901571-3

functional characteristics of viruses and involve of a hydrophobic core and two layers of hydrophilic shells with tumor-targeting ligands. In a study, DOX was elegantly loaded into the hydrophobic core of a virus-based expedient. To mimic the capsid-like structure of viruses, PEG was crosslinked to the core polymer, and bovine serum albumin was assured to the other end of PEG. These formed particles were pH-sensitive as such a pH reduction from a physiological to an endosomal level would convince alterable swelling and a particle size increase of the nanogels. The conversion aided the endosomal escape and release of DOX into the cytosol. Once tumor cells were executed by DOX and split, nanogels would move to neighbouring cells and repeat the cycle, exciting the infection replication of viruses. The talent of pH-sensitive nanogels to escape from endosomes and unremittingly infect adjacent cells is also capable. The above approaches are in infant stages and their aptitudes and mechanisms entail added explication, which determines that radical bioengineering afar simple surface reform of viruses into synthetic provision carriers and uses of these virus-mimicking particles in targeted drug delivery would absolutely extend the clinical choices of cancer therapy in the near prospect [50].

17. Mammalian cell-based drug delivery systems

Mammalian cell-based delivery systems have assembled accumulative devotion, due primarily to their facility to sham many natural properties exposed by their source cells. By merging synthetic NPs with changed cell types such as leukocytes, platelets, and red blood cells, it was potential to cultivate a range of cell membrane-cloaked nanosystems with exceptional features and functions. In a nut shell, living cells not only offer rousing probable in novel drug delivery, but also assist a better indulgent of natural tools upon drug admin [51].

18. Erythrocytes (RBCs)

Red blood cells or erythrocytes are springy and oval biconcave disks with an average diameter of 7–8 μm. They exist as the most profuse cells in mammals, and are the primary transporter of oxygen in human body. RBCs retain high biocompatibility and ample biodegradability without crafting toxic products in vivo, and when equated to synthetic carriers, they exhibit unique features such as sustained circulatory half-life (~120 d in humans and ~40 d in mice) and the facility to exchange certain prodrugs due to enzymatic existence, thus interim as active transporters. In addition, RBCs have vacant volume for drug encapsulation. They also afford ample coating space to bestow diverse carrier functionalities. Some practices can easily alter RBCs without varying their biological properties [52].

Bioinspired Nanomaterials   Materials Research Forum LLC
Materials Research Foundations **111** (2021) 63-95   https://doi.org/10.21741/9781644901571-3

RBCs are acquired from altered mammalian species, and the isolated blood is collected into heparinized tubes by venipuncture formerly being used as carriers. Loading RBCs with the chosen drug molecules can be done during several key means counting electroporation, molecule endocytosis, and hypo-osmotic swelling, which is then monitored by resealing and bestowing cell-penetrating peptides in most cases. The modern design may also employ RBC membranes as a coating for polymer nanoparticles, which has been attested to be popular and hopeful. RBCs will advanced an emergent approach for targeted delivery of anticancer drugs in the near imminent. In a study, Aryal et al. [53] advanced RBC membrane cloaked polymeric nanoparticles (RBCm-cloaked NPs) to deliver DOX. This system seemingly delivered a pooled benefit of both a long circulation epoch and controlled drug release ascribed to the red blood cells and polymeric particles, separately. Two approaches to load DOX into RBCm-cloaked NPs were paralleled in this study, and the results revealed that RBCm-cloaked NPs could potentiate great assurance as a drug-delivery platform in handling diseases such as blood cancer. In alternative study, a RBC-based micromotor was considered to load an imaging agent, CdTe quantum dot (QD), and DOX, to exhibit the coupling of both therapeutic and imaging modalities within a single vehicle. The multi-cargo-loaded, RBC-based micromotors were primed by concurrently loading water-soluble CdTe QD nanocrystals, DOX and iron oxide magnetic nanoparticles into RBCs using a hypotonic dilution based encapsulation method. The outcomes of this study specified that RBC micromotors were adept of concurrently carrying multiple functional cargos with a minimal injurious effect on its dynamic propulsion behaviour and compatibility, which would consequently expand drug delivery efficacy and disease monitoring. It was determined from this revision that a RBC micromotor could offer a novel platform for concurrently imaging a disease, providing an agent and observing therapeutic retort in future theranostic presentations. Adaptation of RBCs thru the drug loading practice is foreseeable; nevertheless, this strategy may hasten the abstraction of delivery carriers by reticuloendothelial cells. Moreover, unlike constraints such as the source of blood, the apparatus used, and the formulation practice may subsidise to the rapid outflow of specific drug encapsulation after preparation or admin. Hence arduous precautions are prerequisite therefore to accomplish optimal handling and treating of the erythrocyte carriers.

19. Immune cells

Immune cells are a group of cells that elicit retorts against foreign elements in order to protect the body against diseases; these cells hereafter cooperatively form the immune system. There are diverse types of immune cells such as lymphocytes, natural killer cells,

Bioinspired Nanomaterials
Materials Research Foundations **111** (2021) 63-95

Materials Research Forum LLC
https://doi.org/10.21741/9781644901571-3

phagocytes, and neutrophils. Leukocytes, also known as white blood cells (WBCs), are a part of the immune system and play a substantial role in suppressing inflammation, infection, and several disease conditions. WBCs have a shorter circulation time (up to 20 days) than RBCs, but they are still smart for drug delivery presentations, because of their specific features, such as increased cellular interactions and substantial tissue penetration capabilities, in exact across physiological barriers [54]. Additionally, WBCs tend to transport, migrate near inflammatory sites and adhere to endothelial wall tissues with tumor cells, aspects highly seemly and capable for novel drug delivery in cancer treatment. Macrophages are one type of WBCs principally designed in bone marrow and derived from monocytes that separate concluded promonocytes from monoblasts. The practice of phagocytosis by macrophages can overcome and digest cellular debris, microbes, cancer cells and other dangerous substances that entered human body. Macrophages have a distinctive facility to differentiate between spiteful and benign cells, though they someway show constraint in individual between unlike types of bacteria, viruses and other foreign constituents. They also retain a homing property and can transfer to tumor sites across endothelial barriers in retort to cytokine excretion from diseased tissues. The conception of macrophage phagocytosis in close conjunction with improvements in nanotechnology has newly been used to combat diseases where therapeutic nanoparticles are being loaded ex vivo into macrophages. Cell-based delivery systems using macrophages are valuable since the intracellular cargo of the cells would endure dormant and harmless to its host deprived of premature release until reaching tumor cells and carrying a strong drug dose. As such, macrophages would act as Trojan horses to carry drug loading nanoparticles, pass through barriers, and offload them into brain tumor sites [55].

20. Stem cells

Stems cells have also appeared as a impending therapeutic aspirant in cancer therapy in modern years, chiefly for gene therapy. Stem cells organised with leukocytes parade intrinsic tropism on inflammatory and tumor sites; later it is imaginable to exploit genetic engineering as well as tumor tropism to convince stem cells to prompt therapeutic gene products that encode antitumor proteins. For example, the realisation in incorporating lipid and polymeric nanoparticles into non-transformed adult human mesenchymal stem cells (MSCs), and the outcomes publicized the capacity of MSC nanoparticles to drift to brain tumors without conciliation of their possibility. In alternative study, MSCs loaded with PLGA-DOX nanoparticles were used to treat pulmonary metastases. Exact targeting to tumors and carrier permeation were perceived; the homing and permeability properties of MSCs were not precious by drug loading, and their ability and knack to kill melanoma

Materials Research Forum LLC
https://doi.org/10.21741/9781644901571-3

cells were reliant on drug dosing. The use of stem cells as nanoparticle cellular carriers is both inventive and favourable; advance inquiries are in progress to apprehend and expose the migration mechanisms of MSCs and MSC-nanoparticles, which are presently not well implicit [56].

21. Platelets

A platelet is a necessary element of the bloodstream and plays a key role in numerous physiologic and pathologic developments such as hemostasis and thrombosis by establishing the plugs that seal injured vessels and stop bleeding and conserving the reliability of blood circulation. Newly, the natural affinity of platelets for circulating tumor cells (CTCs) that are shed from the primary tumor into the bloodstream has drained great attention. The average lifespan of circulating platelets is 8 to 9 days, and hence could prominently expand the pharmacokinetics of intravenously injected therapeutics. Furthermore, transfused platelets retain the facility to migrate to the site of surgical wounds, where residual tumors may survive after surgery. This delivery approach took improvement of the empathy of platelets for cancer cells and the complex could excellently deliver to cancer cell membranes and initiate the extrinsic apoptosis indicating lane. Subsequently, it was digested after endocytosis and enriched amassing at the nuclei for initiation of the essential apoptosis pathway; hence a encouraging synergetic antitumor efficiency was accomplished. Lastly, the complex could added be revised to detect and exclude metastatic tumor cells, thus allotment great ability in cancer therapy [57].

22. Concluding remarks and future perspectives

Bioengineering of nanoparticles is a swiftly mounting field, with tremendous improvements in the last decade. Delightful improvement of the unambiguous delivery and translocation mechanisms espoused by pathogens and mammalian cells, the bioinspired nanoparticles have diverse functions, such as extended circulation, enriched amassing at infected sites, and reduced off-target effects in healthy tissues.

Biological complexity affords desired functions to bioinspired nanoplatforms, which concurrently causes encounters for practice control, refining, scaling up production, and reproducible manufacturing in the research stage. On the basis of the quality-by-design principle, the advantage of minimalism should be monitored to expedite translation studies. Nanoparticles engineered by innumerable biomimetic slants, their structural features and molecular constituents are only unwell considered. Impending revisions are vital to expose details concerning the exact constituents as well as the dissemination, arrangement, and orientation of precise biomolecules on the surface of bioinspired nanoparticles, since these constraints are exceptionally essential for their *in vivo* fates and therapeutic efficacies. In this aspect, incipient expertise such as multiplexed protein analysis, proteomics, and imaging mass spectrometry can be used. Institution of the structure–property correlation will be favourable for streamlining or enhancing preparation processes of remaining biomimetic nanoparticles. This can also stimulate the intention and advance of more operative nanoplatforms. Subsequently, the usefulness and long-term safety concerns of newly bioengineered nanotherapies need to be established by comprehensive anti-infective studies. Similarly, mechanistic studies should be conducted to address molecular and cellular events controlling *in vivo* biopharmaceutical and pharmacokinetic profiles of biomimetic nanotherapies presently advanced. Other cutting-edge technologies, such as computational design, materials genome, and artificial intelligence, can be integrated to discover more effective and translational nanoparticles based on the bioengineering strategies. Despite the aforementioned challenges and restrictions of bioinspired nanotherapies for biomedical applications, we may expect that the biomimetic strategy-based nanomedicine field will afford novel therapeutics against infectious diseases in the near future.

In conclusion, with many promising suggestions developed in modern years where new drug delivery and imaging technologies are advanced, there is a positive position for bioinspired nano-theranostics and its clinical rendition could be recognised in the near future to reform cancer therapy.

References

[1] Aricò AS, Bruce P, Scrosati B, Tarascon J-M, Van Schalkwijk W. 2005. Nanostructured materials for advanced energy conversion and storage devices. Nat. Mater. 4, 366-377. https://doi.org/10.1038/nmat1368

[2] Du D, Yang Y, Lin Y. 2012. Graphene-based materials for biosensing and bioimaging. MRS Bull. 37, 1290-1296. https://doi.org/10.1557/mrs.2012.209

[3] Gaharwar AK, Peppas NA, Khademhosseini A. Biotechnol. Bioeng. 2014, 111, 441-453. https://doi.org/10.1002/bit.25160

[4] Walcarius A, Minteer SD, Wang J, Lin Y, Merkoci A. 2013. Nanomaterials for bio-functionalized electrodes: recent trends. J. Mater. Chem. B 1, 4878-4908. https://doi.org/10.1039/c3tb20881h

[5] Zhou H, Fan T, Zhang D. 2011. Biotemplated materials for sustainable energy and environment: current status and challenges. ChemSusChem 4, 1344-1387. https://doi.org/10.1002/cssc.201100048

[6] An FF, Zhang XH. Strategies for Preparing Albumin-based Nanoparticles for Multifunctional Bioimaging and Drug Delivery. Theranostics. 2017.7(15):3667-3689. https://doi.org/10.7150/thno.19365

[7] Sotiropoulou S, Sierra-Sastre Y, Mark SS, Batt CA. 2008. Biotemplated nanostructured materials. Chem. Mater. 20, 821-834. https://doi.org/10.1021/cm702152a

[8] Zhao M, Lei C, Yang Y, et al. Abraxane, the Nanoparticle Formulation of Paclitaxel Can Induce Drug Resistance by Up-Regulation of P-gp. PLoS One. 2015;10(7):e0131429. https://doi.org/10.1371/journal.pone.0131429

[9] Saravanakumar G, Jo DG, Park JH. Polysaccharide-based nanoparticles: a versatile platform for drug delivery and biomedical imaging. Curr Med Chem. 2012;19(19):3212-3229. https://doi.org/10.2174/092986712800784658

[10] Swierczewska M, Han HS, Kim K, Park JH, Lee S. Polysaccharide-based nanoparticles for theranostic nanomedicine. Adv Drug Deliv Rev. 2016; 99(Pt A):70-84. https://doi.org/10.1016/j.addr.2015.11.015

[11] Huh MS, Lee EJ, Koo H, et al. Polysaccharide-based Nanoparticles for Gene Delivery. Top Curr Chem (Cham). 2017;375(2):31. https://doi.org/10.1007/s41061-017-0114-y

[12] Hudak JE, Bertozzi CR. Glycotherapy: new advances inspire a reemergence of glycans in medicine. Chem Biol. 2014;21(1):16-37. https://doi.org/10.1016/j.chembiol.2013.09.010

[13] Agard NJ, Bertozzi CR. Chemical approaches to perturb, profile, and perceive glycans. Acc. Chem. Res. 2009;42:788-797. https://doi.org/10.1021/ar800267j

[14] Arnold JN, Wormald MR, Sim RB, Rudd PM, Dwek RA. The impact of glycosylation on the biological function and structure of human immunoglobulins. Annu. Rev. Immunol. 2007;25:21-50. https://doi.org/10.1146/annurev.immunol.25.022106.141702

[15] Trombino S, Servidio C, Curcio F, Cassano R. Strategies for Hyaluronic Acid-Based Hydrogel Design in Drug Delivery. Pharmaceutics. 2019;11(8):407. https://doi.org/10.3390/pharmaceutics11080407

[16] G.R.M. Jillian, M.E. Van Dyke, A review of keratin-based biomaterials for biomedical applications, Dent. Mater. 3 (2010) 999-1014. https://doi.org/10.3390/ma3020999

[17] Brandelli A. Bacterial keratinases: useful enzymes for bioprocessing agroindustrial wastes and beyond, Food Bioproc. Tech. 1 (2) (2008) 105-116. https://doi.org/10.1007/s11947-007-0025-y

[18] M.A. Khosa, A. Ullah, A sustainable role of keratin biopolymer in green chemistry: a review, J. Food Process. Beverag. 1 (1) (2013) 8.

[19] L. Kreplak, J. Doucet, P. Dumas, F. Briki, New aspects of the α-helix to β-sheet transition in stretched hard α-keratin fibers, Biophys. J. 87 (1) (2004) 640-647. https://doi.org/10.1529/biophysj.103.036749

[20] A. Aluigi, C. Tonetti, C. Vineis, C. Tonin, G. Mazzuchetti, Adsorption of copper (II) ions by keratin/PA6 blend nanofibres, Eur. Polym. J. 47 (9) (2011) 1756-1764. https://doi.org/10.1016/j.eurpolymj.2011.06.009

[21] A. Ghosh, S.R. Collie, Keratinous materials as novel absorbent systems for toxic Pollutants, Def. Sci. J. 64 (3) (2014) 209-221. https://doi.org/10.14429/dsj.64.7319

[22] Verma, V.; Verma, P.; Ray, P.; Ray, A.R. Preparation of scaffolds from human hair proteins for tissue-engineering applications. Biomed. Mater. 2008, 3, 25007. https://doi.org/10.1088/1748-6041/3/2/025007

[23] Dai L, Si C. Recent Advances on Cellulose-Based Nano-Drug Delivery Systems: Design of Prodrugs and Nanoparticles. Curr Med Chem. 2019; 26(14):2410-2429. https://doi.org/10.2174/0929867324666170711131353

[24] Esposito A, Sannino A, Cozzolino A, Nappo QS, Lamberti M, Ambrosio L, Nicolais L (2005) Response of intestinal cells and macrophages to an orally administered cellulose-PEG based polymer as a potential treatment for intractable edemas. Biomaterials 26:4101-4110 https://doi.org/10.1016/j.biomaterials.2004.10.023

[25] Sezer S., Şahin İ., Öztürk K., Şanko V., Koçer Z., Sezer Ü.A. (2019) Cellulose-Based Superabsorbent Hydrogels. Polymers and Polymeric Composites: Cellulose-Based Superabsorbent Hydrogels. 1177-1203. https://doi.org/10.1007/978-3-319-77830-3_40

[26] L. Bedian, A.M.V. Rodríguez, G.H. Vargas, R. Parra-Saldivar, H.M.N. Iqbal, Bio-based materials with novel characteristics for tissue engineering applications-a review, Int. J. Biol. Macromol. 98 (2017) 837-846. https://doi.org/10.1016/j.ijbiomac.2017.02.048

[27] N. Singh, J. Chen, K.K. Koziol, K.R. Hallam, D. Janas, A.J. Patil. Chitin and carbon nanotube composites as biocompatible scaffolds for neuron growth, Nanoscale. 8 (15); 2016: 8288-8299. https://doi.org/10.1039/C5NR06595J

Bioinspired Nanomaterials   Materials Research Forum LLC
Materials Research Foundations **111** (2021) 63-95   https://doi.org/10.21741/9781644901571-3

[28] S. Cammas, M.-M. Béar, L. Moine et al., "Polymers of malic acid and 3-alkylmalic acid as synthetic PHAs in the design of biocompatible hydrolyzable devices," International Journal of Biological Macromolecules. 1999; 25: 273-282. https://doi.org/10.1016/S0141-8130(99)00042-2

[29] G.-Q. Chen, "A microbial polyhydroxyalkanoates (PHA) based bio- and materials industry," Chemical Society Reviews. 2009; 38(8): 2434-2446. https://doi.org/10.1039/b812677c

[30] R. Acharya, S. Saha, S. Ray, S. Hazra, M.K. Mitra, J. Chakraborty. siRNA-nanoparticle conjugate in gene silencing: a future cure to deadly diseases? Mater. Sci. Eng. C Mater. Biol. Appl., 76 (2017), 1378-1400. https://doi.org/10.1016/j.msec.2017.03.009

[31] Caitriona M.O'D, Andreas BS, Julian DF, Véronique P, Vincent J. Oral delivery of non-viral nucleic acid-based therapeutics - do we have the guts for this? European Journal of Pharmaceutical Sciences. 2019; 133: 190-204. https://doi.org/10.1016/j.ejps.2019.03.027

[32] Weng T, Qi J, Lu Y, et al. The role of lipid-based nano delivery systems on oral bioavailability enhancement of fenofibrate, a BCS II drug: comparison with fast-release formulations. J Nanobiotechnology. 2014;12:39. https://doi.org/10.1186/s12951-014-0039-3

[33] Hua S. Lipid-based nano-delivery systems for skin delivery of drugs and bioactives. Front Pharmacol. 2015; 6: 219. https://doi.org/10.3389/fphar.2015.00219

[34] Sonaje K, Lin KJ, Tseng MT, Wey SP, Su FY, Chuang EY, Hsu CW, Chen CT, Sung HW. Effects of chitosan-nanoparticle-mediated tight junction opening on the oral absorption of endotoxins. Biomaterials. 2011;32:8712-8721. https://doi.org/10.1016/j.biomaterials.2011.07.086

[35] Beuttler J, Rothdiener M, Muller D, Frejd FY, Kontermann RE. Targeting of epidermal growth factor receptor (EGFR)-expressing tumor cells with sterically stabilized affibody liposomes (SAL) Bioconjug Chem. 2009;20(6):1201-1208. https://doi.org/10.1021/bc900061v

[36] Wissing SA, Muller RH. The influence of solid lipid nanoparticles on skin hydration and viscoelasticity - in vivo study. Eur J Pharm Biopharm. 2003; 56(1): 67-72. https://doi.org/10.1016/S0939-6411(03)00040-7

[37] Schubert MA, Muller-Goymann CC. Characterisation of surface-modified solid lipid nanoparticles (SLN). Influence of lecithin and nonionic emulsifier. Eur J Pharm Biopharm. 2005;61(1-2):77-86. https://doi.org/10.1016/j.ejpb.2005.03.006

[38] Muller RH, Radtke M, Wissing SA. Nanostructured lipid matrices for improved microencapsulation of drugs. Int J Pharm. 2002;242(1-2):121-128. https://doi.org/10.1016/S0378-5173(02)00180-1

[39] Teeranachaideekul V, Muller RH, Junyaprasert VB. Encapsulation of ascorbyl palmitate in nanostructured lipid carriers (NLC) - Effects of formulation parameters on physicochemical stability. Int J Pharm. 2007; 340(1-2):198-206. https://doi.org/10.1016/j.ijpharm.2007.03.022

[40] Sunderland K.S., Yang M., Mao C. Phage-Enabled Nanomedicine: From Probes to Therapeutics in Precision Medicine. Angew. Chem. 2017; 56:1964-1992. https://doi.org/10.1002/anie.201606181

[41] Banwell EF, Abelardo ES, Adams DJ, et al. Rational design and application of responsive alpha-helical peptide hydrogels. Nat Mater. 2009; 8(7): 596-600. https://doi.org/10.1038/nmat2479

[42] Sokullu E, Soleymani Abyaneh H, Gauthier MA. Plant/Bacterial Virus-Based Drug Discovery, Drug Delivery, and Therapeutics. Pharmaceutics. 2019; 11(5):211. https://doi.org/10.3390/pharmaceutics11050211

[43] Koutsopoulos S. Self-assembling peptide nanofiber hydrogels in tissue engineering and regenerative medicine: Progress, design guidelines, and applications. J Biomed Mater Res A. 2016; 104(4):1002-1016. https://doi.org/10.1002/jbm.a.35638

[44] Aggeli A., Bell M., Carrick L.M., Fishwick C.W.G., Harding R., Mawer P.J., Radford S.E., Strong A.E., Boden N. pH as a trigger of peptide β-sheet self-assembly and reversible switching between nematic and isotropic phases. J. Am. Chem. Soc. 2003;125:9619-9628. https://doi.org/10.1021/ja021047i

[45] Van Hell AJ, Costa CICA, Flesch FM, Sutter M, Jiskoot W, Crommelin DJA, Hennink WE, Mastrobattista E. Self-Assembly of Recombinant Amphiphilic Oligopeptides into Vesicles. Biomacromolecules. 2007; 8: 2753-2761. https://doi.org/10.1021/bm0704267

[46] Thomas CE, Ehrhardt A, Kay M. Progress and Problems with the Use of Viral Vectors for Gene Therapy. Nat Rev Genet. 2003;4:346-358. https://doi.org/10.1038/nrg1066

[47] Garg, H., Mehmetoglu-Gurbuz, T. & Joshi, A. Virus Like Particles (VLP) as multivalent vaccine candidate against Chikungunya, Japanese Encephalitis, Yellow Fever and Zika Virus. Sci Rep 10, 4017 (2020). https://doi.org/10.1038/s41598-020-61103-1

[48] Kaneda Y. Virosomes: evolution of the liposome as a targeted drug delivery system. Adv Drug Deliv Rev. 2000;43(2-3):197-205. https://doi.org/10.1016/S0169-409X(00)00069-7

[49] Somiya M, Liu Q, Kuroda S. Current Progress of Virus-mimicking Nanocarriers for Drug Delivery. Nanotheranostics. 2017;1(4):415-429. https://doi.org/10.7150/ntno.21723

[50] Bungener L, Serre K, Bijl L. et al. Virosome-Mediated Delivery of Protein Antigens to Dendritic Cells. Vaccine. 2002;20:2287-2295 https://doi.org/10.1016/S0264-410X(02)00103-2

[51] Tan S, Wu T, Zhang D, Zhang Z. Cell or cell membrane-based drug delivery systems. Theranostics. 2015;5(8):863-881. https://doi.org/10.7150/thno.11852

[52] Patel PD, Dand N, Hirlekar RS, Kadam VJ. Drug loaded erythrocytes: as novel drug delivery system. Curr Pharm Des. 2008;14(1):63-70. https://doi.org/10.2174/138161208783330772

[53] Aryal S, Hu CM, Fang RH, et al. Erythrocyte membrane-cloaked polymeric nanoparticles for controlled drug loading and release. Nanomedicine (Lond). 2013;8(8):1271-1280. https://doi.org/10.2217/nnm.12.153

[54] Lim S, Park J, Shim MK, et al. Recent advances and challenges of repurposing nanoparticle-based drug delivery systems to enhance cancer immunotherapy. Theranostics. 2019;9(25):7906-7923. https://doi.org/10.7150/thno.38425

[55] Liu X, Li Y, Sun X, Muftuoglu Y, Wang B, Yu T. et al. Powerful anti-colon cancer effect of modified nanoparticle-mediated IL-15 immunogene therapy through activation of the host immune system. Theranostics. 2018;8:3490-503. https://doi.org/10.7150/thno.24157

[56] Orive G, Cobos R, Gorriti J, Pedraz JL, Meregalli M, Torrente Y. Drug delivery technologies and stem cells for tissue repair and regeneration. Curr Pharm Biotechnol. 2015;16(7):646-654. https://doi.org/10.2174/1389201016071504271124457

[57] Shi Q, Montgomery RR. Platelets as delivery systems for disease treatments. Adv Drug Deliv Rev. 2010;62(12):1196-1203. https://doi.org/10.1016/j addr.2010.06.007

Bioinspired Nanomaterials

Materials Research Foundations **111** (2021) 96-117

Materials Research Forum LLC

https://doi.org/10.21741/9781644901571-4

Chapter 4

Bio-Mediated Synthesis of Nanomaterials for Packaging Applications

P. Sivaranjana*[,1], N. Rajini[2], V. Arumugaprabu[3], S.O. Ismail[4]

[1]Department of Chemistry, Kalasalingam Academy of Research and Education, Krishnankoil 626126, Tamil Nadu, India

[2]Centre for Composite Materials, Department of Mechanical Engineering, Kalasalingam Academy of Research and Education, Krishnankoil 626126, Tamil Nadu, India

[3]Department of Mechanical Engineering, Kalasalingam Academy of Research and Education, Krishnankoil 626126, Tamil Nadu, India

[4]Department of Engineering, School of Physics, Engineering and Computer Science, University of Hertfordshire, Hatfield, AL10 9AB, Hertfordshire, England, United Kingdom

* psivaranjana@gmail.com

Abstract

Change in lifestyle of humans in this present generation with huge dependence on packaging materials has encouraged several studies on development of new variety of packaging materials. Emphasis on replacement of existing non-biodegradable packaging materials with biodegradable materials paved the way for the use of biopolymers. Lack of properties, such as thermal stability and mechanical strength in biopolymers led to the development of biopolymer nanocomposites by adding metal/metal oxide nanoparticles as fillers into the biopolymers. Metal/metal oxide nanoparticles improve mechanical/tensile strength, thermal stability as well as antimicrobial properties of the binding and receiving polymer matrix. Bio-mediated synthesis of metal/metal oxide nanoparticles result in the development of novel packaging materials at a low cost and without releasing hazardous wastes into the environments. Novel packaging materials with metal/metal oxide nanoparticles as additives are capable of increasing the shelf life of food, in certain cases they act as indicators of quality food inside the package. Summarily, this present chapter focuses on bio-mediated synthesis of various metal/metal oxide nanoparticles and their applications in food packaging.

Keywords

Biosynthesis, Filler, Additive, Packaging Material, Antimicrobial, Metal Nanoparticles

Bioinspired Nanomaterials

Materials Research Foundations **111** (2021) 96-117

Materials Research Forum LLC

https://doi.org/10.21741/9781644901571-4

Contents

1. Introduction

The use of non-biodegradable packaging materials has been restricted in several countries to ensure environmental safety. Therefore, there is a need for development of alternative packaging materials with enhanced properties. The development of eco-friendly green packaging materials has potentials to reduce environmental impacts caused by synthetic packaging materials [1]. The biopolymers have been used as new alternative packaging

Bioinspired Nanomaterials   Materials Research Forum LLC
Materials Research Foundations **111** (2021) 96-117   https://doi.org/10.21741/9781644901571-4

materials, they include, but are not limited to, polysaccharides (starch and cellulose derivatives, chitosan and alginates), lipids (bees and carnauba wax, and free fatty acids), proteins (casein, whey and gluten), poly hydroxyl butyrate (PHB), polylactic acid (PLA), poly caprolactone (PCL), polyvinyl alcohol (PVA), poly butylene succinate and their biopolymer blends [2]. Considering environmental safety, the biodegradable natural or biopolymers are preferred. However, the inherent properties and mechanical strength of natural polymers limit them from being utilized as packaging materials, especially on an industrial scale. It is often essential to have their surface modification or physical cross linking or even modified as a composite material [3].

The nanometals have potential to overcome certain limitations in properties of the biodegradable materials. The nanometals have exhibited preferable properties, such as high surface area, fine particle size, high reactivity, high strength and ductility, which make them suitable to be employed frequently in a diversified range of industrial fields [4]. The available vacuum sealed food packaging polymeric materials are permeable to oxygen and moisture, which leads to spoiling of foods. To prevent this situation, polymer packaging materials are coated with nanometal, which inhibits the penetration of oxygen and moisture, thereby preserves the food materials from spoiling [5]. These nanometals also act as antimicrobial agents by preventing growth of harmful microbes, thereby prevent food spoilage and extend their shelf life[6]. Beyond these, nanometals in food packaging also act as smart indicators of change in the chemical composition, pH, gas composition, among others, inside the package that contains food stuff. These changes are communicated to the consumers by change in colour of the tag attached to them. Therefore, the role of nanometals in the food packaging include prevention of shelf life, fighting against the microbial growth and acting as smart indicators for the exact condition of the food stuff to the consumers. Metal and metal oxide (silver, gold, zinc oxide, silica, titanium dioxide, alumina and iron oxides), carbon based nanometals and nano-sized polymers are most commonly used as nanofillers in food packaging applications[7].

Furthermore, the nanometals improves functionalities, such as durability, flexibility, temperature and flame resistance, barrier properties, optical and recycling properties of the recipient packaging materials [8]. Nanometals are added to polymeric packaging materials, such as polyamides (PA), nylons, polyolefin, ethylene-vinyl acetate copolymer, polystyrene (PS), epoxy resins, polyurethane, polyvinyl chloride (PVC) and polyethylene terephthalate (PET) [9]. Development of packaging materials using nanoparticles can be broadly classified into two categories/types: (I) biocomposite materials with inorganic metal nanoparticles as fillers [10, 11], (II) biocomposite materials with plant extracts/agro wastes as fillers [12]. Type I materials possess appreciable strength and shelf life, and

Bioinspired Nanomaterials
Materials Research Foundations **111** (2021) 96-117

Materials Research Forum LLC
https://doi.org/10.21741/9781644901571-4

they are best suitable for packaging applications, whereas type II lacks shelf life and strength. The responses of the type II materials towards the environmental factors, such as humidity, temperature and pressure are higher, when compared with type I. The conventional methods of synthesis and distribution of nanometals into the matrix of the packaging materials results to generation and deposition of hazardous chemicals into the environments. In recent years, studies have been focusing on bioreduction process for the synthesis of nanometals. This is very simple, efficient and cost effective, but time consuming [13]. In this method, the choice of bioreductant includes plant extracts, which contains flavonoids capable of reducing the metal to its nanostate [14]. In this chapter, specific discussion on nanometals synthesized via bioreduction process and utilized as nanofiller in packaging materials is presented.

In spite of the advantages of utilization of nanometals in food packaging materials, most of studies on these materials are still on the stage of demonstration, and their real time applications are yet to receive approval concerning their safety issues, which could be caused by the migrations of nanomaterials from packaging to food stuffs [15]. Additionally, the absorption, distribution, metabolism, excretion and toxicological assessment of nanoparticles in food in humans are to be assessed [16]. Therefore, the use of nanometals in food industry provokes various assessment methodologies to ensure usage as well as the real-time analysis of nanometals in the environment and their impact on various levels of organisms [16]. This is an emerging and evolutionary area involving multidisciplinary studies and provides scope for interdisciplinary research.

2. Metal nanoparticles as fillers in packaging materials

The unique behaviours of metal nanoparticles have attracted researchers to employ them in many different types of matrices [17]. Though, these nanometal particles are added as nano fillers in a minimum quantity, they do not lose their properties, such as antimicrobial nature, thermal resistivity and/or tensile strength [18]. They usually enrich the properties the matrices to which they are added. The antimicrobial activity of metal nanoparticles varies accordingly with their methods of synthesis; either physical, chemical or biological [19]. Interest on bio-synthesized metal nanoparticles increases incredibly, owing to their reliability and cost effectiveness [20]. The most preferred and commonly used metal nanoparticles are zinc, iron, copper, gold, aluminium, nickel and silver. In addition, certain metal oxides include titanium, zirconium, iron and zinc are also used as nanofillers in packaging applications. Among various aforementioned metal nanoparticles (MNPs), silver nanoparticles (AgNPs) is observed to be most appropriately employed in packaging materials, biomedical appliances, cosmetics, pharmaceuticals and textile sectors [21]. The toxicity of AgNPs is found to be very minimum in animal cells.

Bioinspired Nanomaterials   Materials Research Forum LLC
Materials Research Foundations **111** (2021) 96-117   https://doi.org/10.21741/9781644901571-4

Certain inorganic metal nanoparticles are also recognized as safe materials by US Food and Drug Administration (FDA).

3.   Biosynthesized AgNPs

3.1   Biological synthesis of AgNPs using bacterial strains

Biosynthesis of AgNPs includes bioreduction of silver salts with various plant extracts and biological synthesis of AgNPs with bacteria, fungi and biomolecules [22]. Biological synthesis of AgNPs can be via either intracellular or extracellular mechanisms [23]. Comparatively, the extracellular mechanism is observed to be simpler and more economical. Though, the mechanism is not yet completely understood, it is predicted to take place by the presence of nitrate reductase enzymes released by microorganisms, which reduces metallic ions to metallic nanoparticles. Among the biological syntheses of AgNPs, the bacterial synthesis has been identified to be most convenient [24],owing to its availability and vulnerability for genetic modification. The shape, size and nature of AgNPs determine method of synthesis. Hence, many variety of bacterial strains are used for the biological synthesis of AgNPs. These include *Bacillus spp.* [25], *Streptomyces spp.* [26], *Acinetobacter spp.* [27], *Pseudomonas spp.* [28], to mention but a few. Minimum quantity of AgNPs synthesized by biological method using bacterial strains shows a good activity towards major strains of microbes from attacking food stuffs. Amongst them are *S. aureus, P. aeruginosa, B. subtilis* and *E. coli.*

3.2   Biological synthesis of AgNPs using fungi species

The biological synthesis of AgNPs using fungal species is observed to be efficient with extracellular mechanism. The fungal species extracted from plants are more efficient than other species [29]. The following species of fungi are most prominently tested for the biological synthesis of AgNPs, *Fusarium oxysporum* [30], *Guignardia spp.* [31], *Penicillium aculeatum* [32], *Alternaria spp.* [33], *Phenerochaete chrysosporium* [34]. AgNPs synthesized via biological method using fungi have antimicrobial activity against a variety of fungal species as well as bacterial strains; both Gram-positive and negative. Therefore, the AgNPs synthesis using fungal species is an effective process to be employed in packaging material, since it effectively attacks microbes attacking the food stuffs.

4.   Bioreduction of AgNPs using plant extracts

The plant extracts with flavonoid contents are mainly reported bioreductants for the synthesis of AgNPs [35]. Almost all the parts of a plant with flavonoids have been used.

Bioinspired Nanomaterials

Materials Research Foundations **111** (2021) 96-117

Materials Research Forum LLC

https://doi.org/10.21741/9781644901571-4

The experimental condition and active flavonoid contents play major roles in geometry and activity of the AgNPs. In most cases, capping agents are not used during bioreduction of AgNPs with plant extracts, which results in aggregation of the reduced AgNPs. The agglomeration of AgNPs may lead to reduced behavior of AgNPs. In this case, certain researchers generated the AgNPs *in-situ*; inside the polymer films directly, using plant extract as a reducing agent [36-38].

5. Anti-microbial activity of AgNPs

AgNPs have a wide spectrum of antimicrobial activity, including Gram-positive and negative bacteria, fungi and viruses. The mechanism of antibacterial activity of AgNPs have been studied based on its ability to release Ag^+ ions [39] and their potential to inhibit the growth metabolism of bacterial cells [40]. The mechanism of action of AgNPs against bacteria is proposed in three ways [41]: (1) penetration of AgNPs in the range of 1 to 10nm into the cell membrane and interfere with its respiration, (2) interaction of AgNPs with the compounds of sulphur and phosphorus, such as DNA and (3) release of active silver (Ag^+) ions, which reacts with negatively charged cell membrane and damages them.

Moreover, the release of Ag^+ ions are influenced by availability of oxygen atmosphere. In anaerobic condition, the antibacterial effect is almost nil even with higher concentration [40]. The antimicrobial activity of AgNPs is also influenced by the shape and size of the synthesized AgNPs, the symmetry of the particles which offers greater contact surface have greater antimicrobial activity.

6. Packaging materials with AgNPs

The addition of nanoparticles into the already existing packaging materials provides improved physical-chemical properties, reduces hydrophilic behavior and induces biodegradability and antimicrobial activity. AgNPs are added as fillers in both biodegradable and non-biodegradable polymers. Few examples of food packaging material with AgNPs are presented in Table 1.

Table 1: *Food packaging materials with AgNPs as additives.*

Packaging material	Type of food material	Preferred storage period	References
PVC/AgNPs	Beef	Upto 14 days; 4± 2°C	[42]
	Dried fruits and nuts	Upto 21 days; 4± 2°C	[43]
LDPE/AgNPs	Meat, pork and chicken	Upto 21 days; 4± 2°C	[44]
	Cheese	Upto 28 days; 4± 2°C	[45]
PLA/AgNPs	Cheese	Upto 25 days; 4± 2°C	[46]
	Mangoes	Upto 15 days; room temperature	[47]

7. Biosynthesized ZnNPs/ ZnONPs

Among various inorganic metal oxide nanoparticles available, Zinc oxide nanoparticles (ZnONPs)are more preferred, because they are inexpensive and easy to prepare. The US FDA has enlisted ZnONPs as part of safe materials [48]. As mentioned in the previous sub-sections, similar to AgNPs, the ZnNPs/ZnONPs can also be synthesized via bioreduction process, using plant extracts as well as biological synthesis with microorganisms. The following strains of bacteria have been effectively utilized for the biological synthesis of Zn/ZnONPs: *Aspergillus strain, Aspergillus fumigatus, Aspergillus terreus* [49], *Candida albicans* [50]. In addition, few used algal strains include *Chlamydomonas reinhardtii* [51], *Sargassum muticum* [52] and *S. myriocystum.*

Moving forward, biosynthesized Zn/ZnONPs possess appreciable antimicrobial activity against broad spectrum of bacteria and fungi.ZnNPs/ZnONPs also possess antitumor activity, it is cytotoxic and genotoxic towards certain types of human cells [53]. Further large scale production of ZnONPs by bioreduction process is possible [54], by coating the synthesized ZnONPs over cotton fabric, which possesses antimicrobial activity and washing durability. The biosynthesized ZnNPs/ZnONPs show a better catalytic activity than those ZnNPs/ZnONPs obtained from chemical methods. ZnONPs synthesized from the fungal strain, *Aspergillus fumigatus* is stable up to 90 days, while ZnONPs obtained from seaweed, *Sargassum myriocystum* is stable up to 6 months [55]. In case of bioreduction of Zn/ZnONPs using plant extracts, the carboxylic and phenolic groups present in the extract act as bioreductants as well as capping agents [56].

Bioinspired Nanomaterials   Materials Research Forum LLC
Materials Research Foundations **111** (2021) 96-117   https://doi.org/10.21741/9781644901571-4

8. Packaging materials with ZnNPs/ ZnONPs

Zn/ZnONPs are well known for their stability under extreme conditions. They are effective against a wide spectrum of microbial strains, more specifically the food born microbes at a lower concentration [53, 57]. ZnONPs possess a unique property of filtering ultraviolet (UV) rays [58]. Addition of Zn/ZnONPs to both biodegradable or non-biodegradable polymer matrices increases the mechanical properties of the polymer, such as tensile strength, Young's modulus and thermal resistance [59]. ZnONPs coated packaging materials are suitable for storing fish samples [60]. Zn/ZnONPs coated packaging materials also act as a good scavenger of oxygen. All the aforementioned characteristics qualify Zn/ZnONPs to be used as potential additives for packaging applications [61].

Table 2: Food packaging materials with ZnNPs as additives.

Packaging Material	Type of food material	Preferred storage period	References
PVC/ZnNPs	Cheese	30-40 days	[62]
	Sliced apples	Decay rate decreased by 60%	[63]
PLA/ZnONPs	Sliced apples	Decay rate decreased by 65%	[64]
OBG/ZnONPs	Spinach	Upto 7 days	[65]
PU/ZnONPs	Sliced carrot	Upto 9 days	[66]
PVA/ZnONPs	Aqueous food stuffs	pH indicator	[67]

9. Biosynthesized CuNPs/ CuONPs

Copper metal is well known for its antimicrobial activity from the ancient days. This is the reason behind the use of copper vessels for storing water. Even now, copper nanoparticles (CuNPs) are used for water treatment. CuNPs/CuONPs have commendable antimicrobial activity against a variety of bacteria, fungi and viruses [68]. The mechanism of interaction of CuNPs/CuONPs with microbes is through their cell membranes, thereby making them inactive. The bioreduction of CuNPs is successful with extracts of various plant species, such as, *Citrus medica Linn.* [69], *Ziziphus spina-christi* [70], *Asparagus adscendens Roxb.* Used root and leaf [71] include *Eclipta prostrata* leaves [72], *Ginkgo biloba Linn.* [73], *Plantago asiatica* leaf [74], *Thymus vulgarisL.* [75], black tea leaves [76], to mention but a few. Biological synthesis of CuNPs can be carried out with the following bacterial strains: *Escherichia coli, M.psychrotolerans* and *M.morganiiRP42.* Biomediated synthesis of CuNPs/CuONPs is an economical as well as

Bioinspired Nanomaterials
Materials Research Foundations **111** (2021) 96-117

Materials Research Forum LLC
https://doi.org/10.21741/9781644901571-4

simple method, which can be executed with minimum infrastructures. In certain cases, CuNPs synthesized are observed to be more efficient than those synthesized via commercial chemical methods.

10. Packaging materials with CuNPs/ CuONPs

CuNPs are effective materials against both Gram-positive and negative organisms [77], which make them more suitable for packaging applications. The antimicrobial effect of CuNPs/ CuONPs more specifically towards *E.coli* in food stuffs and ability of CuNPs/CuONPs to be blended into polymer suggest them to be good additives in food packaging materials [78]. Impregnation of CuNPs/CuONPs into polymer films or biopolymer films [79] enhances their tensile strengths, transparency, thermal stability, mechanical strengths [80]. More, it adds to the antimicrobial activity, UV barrier property and prolonged shell life, which make the material suitable for food packaging applications [81]. CuNPs coated packaging materials act as freshness indicator in case of meat, by turning dark by reacting with the volatile sulfide released during the spoilage of meat [82]. Currently, researchers are interested in examining the CuNPs/CuONPs impregnated films for electronic as well as catalytic applications.

Table 3: *Food packaging materials with CuNPs/CuONPs as additives.*

Packaging Material	Type of food material	Preferred storage period	References
Cellulose/CuNPs	Fruit juices	Decay rate decreased by 60%	[83]
Hydroxypropyl methylcellulose/CuNPs	Meat	Upto 15 days	[84]
Polylactic acid/CuNPs	All types of food stuffs	Expected to decrease the decay rate by 45%	[85]
Agar/CuNPs	All types of food stuffs	Expected to decrease the decay rate by 40%	[86]
HDPE/CuNPs	All types of food stuffs	Expected to decrease the decay rate by 30%	[87]

11. Titanium dioxide nanoparticles used as additives in food packaging

Titanium dioxide nanoparticles (TiO_2NPs) are effective additives in food packaging, due to their thermal stability, economical, non-toxic and stability properties towards UV light. TiO_2NPs are extensively employed as promising photo catalysts in various types of reactions [88]. They are also used in water treatment and self-cleaning applications. Due to the antimicrobial property of TiO_2NPs, the interest of several researchers has increased

Bioinspired Nanomaterials

Materials Research Forum LLC

Materials Research Foundations **111** (2021) 96-117

https://doi.org/10.21741/9781644901571-4

towards using them in food packaging. Unlike previously discussed metal nanoparticles,TiO_2NPs are more prone to agglomeration; therefore, ionic surfactants are used as capping agents to prevent agglomeration. TiO_2NPsare also synthesized from various plant extracts [89] and microbes [90]. TiO_2NPscan be fabricated with biodegradable fish skin gelatin [91], potato starch [92], pectin [93], super hydrophobic paper [94], poly vinyl chloride [95], polyethylene [96] and PLA [97]. Besides, TiO_2NPs as additives in packaging materials improve the mechanical strength, tensile strength, hydrophobicity, thermal stability, transparency, water vapour permeability, UV transmittance and antimicrobial properties. More specifically, application of TiO_2NPs has been effectively analyzed for foodstuffs, such as bread [98] and strawberry [99] for their nutritional values and decay periods. Also, TiO_2NPsare used for controlling hematophagous fly and sheep-biting louse [100]. Biosynthesized TiO_2NPs are employed in solar cells [101].

12. Other metal and metal oxide nanoparticles used as additives in packaging

The magnesium oxide nanoparticles (MgONPs) are recognized as safe materials by the US FDA. MgONPs are added as additives to PLA, polyethylene (PE) and biodegradable polymers. MgONPs improve oxygen barrier, tensile strength and antibacterial property of polymer packaging materials [102], but they have poor water barrier property. Similarly, iron oxide nanoparticles (FeONPs) additives to polypropylene (PP), PVA and biopolymers show good gas barrier and improved thermal stability [103]. Also, silicon dioxide nanoparticles (SiO_2NPs) provide good insulation, low toxicity and stability to polymer, when they are used in packaging materials[104].

Conclusions

The use of biosynthesized metal/metal oxide nanoparticles as additives in packaging materials has been elucidated. It is evident that they increase shelf life of foodstuffs by inducing water vapour permeability, UV barrier, gas barrier and antimicrobial properties. Metal/metal oxide nanoparticles also increase thermal stability and mechanical/tensile strength of the receiving polymers used in packaging materials.

Besides, metal/metal oxide nanoparticles are found to be compatible with both synthetic as well as biopolymers. Biomediated synthesis of metal/metal oxide nanoparticles result to manufacturing of intelligent packaging materials in an economical and eco-friendly manners. Most metal/metal oxide nanoparticles are recognized as safe materials by US FDA. Therefore, biomediated synthesis of metal/metal oxide nanoparticles and their application in food packaging are important and promising field to satisfy our day-to-day needs.

List of abbreviations

AgNPs	Silver nanoparticles
CuNPs	Copper nanoparticles
CuONPs	Copper oxide nanoparticles
FDA	Food and Drug Administration
FeONPs	Iron oxide nanoparticles
HDPE	High density polyethylene
LDPE	Low density polyethylene
MgONPs	Magnesium oxide nanoparticles
MNPs	Metal nanoparticles
NPs	Nanoparticles
OBG	Olive flounder bone gelatine
PA	Polyamides
PE	Polyethylene
PET	Polyethylene terephthalate
PHB	Poly hydroxyl butyrate
PLA	Poly lactic acid
PP	Polypropylene
PS	Polystyrene
PU	Polyurethane
PVA	Poly vinyl alcohol
PVC	Poly vinyl chloride
SiO_2NPs	Silicon oxide nanoparticles
TiO_2NPs	Titanium oxide nanoparticles
UV	Ultraviolet
ZnNPs	Zinc nanoparticles
ZnONPs	Zinc oxide nanoparticles

Materials Research Forum LLC
https://doi.org/10.21741/9781644901571-4

References

[1]   J.-W. Han, L. Ruiz-Garcia, J.-P. Qian, and X.-T. Yang, "Food Packaging: A Comprehensive Review and Future Trends," *Compr. Rev. Food Sci. Food Saf.*, vol. 17, no. 4, pp. 860–877, Jul. 2018. https://doi.org/10.1111/1541-4337.12343

[2]   O. M. Koivistoinen, "Catabolism of biomass-derived sugars in fungi and metabolic engineering as a tool for organic acid production." 2013.

[3]   J. Vartiainen, M. Vähä-Nissi, and A. Harlin, "Biopolymer Films and Coatings in Packaging Applications—A Review of Recent Developments," *Mater. Sci. Appl.*, vol. 05, no. 10, pp. 708–718, Aug. 2014. https://doi.org/10.4236/msa.2014.510072

[4]   "Nanomaterial advantage | Nature." [Online]. Available: https://www.nature.com/articles/419887a. [Accessed: 01-May-2020].

[5]   N. Peelman *et al.*, "Application of bioplastics for food packaging," *Trends in Food Science and Technology*, vol. 32, no. 2. Elsevier, pp. 128–141, 01-Aug-2013. https://doi.org/10.1016/j.tifs.2013.06.003

[6]   M. Pal, "Nanotechnology: A New Approach in Food Packaging," 2017. https://doi.org/10.4172/2476-2059.1000121

[7]   "Nanomaterials | Food Packaging Forum." [Online]. Available: https://www.foodpackagingforum.org/food-packaging-health/nanomaterials. [Accessed: 01-May-2020].

[8]   C. Buzea, I. I. Pacheco, and K. Robbie, "Nanomaterials and nanoparticles: Sources and toxicity," *Biointerphases*, vol. 2, no. 4, pp. MR17–MR71, Dec. 2007. https://doi.org/10.1116/1.2815690

[9]   M. Cushen, J. Kerry, M. Morris, M. Cruz-Romero, and E. Cummins, "Migration and exposure assessment of silver from a PVC nanocomposite," *Food Chem.*, vol. 139, no. 1–4, pp. 389–397, Aug. 2013. https://doi.org/10.1016/j.foodchem.2013.01.045

[10]   E. Fortunati *et al.*, "Multifunctional bionanocomposite films of poly(lactic acid), cellulose nanocrystals and silver nanoparticles," *Carbohydr. Polym.*, vol. 87, no. 2, pp. 1596–1605, Jan. 2012. https://doi.org/10.1016/j.carbpol.2011.09.066

[11]   H. Y. Yu, X. Y. Yang, F. F. Lu, G. Y. Chen, and J. M. Yao, "Fabrication of multifunctional cellulose nanocrystals/poly(lactic acid) nanocomposites with silver nanoparticles by spraying method," *Carbohydr. Polym.*, vol. 140, pp. 209–219, Apr. 2016. https://doi.org/10.1016/j.carbpol.2015.12.030

[12]   B. Iamareerat, M. Singh, M. B. Sadiq, and A. K. Anal, "Reinforced cassava starch based edible film incorporated with essential oil and sodium bentonite nanoclay as

food packaging material," *J. Food Sci. Technol.*, vol. 55, no. 5, pp. 1953–1959, May 2018. https://doi.org/10.1007/s13197-018-3100-7

[13]   N. Mude, A. Ingle, A. Gade, and M. Rai, "Synthesis of silver nanoparticles using callus extract of Carica papaya - A first report," *J. Plant Biochem. Biotechnol.*, vol. 18, no. 1, pp. 83–86, Jan. 2009. https://doi.org/10.1007/BF03263300

[14]   K. Vasudeo, S. Sampat, and K. Pramod, "Biosynthesis of copper nanoparticles using aqueous extract of Eucalyptus sp . plant leaves," *Curr. Sci.*, vol. 109, no. 2, pp. 255–257, 2015.

[15]   "Nanomaterials in food contact materials; considerations for risk assessment - Annals of the National Institute of Hygiene - Volume 68, Number 4 (2017) - AGRO - Yadda." [Online]. Available: http://agro.icm.edu.pl/agro/element/bwmeta1.element.agro-ed369718-d0fb-47ff-a249-10e3a56ee46b. [Accessed: 01-May-2020].

[16]   Z. Piperigkou *et al.*, "Emerging aspects of nanotoxicology in health and disease: From agriculture and food sector to cancer therapeutics," *Food and Chemical Toxicology*, vol. 91. Elsevier Ltd, pp. 42–57, 01-May-2016. https://doi.org/10.1016/j.fct.2016.03.003

[17]   C. M. Ramakritinan *et al.*, "Synthesis of chitosan mediated silver nanoparticles (Ag NPs) for potential antimicrobial applications," *Front. Lab. Med.*, vol. 2, no. 1, pp. 30–35, 2018. https://doi.org/10.1016/j.flm.2018.04.002

[18]   M. Cruz-Romero, "Crop-based biodegradable packaging and its environmental implications.," *CAB Rev. Perspect. Agric. Vet. Sci. Nutr. Nat. Resour.*, vol. 3, no. 074, 2009. https://doi.org/10.1079/PAVSNNR20083074

[19]   N. Durán, P. D. Marcato, R. De Conti, O. L. Alves, F. T. M. Costa, and M. Brocchi, "Potential use of silver nanoparticles on pathogenic bacteria, their toxicity and possible mechanisms of action," *Journal of the Brazilian Chemical Society*, vol. 21, no. 6. Sociedade Brasileira de Quimica, pp. 949–959, 2010. https://doi.org/10.1590/S0103-50532010000600002

[20]   X. Wei *et al.*, "Synthesis of silver nanoparticles by solar irradiation of cell-free Bacillus amyloliquefaciens extracts and AgNO 3," *Bioresour. Technol.*, vol. 103, no. 1, pp. 273–278, Jan. 2012. https://doi.org/10.1016/j.biortech.2011.09.118

[21]   K. AbdelRahim, S. Y. Mahmoud, A. M. Ali, K. S. Almaary, A. E. Z. M. A. Mustafa, and S. M. Husseiny, "Extracellular biosynthesis of silver nanoparticles using Rhizopus stolonifer," *Saudi J. Biol. Sci.*, vol. 24, no. 1, pp. 208–216, Jan. 2017. https://doi.org/10.1016/j.sjbs.2016.02.025

Bioinspired Nanomaterials
Materials Research Foundations **111** (2021) 96-117

Materials Research Forum LLC
https://doi.org/10.21741/9781644901571-4

[22] T. Yurtluk, F. A. Akçay, and A. Avcı, "Biosynthesis of silver nanoparticles using novel Bacillus sp. SBT8," *Prep. Biochem. Biotechnol.*, vol. 48, no. 2, pp. 151–159, Feb. 2018. https://doi.org/10.1080/10826068.2017.1421963

[23] H. Singh, J. Du, P. Singh, and T. H. Yi, "Extracellular synthesis of silver nanoparticles by Pseudomonas sp. THG-LS1.4 and their antimicrobial application," *J. Pharm. Anal.*, vol. 8, no. 4, pp. 258–264, Aug. 2018. https://doi.org/10.1016/j.jpha.2018.04.004

[24] E. K. F. Elbeshehy, A. M. Elazzazy, and G. Aggelis, "Silver nanoparticles synthesis mediated by new isolates of Bacillus spp., nanoparticle characterization and their activity against Bean Yellow Mosaic Virus and human pathogens," *Front. Microbiol.*, vol. 6, no. MAY, 2015. https://doi.org/10.3389/fmicb.2015.00453

[25] V. L. Das, R. Thomas, R. T. Varghese, E. V. Soniya, J. Mathew, and E. K. Radhakrishnan, "Extracellular synthesis of silver nanoparticles by the Bacillus strain CS 11 isolated from industrialized area," *3 Biotech*, vol. 4, no. 2, pp. 121–126, Apr. 2014. https://doi.org/10.1007/s13205-013-0130-8

[26] D. Manikprabhu and K. Lingappa, "Antibacterial activity of silver nanoparticles against methicillin-resistant Staphylococcus aureus synthesized using model Streptomyces sp. pigment by photo-irradiation method," *J. Pharm. Res.*, vol. 6, no. 2, pp. 255–260, Feb. 2013. https://doi.org/10.1016/j.jopr.2013.01.022

[27] B. A. Chopade *et al.*, "Synthesis, optimization, and characterization of silver nanoparticles from Acinetobacter calcoaceticus and their enhanced antibacterial activity when combined with antibiotics," *Int. J. Nanomedicine*, p. 4277, Nov. 2013. https://doi.org/10.2147/IJN.S48913

[28] V. Gopinath *et al.*, "Biogenic synthesis, characterization of antibacterial silver nanoparticles and its cell cytotoxicity," *Arab. J. Chem.*, vol. 10, no. 8, pp. 1107–1117, Dec. 2017. https://doi.org/10.1016/j.arabjc.2015.11.011

[29] L. Devi and S. Joshi, "Ultrastructures of silver nanoparticles biosynthesized using endophytic fungi," *J. Microsc. Ultrastruct.*, vol. 3, no. 1, p. 29, Mar. 2015. https://doi.org/10.1016/j.jmau.2014.10.004

[30] S. M. Husseiny, T. A. Salah, and H. A. Anter, "Biosynthesis of size controlled silver nanoparticles by Fusarium oxysporum, their antibacterial and antitumor activities," *Beni-Suef Univ. J. Basic Appl. Sci.*, vol. 4, no. 3, pp. 225–231, Sep. 2015. https://doi.org/10.1016/j.bjbas.2015.07.004

[31] M. D. Balakumaran, R. Ramachandran, and P. T. Kalaichelvan, "Exploitation of endophytic fungus, Guignardia mangiferae for extracellular synthesis of silver

nanoparticles and their in vitro biological activities," *Microbiol. Res.*, vol. 178, pp. 9–17, Sep. 2015. https://doi.org/10.1016/j.micres.2015.05.009

[32]   L. Ma *et al.*, "Optimization for extracellular biosynthesis of silver nanoparticles by Penicillium aculeatum Su1 and their antimicrobial activity and cytotoxic effect compared with silver ions," *Mater. Sci. Eng. C*, vol. 77, pp. 963–971, Aug. 2017. https://doi.org/10.1016/j.msec.2017.03.294

[33]   K. Siegel-Hertz, V. Edel-Hermann, E. Chapelle, S. Terrat, J. M. Raaijmakers, and C. Steinberg, "Comparative microbiome analysis of a Fusarium wilt suppressive soil and a Fusarium wilt conducive soil from the Châteaurenard region," *Front. Microbiol.*, vol. 9, no. APR, p. 568, Apr. 2018. https://doi.org/10.3389/fmicb.2018.00568

[34]   M. Saravanan, S. Arokiyaraj, T. Lakshmi, and A. Pugazhendhi, "Synthesis of silver nanoparticles from Phenerochaete chrysosporium (MTCC-787) and their antibacterial activity against human pathogenic bacteria," *Microb. Pathog.*, vol. 117, pp. 68–72, Apr. 2018. https://doi.org/10.1016/j.micpath.2018.02.008

[35]   S. Kumar, A. Shukla, P. P. Baul, A. Mitra, and D. Halder, "Biodegradable hybrid nanocomposites of chitosan/gelatin and silver nanoparticles for active food packaging applications," *Food Packag. Shelf Life*, vol. 16, 2018. https://doi.org/10.1016/j.fpsl.2018.03.008

[36]   S. Paramasivan, N. E.R., R. Nagarajan, V. R. Anumakonda, and H. N., "Characterization of cotton fabric nanocomposites with in situ generated copper nanoparticles for antimicrobial applications," *Prep. Biochem. Biotechnol.*, vol. 0, no. 0, pp. 1–8, 2018.

[37]   V. Sadanand, N. Rajini, B. Satyanarayana, and A. V. Rajulu, "Preparation and properties of cellulose/silver nanoparticle composites with in situ-generated silver nanoparticles using Ocimum sanctum leaf extract," *Int. J. Polym. Anal. Charact.*, vol. 21, no. 5, pp. 408–416, 2016. https://doi.org/10.1080/1023666X.2016.1161100

[38]   P. Sivaranjana, E. R. Nagarajan, N. Rajini, M. Jawaid, and A. V. Rajulu, "Cellulose nanocomposite films with in situ generated silver nanoparticles using Cassia alata leaf extract as a reducing agent," *Int. J. Biol. Macromol.*, vol. 99, pp. 223–232, 2017. https://doi.org/10.1016/j.ijbiomac.2017.02.070

[39]   A. C. S. Almeida, E. A. N. Franco, F. M. Peixoto, K. L. F. Pessanha, and N. R. Melo, "Aplicação de nanotecnologia em embalagens de alimentos," *Polimeros*, vol. 25, no. spe, pp. 89–97, Dec. 2015. https://doi.org/10.1590/0104-1428.2069

[40]   A. S. Asger, K. Jannick, K. Jørgensen, M.-L. Knop, L. Martin, and O. Mikkelsen, "Bactericidal Effect of Silver Nanoparticles Determination of size and shape of

Bioinspired Nanomaterials
Materials Research Foundations **111** (2021) 96-117

Materials Research Forum LLC
https://doi.org/10.21741/9781644901571-4

triangular silver nanoprisms and spherical silver nanoparticles and their bactericidal effect against Escherichia coli and Bacillus subtilis."

[41]   J. R. Morones *et al.*, "JN2005," *Nanotechnology*, vol. 16, no. 10, pp. 2346–53, 2005. https://doi.org/10.1088/0957-4484/16/10/059

[42]   S. Soltani and R. Nourdahr, "Study on the Antimicrobial Effect of Nanosilver Tray Packaging of Minced Beef at Refrigerator Temperature," *Glob. Vet.*, vol. 9, no. 3, pp. 284–289, 2012.

[43]   H. Tavakoli, H. Rastegar, M. Taherian, M. Samadi, and H. Rostami, "The effect of nano-silver packaging in increasing the shelf life of nuts: An in vitro model," *Ital. J. Food Saf.*, vol. 6, no. 4, pp. 156–161, Jan. 2017. https://doi.org/10.4081/ijfs.2017.6874

[44]   L. Kuuliala *et al.*, "Preparation and antimicrobial characterization of silver-containing packaging materials for meat," *Food Packag. Shelf Life*, vol. 6, pp. 53–60, Dec. 2015. https://doi.org/10.1016/j.fpsl.2015.09.004

[45]   F. Beigmohammadi, S. H. Peighambardoust, J. Hesari, S. Azadmard-Damirchi, S. J. Peighambardoust, and N. K. Khosrowshahi, "Antibacterial properties of LDPE nanocomposite films in packaging of UF cheese," *LWT - Food Sci. Technol.*, vol. 65, pp. 106–111, 2016. https://doi.org/10.1016/j.lwt.2015.07.059

[46]   W. Li, L. Li, H. Zhang, M. Yuan, and Y. Qin, "Evaluation of PLA nanocomposite films on physicochemical and microbiological properties of refrigerated cottage cheese," *J. Food Process. Preserv.*, vol. 42, no. 1, p. e13362, Jan. 2018. https://doi.org/10.1111/jfpp.13362

[47]   H. Chi *et al.*, "Effect of PLA nanocomposite films containing bergamot essential oil, TiO 2 nanoparticles, and Ag nanoparticles on shelf life of mangoes," *Sci. Hortic. (Amsterdam).*, vol. 249, pp. 192–198, Apr. 2019. https://doi.org/10.1016/j.scienta.2019.01.059

[48]   J. Pulit-Prociak, J. Chwastowski, A. Kucharski, and M. Banach, "Functionalization of textiles with silver and zinc oxide nanoparticles," *Appl. Surf. Sci.*, vol. 385, pp. 543–553, Nov. 2016. https://doi.org/10.1016/j.apsusc.2016.05.167

[49]   R. Raliya and J. C. Tarafdar, "ZnO Nanoparticle Biosynthesis and Its Effect on Phosphorous-Mobilizing Enzyme Secretion and Gum Contents in Clusterbean (Cyamopsis tetragonoloba L.)," *Agric. Res.*, vol. 2, no. 1, pp. 48–57, Jan. 2013. https://doi.org/10.1007/s40003-012-0049-z

[50]   Shamsuzzaman, A. Mashrai, H. Khanam, and R. N. Aljawfi, "Biological synthesis of ZnO nanoparticles using C. albicans and studying their catalytic performance in the

synthesis of steroidal pyrazolines," *Arab. J. Chem.*, vol. 10, pp. S1530–S1536, May 2017. https://doi.org/10.1016/j.arabjc.2013.05.004

[51]   M. D. Rao and P. Gautam, "Synthesis and characterization of ZnO nanoflowers using *Chlamydomonas reinhardtii* : A green approach," *Environ. Prog. Sustain. Energy*, vol. 35, no. 4, pp. 1020–1026, Jul. 2016. https://doi.org/10.1002/ep.12315

[52]   S. Azizi, M. B. Ahmad, F. Namvar, and R. Mohamad, "Green biosynthesis and characterization of zinc oxide nanoparticles using brown marine macroalga Sargassum muticum aqueous extract," *Mater. Lett.*, vol. 116, pp. 275–277, Feb. 2014. https://doi.org/10.1016/j.matlet.2013.11.038

[53]   A. Król, P. Pomastowski, K. Rafińska, V. Railean-Plugaru, and B. Buszewski, "Zinc oxide nanoparticles: Synthesis, antiseptic activity and toxicity mechanism," *Advances in Colloid and Interface Science*, vol. 249. Elsevier B.V., pp. 37–52, 01-Nov-2017. https://doi.org/10.1016/j.cis.2017.07.033

[54]   R. Yuvakkumar, J. Suresh, A. J. Nathanael, M. Sundrarajan, and S. I. Hong, "Novel green synthetic strategy to prepare ZnO nanocrystals using rambutan (Nephelium lappaceum L.) peel extract and its antibacterial applications," *Mater. Sci. Eng. C*, vol. 41, pp. 17–27, Aug. 2014. https://doi.org/10.1016/j.msec.2014.04.025

[55]   S. Nagarajan and K. Arumugam Kuppusamy, "Extracellular synthesis of zinc oxide nanoparticle using seaweeds of gulf of Mannar, India," *J. Nanobiotechnology*, vol. 11, no. 1, pp. 1–11, Dec. 2013. https://doi.org/10.1186/1477-3155-11-39

[56]   L. Xiao, C. Liu, X. Chen, and Z. Yang, "Zinc oxide nanoparticles induce renal toxicity through reactive oxygen species," *Food Chem. Toxicol.*, vol. 90, pp. 76–83, Apr. 2016. https://doi.org/10.1016/j.fct.2016.02.002

[57]   H. Esmailzadeh, P. Sangpour, F. Shahraz, J. Hejazi, and R. Khaksar, "Effect of nanocomposite packaging containing ZnO on growth of Bacillus subtilis and Enterobacter aerogenes," *Mater. Sci. Eng. C*, vol. 58, pp. 1058–1063, Jan. 2016. https://doi.org/10.1016/j.msec.2015.09.078

[58]   A. Babaei-Ghazvini, I. Shahabi-Ghahfarrokhi, and V. Goudarzi, "Preparation of UV-protective starch/kefiran/ZnO nanocomposite as a packaging film: Characterization," *Food Packag. Shelf Life*, vol. 16, pp. 103–111, Jun. 2018. https://doi.org/10.1016/j.fpsl.2018.01.008

[59]   V. K. Kotharangannagari and K. Krishnan, "Biodegradable hybrid nanocomposites of starch/lysine and ZnO nanoparticles with shape memory properties," *Mater. Des.*, vol. 109, pp. 590–595, Nov. 2016. https://doi.org/10.1016/j.matdes.2016.07.046

[60]   M. Mizielińska, U. Kowalska, M. Jarosz, and P. Sumińska, "A Comparison of the Effects of Packaging Containing Nano ZnO or Polylysine on the Microbial Purity and Texture of Cod (Gadus morhua) Fillets," *Nanomaterials*, vol. 8, no. 3, p. 158, Mar. 2018. https://doi.org/10.3390/nano8030158

[61]   S. V. Calderon, B. Gomes, P. J. Ferreira, and S. Carvalho, "Zinc nanostructures for oxygen scavenging," *Nanoscale*, vol. 9, no. 16, pp. 5254–5262, Apr. 2017. https://doi.org/10.1039/C7NR01367A

[62]   X. Li, Y. Xing, Y. Jiang, Y. Ding, and W. Li, "Antimicrobial activities of ZnO powder-coated PVC film to inactivate food pathogens," *Int. J. Food Sci. Technol.*, vol. 44, no. 11, pp. 2161–2168, Nov. 2009. https://doi.org/10.1111/j.1365-2621.2009.02055.x

[63]   "Effects of Nano-ZnO Power-Coated PVC Film on the Physiological Properties and Microbiological Changes of Fresh-Cut 'Fuji' Apple | Scientific.Net.". https://doi.org/10.4028/www.scientific.net/AMR.152-153.450

[64]   W. Li, L. Li, Y. Cao, T. Lan, H. Chen, and Y. Qin, "Effects of PLA Film Incorporated with ZnO Nanoparticle on the Quality Attributes of Fresh-Cut Apple," *Nanomaterials*, vol. 7, no. 8, p. 207, Jul. 2017. https://doi.org/10.3390/nano7080207

[65]   S. Beak, H. Kim, and K. Bin Song, "Characterization of an Olive Flounder Bone Gelatin-Zinc Oxide Nanocomposite Film and Evaluation of Its Potential Application in Spinach Packaging," *J. Food Sci.*, vol. 82, no. 11, pp. 2643–2649, Nov. 2017. https://doi.org/10.1111/1750-3841.13949

[66]   S. S. K., M. P. Indumathi, and G. R. Rajarajeswari, "Mahua oil-based polyurethane/chitosan/nano ZnO composite films for biodegradable food packaging applications," *Int. J. Biol. Macromol.*, vol. 124, pp. 163–174, Mar. 2019. https://doi.org/10.1016/j.ijbiomac.2018.11.195

[67]   A. Jayakumar *et al.*, "Starch-PVA composite films with zinc-oxide nanoparticles and phytochemicals as intelligent pH sensing wraps for food packaging application," *Int. J. Biol. Macromol.*, vol. 136, pp. 395–403, Sep. 2019. https://doi.org/10.1016/j.ijbiomac.2019.06.018

[68]   H. Almasi, P. Jafarzadeh, and L. Mehryar, "Fabrication of novel nanohybrids by impregnation of CuO nanoparticles into bacterial cellulose and chitosan nanofibers: Characterization, antimicrobial and release properties," *Carbohydr. Polym.*, vol. 186, pp. 273–281, Apr. 2018. https://doi.org/10.1016/j.carbpol.2018.01.067

[69]   S. Shende, A. P. Ingle, A. Gade, and M. Rai, "Green synthesis of copper nanoparticles by Citrus medica Linn. (Idilimbu) juice and its antimicrobial activity,"

Materials Research Forum LLC
https://doi.org/10.21741/9781644901571-4

World J. Microbiol. Biotechnol., vol. 31, no. 6, pp. 865–873, Apr. 2015. https://doi.org/10.1007/s11274-015-1840-3

[70]   R. Khani, B. Roostaei, G. Bagherzade, and M. Moudi, "Green synthesis of copper nanoparticles by fruit extract of Ziziphus spina-christi (L.) Willd.: Application for adsorption of triphenylmethane dye and antibacterial assay," *J. Mol. Liq.*, vol. 255, pp. 541–549, Apr. 2018. https://doi.org/10.1016/j.molliq.2018.02.010

[71]   S. Thakur, S. Sharma, S. Thakur, and R. Rai, "Green Synthesis of Copper Nano-Particles Using Asparagus adscendens Roxb. Root and Leaf Extract and Their Antimicrobial Activities," *Int. J. Curr. Microbiol. Appl. Sci.*, vol. 7, no. 04, pp. 683–694, Apr. 2018. https://doi.org/10.20546/ijcmas.2018.704.077

[72]   I. Chung *et al.*, "Green synthesis of copper nanoparticles using eclipta prostrata leaves extract and their antioxidant and cytotoxic activities," *Exp. Ther. Med.*, vol. 14, no. 1, pp. 18–24, Jul. 2017. https://doi.org/10.3892/etm.2017.4466

[73]   M. Nasrollahzadeh and S. Mohammad Sajadi, "Green synthesis of copper nanoparticles using Ginkgo biloba L. leaf extract and their catalytic activity for the Huisgen [3+2] cycloaddition of azides and alkynes at room temperature," *J. Colloid Interface Sci.*, vol. 457, pp. 141–147, Nov. 2015. https://doi.org/10.1016/j.jcis.2015.07.004

[74]   M. Nasrollahzadeh, S. S. Momeni, and S. M. Sajadi, "Green synthesis of copper nanoparticles using Plantago asiatica leaf extract and their application for the cyanation of aldehydes using K 4 Fe(CN) 6," *J. Colloid Interface Sci.*, vol. 506, pp. 471–477, Nov. 2017. https://doi.org/10.1016/j.jcis.2017.07.072

[75]   Z. Issaabadi, M. Nasrollahzadeh, and S. M. Sajadi, "Green synthesis of the copper nanoparticles supported on bentonite and investigation of its catalytic activity," *J. Clean. Prod.*, vol. 142, pp. 3584–3591, Jan. 2017. https://doi.org/10.1016/j.jclepro.2016.10.109

[76]   M. A. Asghar *et al.*, "Iron, copper and silver nanoparticles: Green synthesis using green and black tea leaves extracts and evaluation of antibacterial, antifungal and aflatoxin B1 adsorption activity," *LWT - Food Sci. Technol.*, vol. 90, pp. 98–107, Apr. 2018. https://doi.org/10.1016/j.lwt.2017.12.009

[77]   A. F. Jaramillo *et al.*, "Comparative Study of the Antimicrobial Effect of Nanocomposites and Composite Based on Poly(butylene adipate-co-terephthalate) Using Cu and Cu/Cu2O Nanoparticles and CuSO4," *Nanoscale Res. Lett.*, vol. 14, no. 1, pp. 1–17, May 2019. https://doi.org/10.1186/s11671-019-2987-x

Materials Research Forum LLC
https://doi.org/10.21741/9781644901571-4

[78]   M. Grigore, E. Biscu, A. Holban, M. Gestal, and A. Grumezescu, "Methods of Synthesis, Properties and Biomedical Applications of CuO Nanoparticles," *Pharmaceuticals*, vol. 9, no. 4, p. 75, Nov. 2016. https://doi.org/10.3390/ph9040075

[79]   V. Sadanand, N. Rajini, A. Varada Rajulu, and B. Satyanarayana, "Preparation of cellulose composites with in situ generated copper nanoparticles using leaf extract and their properties," *Carbohydr. Polym.*, vol. 150, pp. 32–39, Oct. 2016. https://doi.org/10.1016/j.carbpol.2016.04.121

[80]   Y. A. Arfat, J. Ahmed, N. Hiremath, R. Auras, and A. Joseph, "Thermo-mechanical, rheological, structural and antimicrobial properties of bionanocomposite films based on fish skin gelatin and silver-copper nanoparticles," *Food Hydrocoll.*, vol. 62, pp. 191–202, Jan. 2017. https://doi.org/10.1016/j.foodhyd.2016.08.009

[81]   A. Eivazihollagh *et al.*, "One-pot synthesis of cellulose-templated copper nanoparticles with antibacterial properties," *Mater. Lett.*, vol. 187, pp. 170–172, Jan. 2017. https://doi.org/10.1016/j.matlet.2016.10.026

[82]   C. Sharma, R. Dhiman, N. Rokana, and H. Panwar, "Nanotechnology: An untapped resource for food packaging," *Frontiers in Microbiology*, vol. 8, no. SEP. Frontiers Media S.A., 12-Sep-2017. https://doi.org/10.3389/fmicb.2017.01735

[83]   A. Llorens, E. Lloret, P. Picouet, and A. Fernandez, "Study of the antifungal potential of novel cellulose/copper composites as absorbent materials for fruit juices," *Int. J. Food Microbiol.*, vol. 158, no. 2, pp. 113–119, Aug. 2012. https://doi.org/10.1016/j.ijfoodmicro.2012.07.004

[84]   S. Ebrahimiasl and A. Rajabpour, "Synthesis and characterization of novel bactericidal Cu/HPMC BNCs using chemical reduction method for food packaging," *J. Food Sci. Technol.*, vol. 52, no. 9, pp. 5982–5988, Sep. 2015. https://doi.org/10.1007/s13197-014-1615-0

[85]   D. Longano *et al.*, "Analytical characterization of laser-generated copper nanoparticles for antibacterial composite food packaging," in *Analytical and Bioanalytical Chemistry*, 2012, vol. 403, no. 4, pp. 1179–1186. https://doi.org/10.1007/s00216-011-5689-5

[86]   S. Shankar and J. W. Rhim, "Effect of copper salts and reducing agents on characteristics and antimicrobial activity of copper nanoparticles," *Mater. Lett.*, vol. 132, pp. 307–311, Oct. 2014. https://doi.org/10.1016/j.matlet.2014.06.014

[87]   D. N. Bikiaris and K. S. Triantafyllidis, "HDPE/Cu-nanofiber nanocomposites with enhanced antibacterial and oxygen barrier properties appropriate for food packaging applications," *Mater. Lett.*, vol. 93, pp. 1–4, Feb. 2013. https://doi.org/10.1016/j.matlet.2012.10.128

Materials Research Forum LLC
https://doi.org/10.21741/9781644901571-4

[88] H. M. Yadav, J. S. Kim, and S. H. Pawar, "Developments in photocatalytic antibacterial activity of nano TiO2: A review," *Korean Journal of Chemical Engineering*, vol. 33, no. 7. Springer New York LLC, pp. 1989–1998, 01-Jul-2016. https://doi.org/10.1007/s11814-016-0118-2

[89] G. Rajakumar, A. A. Rahuman, B. Priyamvada, V. G. Khanna, D. K. Kumar, and P. J. Sujin, "Eclipta prostrata leaf aqueous extract mediated synthesis of titanium dioxide nanoparticles," *Mater. Lett.*, vol. 68, pp. 115–117, Feb. 2012. https://doi.org/10.1016/j.matlet.2011.10.038

[90] A. Vishnu Kirthi *et al.*, "Biosynthesis of titanium dioxide nanoparticles using bacterium Bacillus subtilis," *Mater. Lett.*, vol. 65, no. 17–18, pp. 2745–2747, Sep. 2011. https://doi.org/10.1016/j.matlet.2011.05.077

[91] Q. He, Y. Zhang, X. Cai, and S. Wang, "Fabrication of gelatin-TiO2 nanocomposite film and its structural, antibacterial and physical properties," *Int. J. Biol. Macromol.*, vol. 84, pp. 153–160, Mar. 2016. https://doi.org/10.1016/j.ijbiomac.2015.12.012

[92] S. A. Oleyaei, Y. Zahedi, B. Ghanbarzadeh, and A. A. Moayedi, "Modification of physicochemical and thermal properties of starch films by incorporation of TiO2 nanoparticles," *Int. J. Biol. Macromol.*, vol. 89, pp. 256–264, Aug. 2016. https://doi.org/10.1016/j.ijbiomac.2016.04.078

[93] A. Nešić *et al.*, "Pectin-based nanocomposite aerogels for potential insulated food packaging application," *Carbohydr. Polym.*, vol. 195, pp. 128–135, Sep. 2018. https://doi.org/10.1016/j.carbpol.2018.04.076

[94] H. Li, J. Yang, P. Li, T. Lan, and L. Peng, "A facile method for preparation superhydrophobic paper with enhanced physical strength and moisture-proofing property," *Carbohydr. Polym.*, vol. 160, pp. 9–17, Mar. 2017. https://doi.org/10.1016/j.carbpol.2016.12.018

[95] C. López de Dicastillo, C. Patiño, M. J. Galotto, J. L. Palma, D. Alburquenque, and J. Escrig, "Novel antimicrobial titanium dioxide nanotubes obtained through a combination of atomic layer deposition and electrospinning technologies," *Nanomaterials*, vol. 8, no. 2, Feb. 2018. https://doi.org/10.3390/nano8020128

[96] Y. Xing *et al.*, "Effect of TiO 2 nanoparticles on the antibacterial and physical properties of polyethylene-based film," *Prog. Org. Coatings*, vol. 73, no. 2–3, pp. 219–224, Feb. 2012. https://doi.org/10.1016/j.porgcoat.2011.11.005

[97] D. Roilo, C. A. Maestri, M. Scarpa, P. Bettotti, and R. Checchetto, "Gas barrier and optical properties of cellulose nanofiber coatings with dispersed TiO2

Materials Research Forum LLC

https://doi.org/10.21741/9781644901571-4

nanoparticles," *Surf. Coatings Technol.*, vol. 343, pp. 131–137, Jun. 2018. https://doi.org/10.1016/j.surfcoat.2017.10.015

[98]   A. Mihaly-Cozmuta *et al.*, "Preparation and characterization of active cellulose-based papers modified with TiO2, Ag and zeolite nanocomposites for bread packaging application," *Cellulose*, vol. 24, no. 9, pp. 3911–3928, Sep. 2017. https://doi.org/10.1007/s10570-017-1383-x

[99]   D. Li, Q. Ye, L. Jiang, and Z. Luo, "Effects of nano-TiO $_2$ -LDPE packaging on postharvest quality and antioxidant capacity of strawberry (*Fragaria ananassa* Duch.) stored at refrigeration temperature," *J. Sci. Food Agric.*, vol. 97, no. 4, pp. 1116–1123, Mar. 2017. https://doi.org/10.1002/jsfa.7837

[100]   K. Velayutham *et al.*, "Evaluation of Catharanthus roseus leaf extract-mediated biosynthesis of titanium dioxide nanoparticles against Hippobosca maculata and Bovicola ovis," *Parasitol. Res.*, vol. 111, no. 6, pp. 2329–2337, Dec. 2012. https://doi.org/10.1007/s00436-011-2676-x

[101]   N. A. Órdenes-Aenishanslins, L. A. Saona, V. M. Durán-Toro, J. P. Monrás, D. M. Bravo, and J. M. Pérez-Donoso, "Use of titanium dioxide nanoparticles biosynthesized by Bacillus mycoides in quantum dot sensitized solar cells," *Microb. Cell Fact.*, vol. 13, no. 1, pp. 1–10, Jul. 2014. https://doi.org/10.1186/s12934-014-0090-7

[102]   C. Swaroop and M. Shukla, "Nano-magnesium oxide reinforced polylactic acid biofilms for food packaging applications," *Int. J. Biol. Macromol.*, vol. 113, pp. 729–736, Jul. 2018. https://doi.org/10.1016/j.ijbiomac.2018.02.156

[103]   M. J. Khalaj, H. Ahmadi, R. Lesankhosh, and G. Khalaj, "Study of physical and mechanical properties of polypropylene nanocomposites for food packaging application: Nano-clay modified with iron nanoparticles," *Trends in Food Science and Technology*, vol. 51. Elsevier Ltd, pp. 41–48, 01-May-2016. https://doi.org/10.1016/j.tifs.2016.03.007

[104]   S. Mallakpour and H. Y. Nazari, "The influence of bovine serum albumin-modified silica on the physicochemical properties of poly(vinyl alcohol) nanocomposites synthesized by ultrasonication technique," *Ultrason. Sonochem.*, vol. 41, pp. 1–10, Mar. 2018. https://doi.org/10.1016/j.ultsonch.2017.09.017

Bioinspired Nanomaterials
Materials Research Foundations **111** (2021) 118-154

Materials Research Forum LLC
https://doi.org/10.21741/9781644901571-5

Chapter 5

Bio-Mediated Synthesis of Metal Nanomaterials for SERS Application

Sangeetha Kumaravel[1,2] and Subrata Kundu[1,2,*]

[1]Electrochemical Process Engineering (EPE) division, CSIR-Central Electrochemical Research Institute (CECRI), Karaikudi-630003, Tamil Nadu, India

[2]Academy of Scientific and Innovative Research (AcSIR), Ghaziabad-201002, India

*skundu@cecri.res.in and kundu.subrata@gmail.com

Abstract

The discovery of nanomaterials (NMs) caused a great revolution in the field of science especially in material science. The highly exotic and tunable size and shape of NMs have devoted more interest due to their unique physiochemical properties. There are various methods and methodologies involved to prepare NMs in a desired morphology. Among these, the fabrication of bio-molecules mediated NMs are highly attractive because their size and shape can be easily tuned by simple, eco-friendly and reliable way. Deoxyribonucleic acid (DNA) is considered to be one of the most promising and well-studied bio-molecule in the fabrication of various types of NMs. The rich functionalities with the double-helix structure of DNA facilitate to accommodate a higher number of metal ions on its surface and results in perfect chain-like nano-assemblies. Moreover, the DNA mediated NMs can be highly useful for the Surface Enhanced Raman Scattering (SERS) studies with appropriate analytes. The SERS technique provides the fingerprint information of the analyte molecules even at very low concentration (such as even in ppm levels). The SERS intensity is greatly influenced by the size and shape of the NMs prepared using DNA scaffolds due to their assembly in a close proximity and generation of higher number of 'hot spots'. In this present book chapter, we elaborated the numerous methodologies involved for the synthesis of DNA-based NMs considering their size, shapes, and also highlighted the possible mechanism involved for their growth with DNA scaffolds. In-addition, the possible application of DNA mediated NMs towards SERS studies has also been detailed in this book chapter.

Bioinspired Nanomaterials
Materials Research Foundations **111** (2021) 118-154

Materials Research Forum LLC
https://doi.org/10.21741/9781644901571-5

Keywords

Nanomaterials, Biomolecules, DNA, Incubation, Photoreduction, Seed-Mediated, Inter-Partical Distance, SPR, Generation, Noble Metals, SERS, Enhancement Factor, Methylene Blue, Rhodamine 6G, Tumor Detection

Contents

Bioinspired Nanomaterials   Materials Research Forum LLC
Materials Research Foundations **111** (2021) 118-154   https://doi.org/10.21741/9781644901571-5

1. Introduction

Currently, the nanomaterials (NMs) reached significant economic impact with the advantage of large tunable property compared to its bulk counterpart [1–4]. Indeed, NMs have existed in nature from the origin of earth, for example NMs from volcanoes and forest fires [5]. Later, the unintentional NMs produced by human being from the process of mining and factory emissions were called as thropogenic were identified [6]. From the last fifty years, particularly after the industrial revolution, NMs engineered for various applications, but comparatively lesser quantity to those naturally existing and unintentional nanomaterials in the earth [6]. Those engineered NMs generally engaged with nanoscience and nanotechnology which applied across physics, chemistry, biology and geology [7]. These interdisciplinary advancement in the NMs development shows rapid changes in the device fabrication [8], drug designs [9] and biomedical applications [10]. Usually, these NMs are having high surface area with flexibility in their electric properties [1]. In 29th December 1959, the first concept about the nanoscience and nanotechnology was explained in a title "There's Plenty of Room at the Bottom" by physicist Richard Feynman in American Physical Society meeting at the California Institute of Technology [11]. Several methodologies were followed to develop such nano-sized (10^{-9} of a meter) material with high selectivity towards their size and shapes [12–15]. Generally, there are two approaches such as top-down and bottom-up methods being used in the NMs synthesis which includes photolithography, reduction and pyrolysis methods [16]. The use of synthetic structural directing agents such as citrates [17,18], carbon nanotubes (CNTs) [19], silica materials [20], polyvinylpyrrolidone [21] (PVP), various surfactants such as cetyltrimethylammonium bromide (CTAB, cationic surfactant) [22,23], tetradecyl phosphate, Hexadecyl ammine (cationic surfactant), dodecyl ammine (neutral surfactant) and zeolites resulting in different morphological NMs such as nano-wires, cubes, spheres, clusters, rods and so on [21,24]. These synthetic structural directing agents can be widely applied in the NMs synthesis although for biological systems it may cause more side effects in practice [9]. The biomedical application of NMs for drug delivery systems need special attention during NMs synthesis because synthesized NMs should be stable, should disperse properly in the proximal fluid (mostly blood) [10]. Therefore, there are increasing attentions towards many scientific communities for designing bio-based NMs.

In this line, naturally existing bio-molecule based synthesis of NMs gain tremendous attention for the advancement in material science [25,26]. The bio-mediated NMs showed advantageous over other methods due to the less-toxicity and requires only of mild synthetic process [26]. Various microorganisms are rich in functional groups with their patterns and size that results in a nanoscale level production of materials in an environment friendly way. These bio-molecules can act as template or capping agent for getting a perfect size and shape selectivity towards NMs [27]. Different microorganisms such as proteins [28], amino acids [29], viruses [30], DNA [31] and enzymes [28], plants [32] and biomass [32] have been utilized for designing NMs. In 2001, Niemeyer detailed a review about the coupling of NMs with the bio-molecules including immunoglobulins, serum albumins, proteins, peptides, DNA, glutathione and hemeproteins [12]. Another main advantages of bio-molecules is that both nanomaterials and most of the bio-molecules are coming under same nanometer scale regime **(Figure 1)**. In this book chapter we have highlighted the use of DNA as scaffolds/template for the design of NMs and the use of DNA based NMs applied for Surface Enhanced Raman Scattering (SERS) studies.

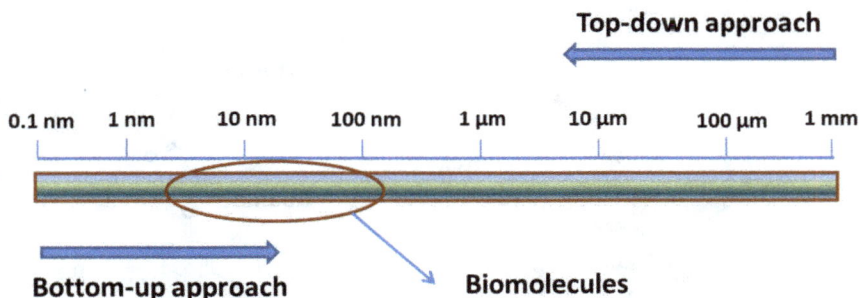

Figure 1. Two different approaches in the nanomaterials fabrication.

2. Styles and advantages of DNA in nanomaterials designs

Deoxyribonucleic acid (DNA) has been used in the architecture of novel NMs by simple synthetic procedures which results in a nanowires and nanostructures [12]. The DNA double helix structure exhibits a diameter ~ 2 nm and a length of 0.34 nm that largely constrains 3.4 nm lengths of ten helix together [33]. DNA structure contains the nucleotide base pairs of adenine (A), guanine (G), thymine (T) and cytosine (C) are interconnected by the hydrogen bonding. Also, the side chain of DNA contains monophosphorylated deoxyribose sugar units **(Figure 2)** [34]. One among the advantages with DNA is it is rich in functionality, flexibility, stability and also the wide angle tolerance makes it more suitable for developing advanced material in the bottom-up

Bioinspired Nanomaterials
Materials Research Foundations **111** (2021) 118-154

Materials Research Forum LLC
https://doi.org/10.21741/9781644901571-5

approach [35]. The Watson and Crick discovery of DNA was passed half the century, the perfect polymeric double-helix structure with high mechanical strength makes the DNA assembly as a better choice in the NMs designing. In 1998, for the first time, Braun *et al.* reported that the DNA metallization used for the construction of functional circuits with silver metal NPs [36]. They reported the construction of DNA encapsulated silver nano-wire between two gold electrodes that bridges *via* connecting the two electrodes to form a complete circuit. Initially, the DNA was templated only for nanowire fabrication. After that, the recent advancement has been done with the tuning of DNA sequence resulting in multiple structures with varying size and length of DNA sequences. The development in the DNA nanotechnology provides several desired shape of 2D or 3D nano-architectures. This structural modification with the Watson-Crick base pairing DNA was first reported by Seeman in 1982. Sticky-ended ligation techniques has been used where the DNA double helix structure are covalently connected with proper space to generate 3D networks of nucleic acid [34].

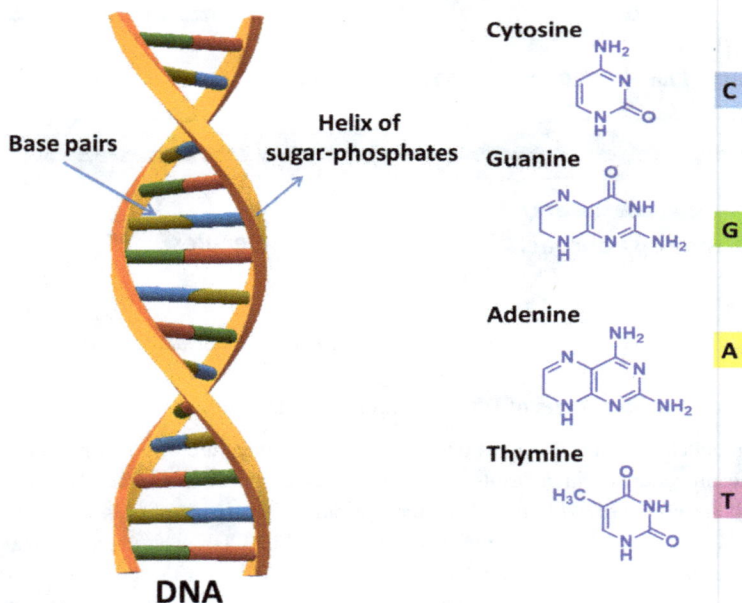

Figure 2. *DNA structure containing based pairs of adenine (A), guanine (G), thymine (T) and cytosine (C).*

DNA can be considered as a 'tool box' where the rigid sequence of DNA can be modified as double and triple crossover tile, helix bundles, cross linked four armed junctions, triangular four armed, cross-shaped junctions **(Figure 3)** [33,37]. This modified construction of DNA sequence can produces 1D, 2D and 3D templates with the increased interparticle space in sub nanometer assemblies. In 2006, the development of DNA origami technique was proposed for the production of 2D and 3D based nano-assemblies as proposed by Paul Rothemund [38]. This technique involves a folding of viral DNA into NM assembly [38]. Sharma *et al.* made an interesting result of making bivalent thiolate-Au linkage between DNA templated discrete Au nanoparticles (NPs) arrays with higher yields [39]. Thus, this benefits with this DNA origami based tuning gives large interspace cavity which readily accommodate other molecular components such as proteins.

While accommodation of other molecules, the exact programmable site assembly of DNA scaffold on the protein is a challenging part. Liu *et al.* have developed periodic linear arrays with the utilization of triple-crossover DNA tiles of aptamer sequence to which thrombin (protein) specifically binds successfully forming linear arrays [40]. There are several reports in the development of DNA based nanomaterials for diverse applications such as in catalysis, sensors, SERS and so on. Our group also highlighted DNA based NMs for multiple applications including SERS, catalysis, electrocatalysis, supercapacitor, and for feedstock conversions [41-44]. We have prepared aquasols and organosols of various noble metals such as Au, Ag, Pd, Pt, Os, Ir, Rh, Ru and Re with their molecular self-assembly with DNA [41–44]. The presence of DNA in the aqueous and organosol results in an increase in stability and long-lasting property of the material [41,43]. Apart from the noble metals, we have also synthesized DNA templated chalcogenide materials such as CoS, CoSe, and NiTe, for water splitting electrocatalysis [45–47].

In this book chapter, we have mainly highlighted the development of DNA based noble metal NMs, their detailed *in-situ* and *ex-situ* properties and possible application in SERS studies as discussed below.

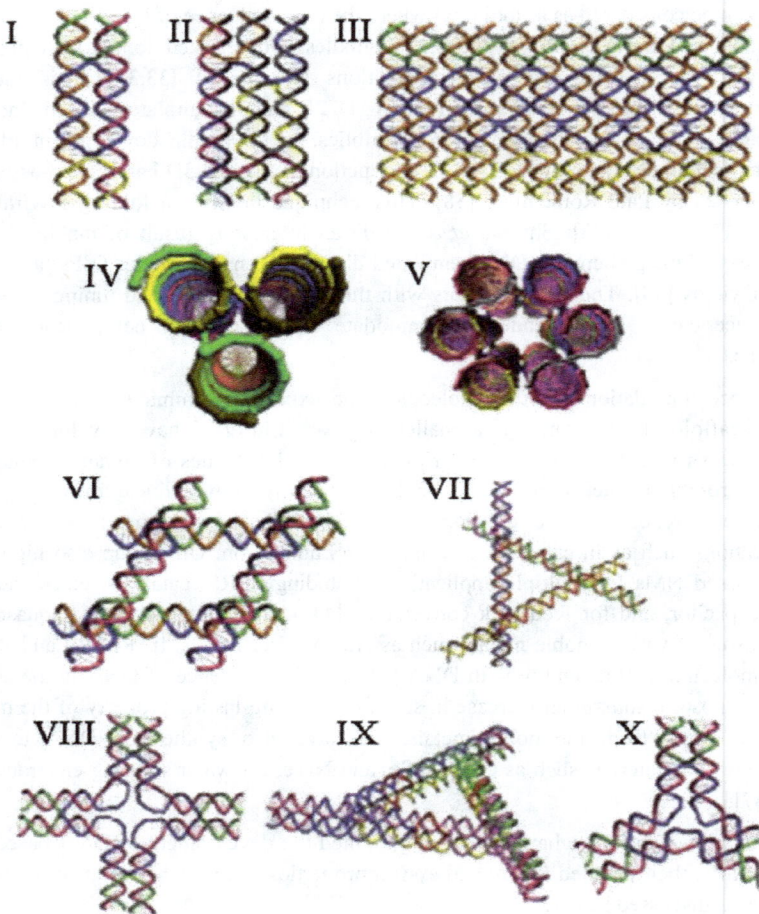

Figure 3. Types of DNA sequences, double and triple cross tile (I and II), 12 set of tile (III), 3 and six bundle tiles (IV and V), cross coupled, triangle tile and double cross tiles (VI - X). Reprinted from ref. 33 with permission. Copyright 2012. American Chemical Society.

Bioinspired Nanomaterials

Materials Research Forum LLC

Materials Research Foundations **111** (2021) 118-154

https://doi.org/10.21741/9781644901571-5

3. Introduction of surface enhanced raman spectroscopy (SERS)

Surface Enhanced Raman Spectroscopy (SERS) also abbreviated as Surface Enhanced Raman Scattering is a powerful surface-sensitive technique which is used for molecular detection and characterization of analyte with enhanced Raman scattering by several order [48]. The SERS happens when the analyte molecule adsorbed or very closes to the metal surface [49,50]. The laser beam with certain wavelength strikes the metal surface and produces the surface plasmons, which then interacts with the analyte molecules significantly enhancing the Raman signals in a few order of magnitude [51–53]. The SERS was first observed by Fleischmann *et al.* in 1974, the Raman signals of pyridine molecule adsorbed on highly roughened silver surface greatly enhanced by many orders of magnitude [54]. But this group has not realized the exact reason behind the enhancement in the Raman signals. Later on Van Duyne *et al.* explained the physical phenomenon with proper calculations of how the Raman signal enhancement happened [51]. After the discovery of SERS, it has been widely used in various applications and it gives information about chemical composition and also structure of target molecule.

3.1 Mechanisms in SERS

The SERS enhancement factor (EF) is used to determine the enhanced Raman intensity, which is written below [55],

$$EF = \frac{I_{SERS}}{N_{SERS}} \times \frac{N_{Raman}}{I_{Raman}}$$

Where N_{Raman} and N_{SERS} signifies the number of probe molecules which contribute to signal intensity of normal and enhanced respectively, while I_{Raman} and I_{SERS} denote the corresponding normal Raman and SERS intensities.

The EF was mainly influenced by two kinds of reaction mechanisms named as electromagnetic enhancement (EM) and chemical enhancement (CHEM) [48]. In both mechanisms, the intensity of Raman signals is boosted by producing more localized surface plasmon resonances (LSPR). Hence, understanding the mechanisms is essential to design NMs for enhancing the Raman signals which were explained briefly as follows.

3.1.1 Electromagnetic enhancement (EM)

In the EM mechanism, the incident laser light interacts with the localized surface plasmon resonance (LSPR) of the plasmonic nanomaterial resulting in the generation of secondary electric field which effectively enhances the electromagnetic field of NMs (Figure 4). This enhanced electromagnetic field further boosts the SERS signal of analyte

Bioinspired Nanomaterials
Materials Research Foundations **111** (2021) 118-154

Materials Research Forum LLC
https://doi.org/10.21741/9781644901571-5

molecules which is adsorbed on the rough surfaced NMs. To profit from this enhancement, analyte molecule must be in close contact (about 10 nm) with the plasmonic nanostructure surfaces. The molecule should be adsorbed on the surface nanostructures either by chemical or physical adsorption. EM contribution is higher in the SERS signal enhancement in the order of 10^4 to 10^7 and the enhancement factor (EF) has been calculated using the given formula above.

3.1.2 Chemical enhancement (CHEM)

Chemical enhancement (CHEM) is another kind of mechanism in SERS which increase the enhancement factor (EF) but comparatively lesser than EM effect. It involves in the enhancement of Raman signals through the charge transfer mechanism between the nanostructured material and analyte molecules (Figure 4). Its contribution in enhancing the EF is much lesser than EM. The charge transfer mechanism is mostly accepted as explanation for CHEM. The analyte molecule should directly make a chemical bonding with the NMs in order to occur CHEM. The EF almost lies in the order of 10 to 100 in the case of CHEM. The exact factor for enhancement with week Raman signal and definition are still subject to argument.

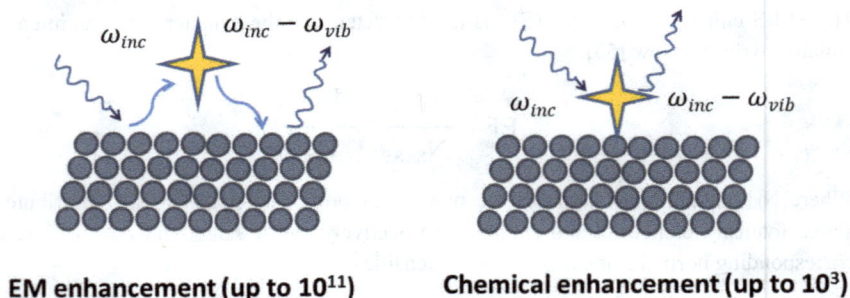

EM enhancement (up to 10^{11})   **Chemical enhancement (up to 10^3)**

Figure 4. The Electromagnetic and Chemical enhancement mechanism involving in SERS.

4.   Factors influencing the SERS enhancement

The wave length, polarization and the angle of incidence laser light is one of the important factors that can influence the EF. The size and shape of metal NPs (MNPs) greatly influence the SERS enhancement. In particular, the nano-chains and nano-wires kind of MNPs shows highest enhancement over other structures. The analyte interaction

Bioinspired Nanomaterials   Materials Research Forum LLC
Materials Research Foundations **111** (2021) 118-154   https://doi.org/10.21741/9781644901571-5

with NMs such as the adsorption efficiency (surface coverage), orientation of analyte is important parameters in SERS enhancement.

4.1 Role of hotspot in SERS

Since the EF is greatly dependence on EM effect, the researchers focus on developing nano-structured materials to enhance such EF [56]. In general, EF changes with the electric field generated at NMs and also it is proportional to 10^4 of the electric field [57]. In plasmonic nanostructures the electromagnetic field is not evenly arranged but located spatially in constricted regions called as hotspots, it plays a vital role in enhancing the Raman signals of analyte molecule by increasing the EF [58]. The hotspots are mostly located at the edges and corners of the NMs. Such, hotspots are classified in to three types called as first, second and third generation categories. The gold (Au) and silver (Ag) NPs with irregular size and shapes comes under the first generation category.

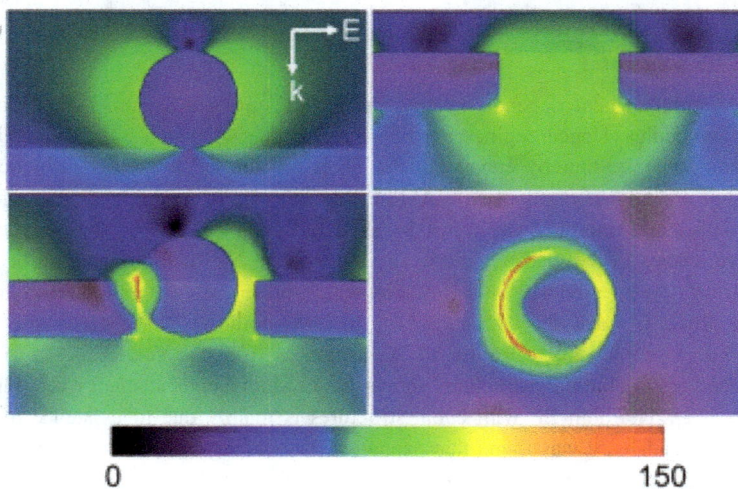

Figure 5. Electric field amplitude of single nanoparticle, single nano-hole and coupled hole–particle structure. Reprinted from ref. 59 with permission. Copyright 2013. Royal Society of Chemistry.

Mostly these are made from homogeneously freely suspended single particles like nano-spheres or nano-cubes [59]. Similar kind of detection observed with varying Au

Bioinspired Nanomaterials
Materials Research Foundations **111** (2021) 118-154

Materials Research Forum LLC
https://doi.org/10.21741/9781644901571-5

nanostructures such as single NPs, single nano-hole and coupled hole–particle **(Figure 5)** [60]. The nano-gaps which are generated between the two coupled nano structures of 1 to 5 nm size particles or nanoaggregates belong to second generation hotspots and show remarkable SERS activity. The hotspots are generated at the edges of the single nanoparticle. This kind of hotspots exhibits almost two to four orders higher EF over the first generation hotspots. The third generation hot spots are most recent findings. These are mostly explored in Tip Enhanced Raman Spectroscopic techniques (TERS) where the hotspots are exploiting the dielectric properties of NMs. Ding *et al.* summarized these three types of hotspots which can be used in trace molecule detection [59]. In the case of second generation category, the hotspots are generated throughout the particles and the observed hotspots are larger than first generation. In the third generation category, hotspots are created due to dielectric properties of nanostructures. Hence, engineering the hotspots in NMs is essential in enhancing the Raman signal intensity to detect the analyte molecules at very low concentration range.

4.2   SERS substrates

The SERS enhancement depends on how the analyte molecules can interact with the substrate molecules. Hence engineering of nanostructures to increase the interaction of analyte and substrate thereby enhancing the EF is crucial. EF is greatly influenced by varying size and shapes of nanostructures. In literature Au, Ag and copper (Cu) metallic nanostructures are emerging substrates in SERS analysis because of their remarkable LSPR frequencies in the visible region. The Au and Ag NPs commonly used in SERS analysis over Cu because Cu is less stable which undergoes air oxidation at normal conditions [61]. Besides, the size and shape of NMs could effectively alter the SERS activity. For example, the SERS EF of Ag increases from the size of 60 to 100 nm particles and also cubic nanostructures shows 10^5 orders of enhancement over spherical structures [62]. The high EF with cubic structures is mainly due to existence of more hotspots at the edges. It was found that nanoaggregates of Ag colloidal particles exhibits 10^{14} orders of enhanced SERS activity because of generation of nano-gaps during aggregation helps to create SERS hotspots throughout the particles. The bi-metallic NMs such as Au-Ag and Au-Pd were also attractive in enhancing the SERS activity by creating more hotspots [63,64]. Apart from noble metals, transition metals such as Rhodium, Ruthenium and Osmium NPs also acts as SERS substrates [65]. In line, DNA mediated NMs are used as emerging SERS substrates and showing its great role in enhancing the EF. The DNA-mediated Au-Ag nano-mushrooms exhibit the SERS enhancement in higher order of 10^9 [52]. The Au NPs assemblies which are functionalized by DNA show remarkable SERS activity [66]. Rhodamine 6G, 4-aminothiophenol, pyridine and 2,4,6-trinitrotoluene molecule were analyzed at very low concentrations in the range of 10^{-5} to

10^{-6} M using SERS analysis by taking DNA-Au NPs of 2-3 nm size [67]. Thus, engineering of nanostructured materials with finite sizes and shapes will help in enhancing the SERS activity.

5. Methodologies in the synthesis of DNA mediated nanomaterials

The biomolecule DNA based NMs synthesis is the bottom-up approach in the development of two dimensional and three dimensional nanostructures. The presence of double helix pattern of DNA facilitates to accommodating the metal ions and controlling the nucleation of NMs [16]. Also, the wide programmable nature with different stable sequences of DNA is the foremost advantage in the NMs fabrication [33]. It has a strong interaction with the metal ions and stops agglomeration of the NMs. The synthesis of DNA based NMs involved wide range of methodology where each method has its own advantage and disadvantage of control over the reaction, morphology, scalability, stability and time. Among the different methods, the incubation method, wet-chemical method, photo reduction method and seed-mediated methods are more effective, stable and widely used in the DNA based NMs synthesis. In addition to this, there are few other methods such as microwave heating, electrochemical deposition and localized DNA medication methods are also in general practice for their synthesis. The methodology explained below precisely showed various works considering DNA based NMs for SERS application.

5.1 Incubation method of preparing DNA based nanomaterials

The incubation method is a simple and effective way of assembling the metal NPs over DNA, especially in the dimer preparations. It involves, the presence of negatively charged phosphate groups and a sugar moiety in the double helix of DNA has been electrostatically interacting with the metal ions [17]. The time for this process of interaction state is called as incubation period. Based on the DNA tile sequence or their nature, the interaction sites may vary which results in a signification tuning in the structures of the NMs. This incubation method is comparatively simple to produce various 2D and 3D nanostructures [68]. The period of incubation and the pH maintenance is the key point in this method. Thacker *et al.* have developed a DNA origami based Au NPs dimer for SERS application [18]. The DNA helps to positioning the NPs into a perfect dimer with inducing the plasmonic coupling effect having an inter-particle gap between the dimer being ~3.3 nm. The incubation method of preparing dimer starts with, the 40 nm Au NPs suspended in a 2.5 mM Bis(p-sulfonatophenyl)phenylphosphine dihydrate dipotassium (BSPP) solution which can act as a coordination complex. Further, the DNA origami has been treated with 5' dithiol was incubated for 30 minutes with the above suspension and the pH were maintained using citrate-HCl buffer solution (pH=3). The presence of DNA brings the perfect and stable attachment of Au dimer NPs. The transmission electron microscopic (TEM) image observed in dimer formation with the inter-distance is 3.3 nm. The intensity analysis was carried out with gel electrophoresis show the free moving dimer NPs and the aggregated NPs with gel.

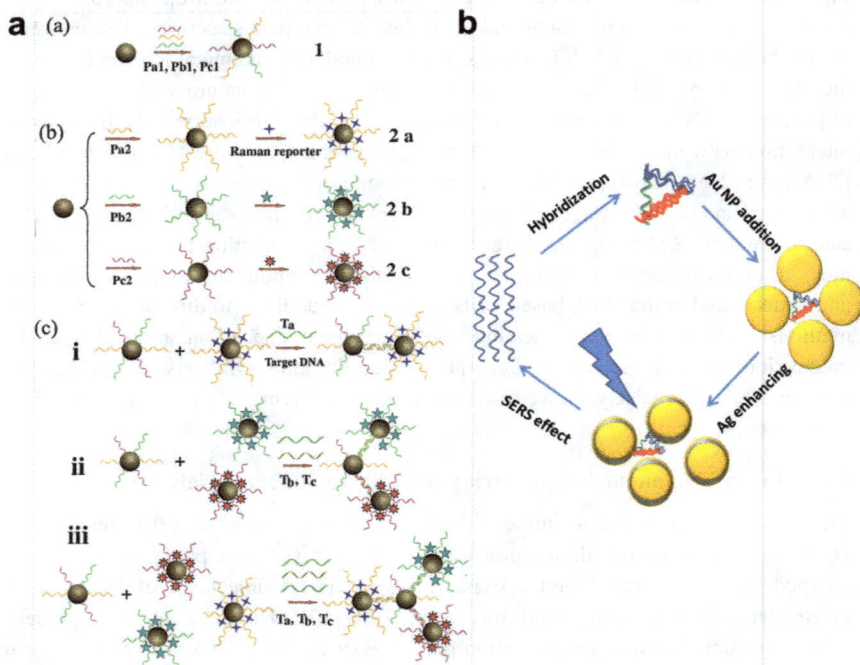

Figure 6. The mixed DNA-fabricated Ag NPs assembly (a) Reprinted from ref. 17 with permission. Copyright 2011. Royal Society of Chemistry; incubated assembly of DNA pyramids (b) and TEM image of Au/ DNA NPs and the Ag@Au/DNA NPs.

They found the maximum yield of 61% dimer from the aggregate bands. This work exhibits the utilization DNA brings the smallest controllable inter-particle gap between the NPs. Zhang *et al.* have prepared a most stable DNA-Ag NPs for multiplex DNA detections (Figure 6a) [17]. The detection of biological analyte was important in the

Bioinspired Nanomaterials   Materials Research Forum LLC
Materials Research Foundations **111** (2021) 118-154   https://doi.org/10.21741/9781644901571-5

medicinal field for diagnosis tests. The most important concern in the diagnosis is the simultaneous detection of multiple analytes. Therefore, DNA with a large number of binding sites in Ag NPs increases the surface area of the NPs. Hence, the mixture of equal molar ratio of probe sequence was effectively detected. The as prepared 30 nm Ag NPs were processed with triple cyclic disulfide-derivate single-stranded oligonucleotide and the pH maintained using 100 mM phosphate buffer (pH=7.4). Then, the NaCl was added up to 1M for 48 h with the maintained pH using phosphate buffer and incubated with methanol. Also, in the human breast cancer cell detection, the DNA–Au nanoaggregates shows excellent increased Raman signal.

Similarly, Keum *et al.* has developed three-dimensional DNA nanostructures with the oligonucleotides [69]. At first, the DNA assembly was hybridized using the buffer containing 2 mM EDTA, 12.5 mM $MgCl_2$ and 20 mM Tris was added and heated up to 95 °C for 5 minutes. After cooled, DNA pyramid was modified with thiol and conjugated with gold NPs. The above mixture was kept for incubation for 2 days under continuous stirring (Figure 6 b). The obtained Au/DNA mixture was combined with stabilizer and 1 mM $AgNO_3$ was incubated to get an Ag NPs over Au/DNA pyramids. The TEM image shows the DNA mediated Au NPs size of 20 nm which can act as a seed and coating with silver on the Au/DNA. Here, the DNA mediation controlled the NPs size and the absence of DNA results an irregular formation of Au NPs. With this advanced DNA material, they have observed the enhanced signal for dye system and also with complementary DNA detection.

Similar to the previous report, the citrate developed gold NPs were incubated with hairpin DNA (5'-thiol) and the 50 µL bio-barcode DNA. Allowed to stirring for 24 h at 37 °C, further it was aged with NaCl for 6 h. Finally, the mixture was washed thoroughly with the pH maintained using a phosphate buffer solution (pH=7) [70]. The 3D chiral nano-structure was also developed in the presence of DNA origami with this incubation method [68]. With the citrate based development of Au NPs conjugated with bis(p-sulfonatophenyl)phenylphosphine (BSPP) dipotassium salt. Further, the thiol modified oligonucleotide was purified with column and the phosphinated Au NPs were incubated for 20 h. This incubation mixture contains all that of pH buffer, 50 mM NaCl, 89 mM Tris, 89 mM boric acid and 2 mM EDTA [71]. This incubation results in a stable formation of 3D DNA origami plasmonic NMs. The key point to get stable DNA based NMs in the incubation method is that a proper maintaining pH, initial processing of DNA with stabilizer and the incubation period.

5.2   Wet-chemical method of preparing DNA based nanomaterials

Wet chemical method of preparing NMs is a well-known technique and it is convenient to do even with lesser equipment. With this method stabilizing NPs in a controlled morphology is somewhat difficult, usually it gets aggregated and settled as bulk aggregates with strong reducing agents. In order to control the NPs size and stabilize in the aqueous or organosol medium can be achieved with the help of bio-molecule DNA assembly as a template. Simply, this wet chemical synthesis method involves reduction of metal ions into metals starting nucleation over DNA and forming a stable nanomaterial solution. Based on the application, the medium where the NPs stabilized can be varied as aqueous and organosol. In our group, we have studied the stabilization of metal NPs having been done in both aqueous and organosol medium in presence of DNA. The following are some of the reported methodologies carried out to prepare DNA mediated NMs synthesis in both aqueous and organic medium.

5.2.1   DNA stabilized nanomaterials in aqueous medium

The wet chemical method for preparing stable metal NPs is a simple and quick process compared to other methods. Other than Ag and Au, several other metals such as Ru, Rh, Pd, Ir, and Re NPs were also prepared in aqueous medium in presence of DNA [41,42,72,73]. The wet-chemical method of preparing metal NPs involves three stages in the formation **(Figure 7a)** [16]. At first, the metal ions electrostatically coordinated with the negative moieties of DNA in its backbone. This interaction stave is simply called as 'activation stage'. Next step is the reduction of metal ions attached with the DNA. The reduced metallic NPs electrostatically sticks over the DNA and forms nano-chain assemblies. The reducing agents such as sodium borohydride ($NaBH_4$), tetrabutyl ammonium borohydride ($TBABH_4$) and hydroquinone etc. were generally used for the reduction of the metal ions to generate metal NPs.

Further, the self-assembled metals start to nucleate over and over on the surface like a nucleation site and form DNA-templated structures. Rhodium NPs were prepared and self-assembled on DNA *via* a wet-chemical method [41]. The reaction was carried out in aqueous medium, where the Rh precursor was initially interacting with the 0.01 M DNA stock solution upon stirring for 20 minutes. Then it undergoes reduction with 0.01 M ice cold $NaBH_4$ which results a stable solution of Rh-DNA nano-chain assembly. The as-prepared Rh NPs was stable more than three months with the help of DNA in the aqueous medium. Similarly, the synthesis of RuO_2 was carried out the same way, that the ruthenium precursor interact with DNA has been reduced as metallic Ru in aqueous medium [42]. In the ambient condition the metallic Ru was converted to RuO_2 NPs stabilized on DNA **(Figure 7b)**. The strong interaction between DNA and the metal NPs

Bioinspired Nanomaterials
Materials Research Foundations **111** (2021) 118-154

Materials Research Forum LLC
https://doi.org/10.21741/9781644901571-5

were identified using various characterization techniques such as, Ultraviolet-Visible spectroscopic analysis (UV-Vis), Fourier Transform Infrared Spectroscopic (FT-IR) analysis, energy dispersive X-ray spectroscopic analysis (EDS) and color mapping analysis. In the UV-Vis study, the presence of aromatic groups in bare DNA solution shows the absorption wavelength of 262 nm [42].

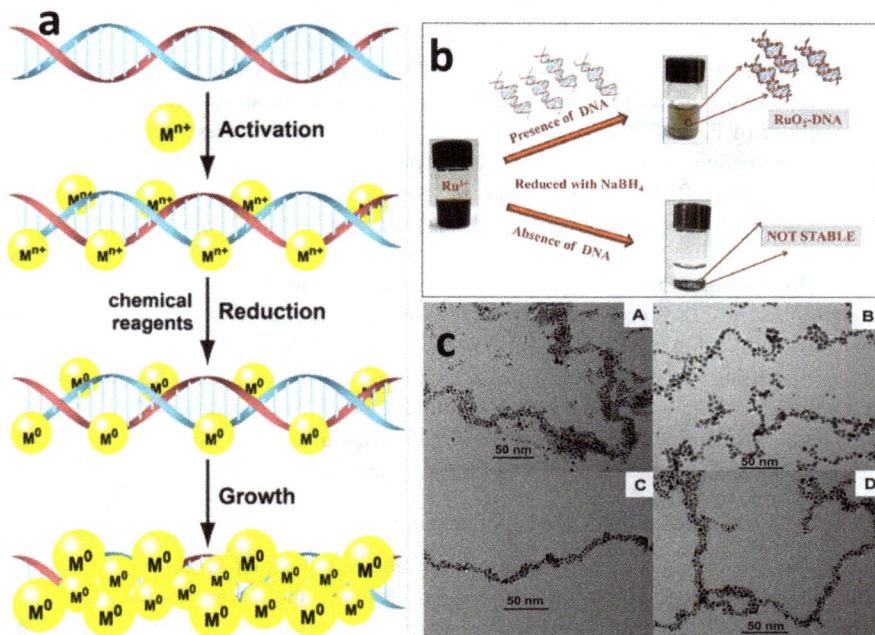

Figure 7. *Mechanism involved in the chemical reduction of metal ions over DNA (a).Reprinted from ref. 16 with permission. Copyright 2018. Royal Society of Chemistry; RuO_2 stabilized on DNA and aggregated in the absence of DNA (not stable) (b) and the TEM image of varying DNA ratio on Ag NPs (A) 4 : 1, (B) 3 : 1, (C) 2 : 1 and (D) 1 : 1 (Au NPs : DNA)(c). Reprinted from ref. 67 with permission. Copyright 2014. Royal Society of Chemistry.*

Once the metal ions get interacted with DNA, the absorption range gets shifted towards longer wavelength. Also, the FT-IR analysis helps to figure out the functional group participation of bare DNA and the DNA based NMs. Major functional group presence in DNA such as –OH group at (3422 cm^{-1}), C=O amide group at 1723 cm^{-1}, C-H bond at 1496 cm^{-1}, 1124 cm^{-1} PO$_2^-$ stretching vibration of DNA and the lower wavenumber of 525 cm^{-1} appears for fundamental deformation of DNA. These peaks were either shifter or disappeared after the NMs assembled over the DNA. This interaction study clearly explains the strong interaction of bio-molecule DNA with the metal NPs and results a perfect nano-chain assembly.

In morphological studies the DNA developed Rh, Ru and Pt NMs shows the particle size was below 5 nm. From the EDS and color mapping analysis, along with the metal NPs the presence of P, N, C and O were observed. These elemental presences are observed from the aromatic group, carbohydrates and a phosphate moiety in the DNA implies the robust binding of DNA in the nanomaterial formation. Zhang *et al.* have developed the Au NPs on DNA in aqueous medium and found the inter particle gap is 2- 3 nm [67]. In the typical synthesis, 20 ml of Au precursor was added with 50 ng mL^{-1} of λ-DNA under stirring for 3 hours. Finally, a color change to wine red indicates the formation of DNA–Au hybrids. The TEM morphology analysis shows the varying DNA concentrated nano-chain assembly of Au NPs **(Figure 7 c)**. This ideal material was testified in SERS analysis by using Rhodamine 6G, 4-aminothiophenol, pyridine and 2,4,6-trinitrotoluene as a substrate. Hong *et al.* have prepared similar kind of Au dimers on DNA for SERS analysis of Rhodamine B. The wet chemical approach of NMs synthesis using the bio-molecule DNA was further studied with organosol medium.

5.2.2 DNA stabilized nanomaterials in organic medium

The utilization of organic solvents instead of aqueous medium is another method of developing DNA based NMs. From the reports observed, the solvent medium is also an important factor to control the nanomaterial size on DNA. The process of metal ions reduction in organic medium is quicker than in aqueous medium. The method of using organic solvents helps to overcome some of the difficulties in the aqueous medium such as washing or removal of excess reagents and stabilizer used. Also, it has more control over size and gets a stable colloidal monodispersed suspension. Generally, organic solvents such as ethanol, acetonitrile, acetone, toluene, CHCl$_3$ and hexane were used in the synthesis. There are very few reports available in the DNA based NMs synthesis in organic medium. Our group was also working in the DNA based metal NPs synthesis such as Re, Ir Rh and Os in organic medium [72, 73, 76]. In the typical synthesis 20 ml DNA and 20 mL of aqueous perrhenate (0.1 M) solution was taken in the organic solvent acetonitrile; further 3 ml of glacial acetic acid was added with 0.25 g of TOAB (Phase transferring agent) and stirred for 20 minutes [73]. Then the mixture was kept in a refrigerator for 12 hours to separate the organic layer from aqueous layer. After the phase transfer process the color of the solution turned from light brown to dark brown which indicates the formation of Re NPs on DNA. From the morphological analysis, the observed particle size is ~ 1.5 ± 0.7 nm. This shows the high controlled over morphology of Re NPs over DNA. Similarly, the synthesis of Rh NPs on DNA was carried out by using organic solvents such as acetone. Here the stock solution of DNA mixed with metal ions gets reduced simultaneous with TBABH$_4$ with continuous stirring. The initial observation of faint violet color was changed to dark violet color is the identification of the formation of Rh NPs over DNA [73]. The morphology shows a perfect long wire like structure with the particle size of ~2.6 ± 0.2 nm. In 2005, Wei *et al.* reported the self-assembly of Ag NPs over DNA and utilized for SERS application with detection of Rhodamine 6G (R6G) and 4-aminothiophenol (4-ATP) [74]. One step synthesis of Ag NPs on DNA was carried out in ethanol/water system of 2 ml of CTAB (surfactant, capping agent) taken in ethanol with 5 mM AgNO$_3$ was kept stirring for 10 minutes. After that, the metal ions were reducing with NaBH$_4$ and the color change observed to greenish yellow. Following the same preparation method, in the year of 2008, the self-assembly of λ-DNA networks/Ag NPs was fabricated [75]. They have varied DNA concentration such as 20, 50, and 100 ng/µl and prepared CTAB-capped Ag NPs. The prepared Ag NPs on DNA were used as a SERS substrate for the detection of methylene blue (MB) [75]. From the above discussion, the advantages with the DNA based NMs wet-chemical synthesis was elaborated using both aqueous and organic solvents.

5.3   DNA stabilized nanomaterials preparation by photoreduction

The photoreduction method has also been widely used in the NMs synthesis. This method of synthesis needs proper circumstance and equipment for photo-irradiation. In most cases, Au and Ag are the well responsive material for photoreduction method of synthesizing NMs. There are several reports to carry photo reduction of such Au and Ag colloidal solutions by losing the electron while ejection to form metallic NPs. In addition with Au and Ag, DNA mediated Rh and Pd NPs can also be prepared by the photoreductive method. The presence of DNA can act as a template as well as mild reducing agent during the synthesis.

The following are some of the reports carried out photoreduction of DNA mediated NMs for SERS applications. In 2013, we have reported the lesser inter-particle distance of ~1.7 ± 0.2 nm of Ag NPs using DNA [76]. The Ag NPs nano-clusters has been prepared by mixing DNA stock solution with silver nitrate solution and kept at stirring for 1 day to get the homogenous aqueous solution. Then the solution was kept under UV light radiation for photo reduction at a wavelength of 260 nm for about 3 hours continuously.

Bioinspired Nanomaterials
Materials Research Foundations **111** (2021) 118-154

Materials Research Forum LLC
https://doi.org/10.21741/9781644901571-5

In that case, the color changes were observed from light yellow to yellowish green color indicates the formation of Ag NPs over DNA. These prepared Ag-NCs were utilized for MB detection in the range of EF - 10^8 to 10^{10}. In the SERS substrate, the inter-particle gap is an important factor to produce high efficiency in the enhancement factor. Therefore to reduce the inter-particle distances by tuning the morphology is the ultimate progress to get high SERS efficiency. Here the advantage with DNA, its diameter is only 2-3 nm therefore the NMs developed on the DNA template have lesser inter particle distance.

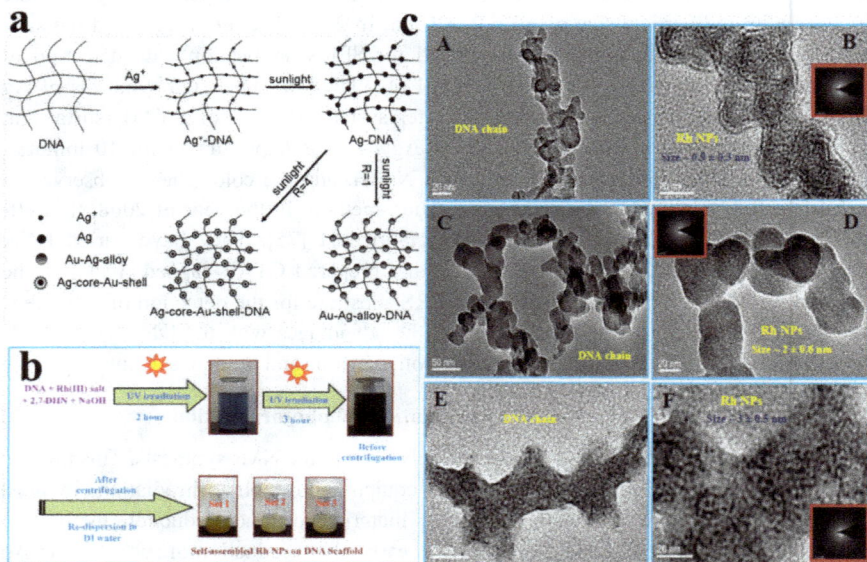

Figure 8. Synthesis scheme of DNA mediated Au-Ag bimetallic NPs (a) Reprinted from ref. 77 with permission. Copyright 2009. Royal Society of Chemistry; the UV-irradiated synthesis scheme of Rh NPs on DNA (b) and corresponding high and low magnified TEM image of the Rh NPs on DNA with below 5 nm (c).

Another interesting report by Yang *et al.* has developed a sunlight-induced formation of bi-metallic SERS substrate for tumor detection [77]. In the synthesis, the silver nitrate solution was mixed with 20 μL of DNA stock solution and kept stirring for 3 hours in RT, and then the mixture was placed under direct sunlight until the colorless solution

Materials Research Forum LLC
https://doi.org/10.21741/9781644901571-5

turns to greyish indicating the formation of Ag seeds. With this Ag seed, the addition of sodium tetrachloroaurate solution was added and irradiates in a direct sunlight at a varying time resulting in silver-core-gold-shell NPs (Figure 8a). The same group reported the photo reduced process of Au NPs synthesis under 16 W UV lamp [78]. Similarly, the synthesis of Rh NPs over DNA has been carried out by simple UV photo-irradiation route and produces large number of active 'hot spots' for effective SERS detection [79]. In the synthesis, 60 ml DNA stock solution mixed with Rh salt solution and 30 ml of 2,7-DHN solution. The above mixture was stirred for 3 minutes and placed under UV radiation for 5 hours (Figure 8b). The formation of black solution was centrifuged and re-dispersed which become yellow in color. From morphological study, low and high magnified TEM analysis had shown the self-assembled Rh NPs over DNA scaffold. The average chain diameter was calculated to be ~ 35 ± 10 nm (Figure 8c). The inset image shows the selected area diffraction (SAED) pattern for the Rh NPs. Similarly, Pd NPs has also been prepared by photoreduction of passing UV irradiated for 4 hours. The obtained SERS substrate was testified with different analyte molecules [80].

5.4   Seed mediated method of preparing DNA based nanomaterials

The seed mediated method of NMs preparation is slightly different from the above methods. In the chemical method, the metal ions electrostatically attached with the DNA were reduced using reducing agents and the NMs starts growing on the surface over and over as catalytic sites. Here, in the seed mediation method, the readily developed metal NPs serves as a seed to mediate the growth of various metals over the surface of the seeds. Sun *et al.* reported a DNA based fabrication of ZnO/Ag composites for SERS application [81]. In this work, they have chosen n-type semiconducting material ZnO as a seed/template for developing Ag as a SERS substrate. Initially, the zinc acetate solution was interacting with DNA *via* stirring and the mixture was drop casted over ITO surface, dried at 350 °C for 20 minutes (Figure 9a). Further, the ZnO NR arrays ITO were dipped in silver nitrate solution and then illumination carried out with UV-light for 10 minutes and thus forms ZnO/Ag composites (Figure 9b).

In SERS, the prepared material shows excellent efficiency in the R6G probe detection and 4-ATP. The EF value was calculated for probe detection of 4-ATP is 7.4×10^8. In the seed mediation process, silicon also plays a very active and reproducible role as a base material for devolving the SERS substrate. Here, AgNPs@Si were prepared by etching of hydrogen fluoride (HF), the Si-H bonds reduces the silver metal ions on the silica wafer [82]. Further, the etched silica wafer was immersed into the AgNO₃ solution and stirred continually for 90 s. Finally, it forms AgNPs@Si NMs for effective detection of DNA. In 2014, the DNA mediated bimetallic nano mushrooms were effectively prepared for SERS amplification [52]. They critically developed Au head and the silver cap nano-gaps for SERS to achieve the EF value ~1.0×10^9. Firstly, the Au NPs were prepared with sodium citrate with boiling and stirring. Then, 3 μL of a 100 μM DNA solution were added with Au NPs and 100 mM phosphate buffer was added further. With this modified Au NPs, the L-sodium ascorbate and silver nitrate solution were mixed vigorously for 1 minute. Finally, the Au-Ag mixture was washed twice with ultra-fast centrifugation. The TEM image shows the nano mushroom structure with lesser inter particle gap. The creation of minimum nano-gaps increases the SERS efficiency. The developed SERS active NMs are promising in the detection of molecular diagnostics and cellular imaging. The seed mediate method involves several other metals were also reported to develop SERS active substrates. Thus, it is well-discussed that the designing of stable SERS substrate is also critical important for studying SERS enhancement. The number of hot-spots and the reducing inter-particle distance using biomolecule DNA is a better approach to increase the SERS efficiency. Further, the detailed discussion of SERS enhancement with DNA based NMs for probe detections are given in the following parts.

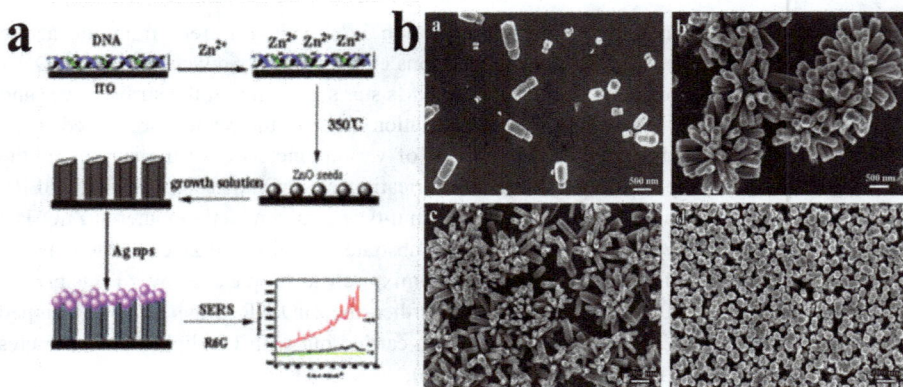

Figure 9. *Synthesis scheme of DNA mediated ZnO/Ag composites (a) and FE-SEM image of ZnO nano array assembled over ITO surface (b). Reprinted from ref. 81 with permission. Copyright 2011. Royal Society of Chemistry.*

6.   Surface enhanced raman scattering (SERS) applications

The SERS have been employed in various kinds of applications for their enhanced Raman signals. It has been widely used in biological and biomedical applications, intracellular analysis, immune nano-sensor, DNA analysis, microorganism detection, defense and sensor applications [83]. For examples, the core-shell morphology of silica-Au NPs modified with redox-active molecules has been used to analyze the intracellular process [84]. The terror/criminal substances malathion, chorpyrifos and trinitrotoluene (TNT) were detected at very low concentrations using Ag NPs as SERS substrate [85]. The biologically important molecules like DNA [44], RNA [86] and proteins [87] and microorganisms such as virus [88], bacteria [89] and yeast [90] can also be detected by using SERS analysis. The SERS method also useful in detecting of cancer markers, angiogenin (ANG), and alpha-fetoprotein (AFP) with minimum detection level [91].

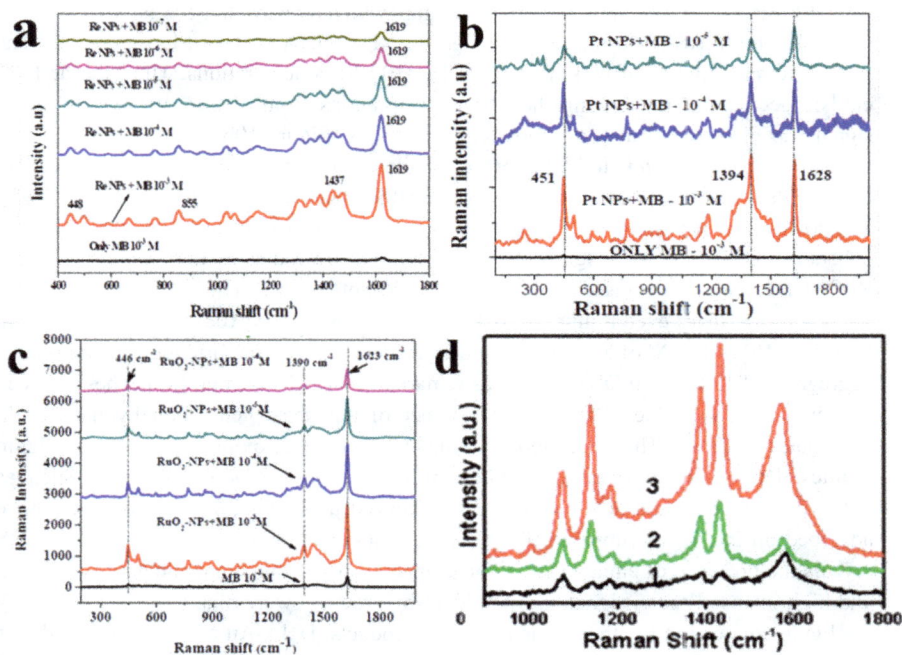

Figure 10. *SERS signals of different concentrated MB molecule on 20 nM size Re NPs (a); SERS performance of Pt NPs on DNA with varying MB concentration (b); on RuO₂ (c) and SERS of 4-ATP on DNA-Au substrate (d) Reprinted from ref. 67 with permission. Copyright 2014. Royal Society of Chemistry.*

Bioinspired Nanomaterials
Materials Research Foundations **111** (2021) 118-154

Materials Research Forum LLC
https://doi.org/10.21741/9781644901571-5

SERS application using different NMs synthesized by wet aqueous method has been discussed below. The SERS activity depends on the size of the substrate and analyte concentration can be proven by the following examples. The SERS spectra of Rhodamine B (RB) on 30 nm Au NPs at various distance between RB and Au NPs were reported [92]. The intensity of major Raman signals at 718 cm^{-1}, 878 cm^{-1}, 1095 cm^{-1}, 1300 cm^{-1}, 1440 cm^{-1} and 1650 cm^{-1} is low when the distance between RB and Au NPs is 0 nm. Further, Raman signal intensity gradually increased during the increasing of distance between RB and Au NPs. The overall study reveals that SERS of RB is dependence on Au NPs size and the location point/interparticle distance. Another example of enhanced SERS activity is shown by DNA supported Re NPs substrate [73]. Here, the small size of Re NPs offers the required inter-particle gap which creates huge number of SERS hotspots. The SERS performance of MB on Re NPs@DNA is shown in Figure 10a. The intense Raman signals of MB are obtained at 10^{-3} M concentrations. The peak at 1610 cm^{-1} is decreased by changing the MB concentrations from 10^{-3} to 10^{-7} M and weak signal was observed even at very low concentrations (Figure 10b). The calculated EF is very close to 10^7 which is in higher order. Similarly, by varying the DNA concentration, the Pt NPs stabilized DNA shows excellent SERS efficiency over MB (10^{-3} – 10^{-5}) [93]. The lowest concentration of MB was detected with SERS substrate Pt NPs with the enhancement 2.52×10^5 (Figure 10b). The Figure 10c indicates the SERS of MB on RuO$_2$-DNA nano-chain aggregates [42]. The major intense peaks of MB at 445, 1391 and 1620 cm^{-1} are sharp even minimum MB concentrations (Figure 10c). The EF is found to be 1.87×10^5 at 10^{-3} M of MB. SERS studies evidencing that the formation of nano-chain aggregates of RuO$_2$ with DNA makes as remarkable SERS substrate [42]. The enhanced Raman intensity is due to the higher number of hot spots generation by nano-chain aggregates on DNA. The SERS tool is utilized to detect the oxidized products of lignin mimics. During the biomass conversion of lignin, the byproducts obtained such as aldehydes, alcohols and acids were *in-situ* analyzed using SERS. These products were adsorbed on the SERS substrate (RuO$_2$-DNA nano-chain aggregates) and *in-situ* SERS reveals the presence of cinnamyl alcohol, cinnamaldehyde and cinnamic acid. The sharp and high intense Raman signals of individual by products were obtained and evidencing the lignin conversion to other value added by products. DNA-Au NPs assemblies shows strong C-C bending and stretching at 1005 and 1570 cm^{-1} and C-H bending peak at 1185 cm^{-1} attributes to the 4-ATP. The sharp Raman signals of 4-ATP exhibits at 10^{-5} M concentrations which is a very minimum level (Figure 10d). The substrate also used to detect Rhodamine 6G, the analytes such as pyridine and trinitrotoulene at very low concentrations. Organic solvents have been used to synthesis nanostructured materials in confined shape and structures using DNA as stabilizer and studied the SERS applications. The Rh NPs-DNA colloidal NPs prepared by wet chemical organosol method used as

SERS substrate to analyze the MB molecule [79]. As seen in Figure 11a, the Raman characteristic signals of MB have appeared in the region of 300 to 1800 cm^{-1} with low intensity (without SERS substrate). In the presence of SERS substrate, the Raman intensity is enhanced at various MB concentrations (10^{-3} - 10^{-6} M). The maximum peak intensity was observed for 10^{-3} M of MB, upon decreasing the signals from 10^{-3} to 10^{-6} M (Figure 11a). The highest EF value is obtained about 1.53×10^2 orders at the lowest MB concentration of 10^{-6} M. The superior SERS activity of Rh NPs-DNA is assigned to be small sized Rh NPs particles (below 5 nm) formation during DNA mixing, which allows more SERS hotspots. The presence of DNA controls the Rh NPs in an ultra-small size. Three different sizes of Rh NPs such as small, medium, and large were prepared and SERS studies are investigated by using MB as SERS probe (Figure 11b).

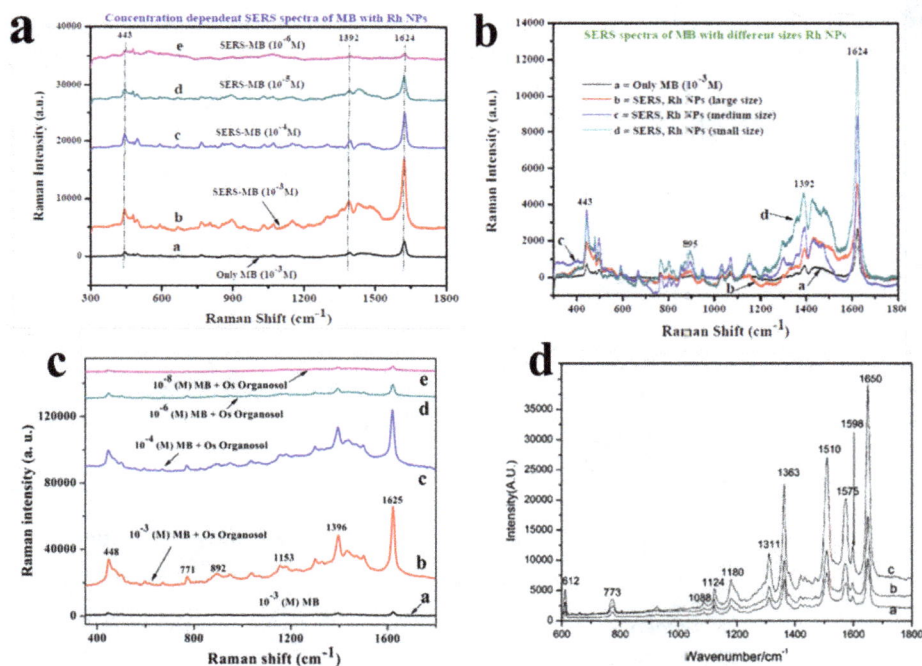

Figure 11. *SERS of MB on Rh@DNA (a); MB on different sized RhNPs-DNA (b); MB on Os organosol-DNA (c), and SERS of Rhodamine 6G on CTAB stabilized AgNPs-DNA substrate (d). Reprinted from ref. 74 with permission. Copyright 2005. American Chemical Society.*

Rh NPs used are the negatively charged whereas MB dye is positively charged. Due to opposite charge, there will be an attraction between dye and substrate which will results in better SERS activity. The size effect plays significant role in enhancing the Raman signals of MB. As seen in Figure 11b, the intensity was enhanced in the case of small size Rh NPs over the medium and large particles. In another reports, Osmium (Os) organosol on DNA used to record the SERS of MB molecule [72]. Figure 11c depicts the SERS of different concentrations of MB on Os organosol-DNA substrate. In this work, the Os organosol of aggregated short wires-like morphological substrate shows highest EF value for MB dye (4.4×10^5) over the organosol of aggregated long wires (3.2×10^2). This results also one of the evidence that morphology of SERS substrate is crucial in enhancing the Raman signals of analytes. Similar kind of SERS results were observed for 4-ATP on CTAB stabilized Ag NPs. The SERS activity is increased by allowing the incubation time of DNA with Ag NPs (Figure 11d) [74]. In similar way iridium organosol on DNA, Rhenium organosol on DNA were used as SERS substrate to study the SERS activity of MB dye and obtained enhanced EF value. The overall discussions imply that substrate morphology is an important factor in enhancing the SERS activity.

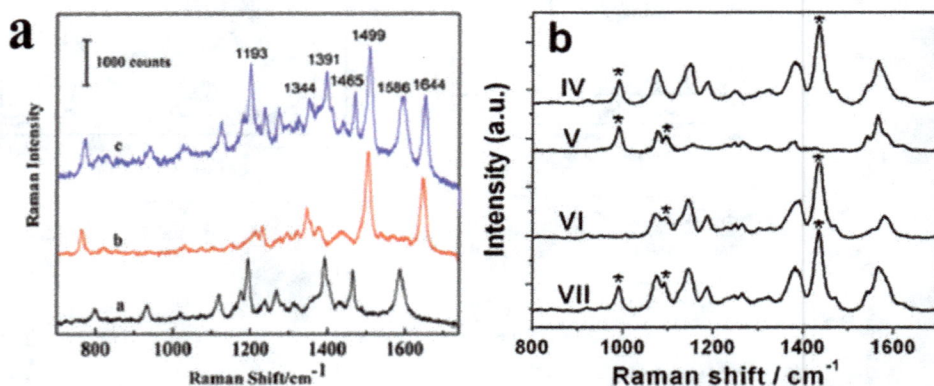

Figure 12. *SERS signal of different biomarkers probes Cy3, Rox and mixture of Rox and Cy3 (a) Reprinted from ref. 70 with permission. Copyright 2014. Royal Society of Chemistry. and different DNA sequence detection with Ag NPs (b). Reprinted from ref. 17 with permission. Copyright 2011. Royal Society of Chemistry.*

In the incubation method of preparing DNA based NMs, the high control over the inter particle gaps is possible. The least inter-particle distance significantly induces the SERS signal by a few orders. The Au NPs with the inter-particle distance of 3.3 ± 1 nm shows red shift in the far field scattering measurement [18]. The enhancement in the SERS signal was testified with DNA sequence detection and dye molecule Rhodamine 6G at a lowest concentration. A similar kind of DNA based Au nano aggregates has been studied for the detection of human breast cancer cell (MCF-7) [70].

The biomarkers on cancer cell with probe was analyzed and shown in Figure 12a. The Cy3 probe peaks at 1193 cm^{-1}, 1391 cm^{-1}, 1465 cm^{-1} and 1586 cm^{-1} and Rox peaks at 1499 cm^{-1}, 1344 cm^{-1} and 1644 cm^{-1} were detected with the DNA-Au nanoaggregates. Li *et al.* reported the SERS performance of silver nanoparticle on oligonucleotide DNA [70].

The multi target on detection of different sequence of DNA named as Ta, Tb and Tc shown in Figure 12b. The signal at 1098 cm^{-1} and at 991 cm^{-1} clearly determined the mixture of DNA with the SERS substrate. The SERS effect on photo-reduced DNA mediated nanomaterial synthesis is also an interesting way of increasing the SERS efficiency. The report carries the lowest concentration of tumor detection (1×10^{-15} M) with the SERS substrate Au–Ag–DNA mixture solution [77]. The highest enhancement factor calculated for 2,4,6-Trinitrotoluene (TNT) is 1.5×10^{12}. Our group also developed the Ag NPs on DNA results a mediated wire-like clusters formation for SERS analysis [94]. As we discussed before, the inter-particle gap is a crucial player in SERS enhancement, here the inter particle distance was reduced up to $\sim 1.7 \pm 0.2$ nm using DNA scaffold. The SERS performance of this wire cluster was analyzed with the probe molecule methylene blue. This report proves that the Raman signal enhancement is not only depending on the excitation laser but also inter particle gaps. The EF obtained is 10^8 to 10^{10} for the probe MB detection. It is conformed that the increasing number of hot spots on the wire or chain-like morphological NMs enhances the SERS signal at a single molecule levels [78,80]. Therefore, the control over size and shapes in the NMs fabrication brings huge variation in the SERS enhancement [95].

7.   Conclusions of DNA based metal nanostructures for SERS studies

In the wide range of usage with bio-mediated NMs, SERS plays an important role in the bio- medicinal field. Among the various bio-molecules, DNA is a highly preferable and more utilized one in the NMs synthesis. In this book chapter, we have highlighted the various methods involved in the synthesis of DNA based nanomaterials; the types of DNA sequences or tiles modulated with computational methods; the morphological flexibility of various nanostructures such as nano-chains, clusters, wires and the sequential 2D and 3D structures with DNA scaffolds. We also detailed the use of prepared DNA based nanomaterial for SERS studies. The mechanism involved in the SERS and the places where the SERS techniques can be widely applied is also highlighted. Finally, the specific role of size and shape effect of metal NPs in the SERS was also discussed.

- The bio-molecule DNA has been used as a greener approach towards the NMs synthesis.

- The various methodologies in the DNA mediated NMs preparation, the electrostatic interaction of DNA with the metal ions and the controlled over nucleation was studied.

- Among the methods, incubation and photo-reductive DNA based NMs synthesis is an effective and stable way to prepare the DNA based NMs.

- The synthesized DNA based NMs have been used for SERS studies and found high reproducibility, selectivity and sensitivity for SERS probe detection.

- The plasmonic effects with the noble metals were increased by reducing the NPs distance *via* high functionality DNA was observed with increased SERS efficiency.

- The DNA mediated NMs with less inter-particle distance produces more number of 'hot spots' which induces the localized surface charges results increased Raman signal of the analyte molecule adsorbed very closes to the metal surface and SERS EF increases in a few orders of magnitude.

- The major concern with this SERS study is the reproducibility and the exact mechanism associate with this Raman signal enhancement still needs to be developed for better understanding.

The observed increment of week Raman signal with the introduction of plasmonic NMs is revolutionary findings to make the Raman analysis for future technique to detect ultra-low concentrated analyte molecules. The mechanism of SERS efficiency includes electromagnetic and chemical effect was observed with various noble metals and few semi-conducting materials. The fact of reducing inter-particle distance to increase the SERS efficiency was interesting with various size and shape-selective NMs. In this line, the development of DNA based NMs is a great opportunity to reduce the inter-particle distance below 5 nm and their chain-like/wire-like morphology increases the stability of 'hot spots' for effective SERS analysis even at very low concentrations. In future, such DNA based NMs will be effective for studying many biological applications at a trace detection level in our day-today-life.

References

[1]   K.D. Gilroy, A. Ruditskiy, H. Peng, D. Qin, Y. Xia, Bimetallic Nanocrystals: Syntheses , Properties , and Applications, Chem. Rev. 116 (2016) 10414−10472. https://doi.org/10.1021/acs.chemrev.6b00211

[2]   J.R. Heath, Nanoscale Materials, Acc. Chem. Res. 32 (1999) 990059

[3]   X. Peng, Q. Pan, G.L. Rempel, Bimetallic dendrimer-encapsulated nanoparticles as catalysts: a review of the research advances, Chem.Soc.Rev. 37 (2008) 1619–1628. https://doi.org/10.1039/b716441f

[4]   R.J. White, R. Luque, V.L. Budarin, J.H. Clark, D.J. Macquarrie, Supported metal nanoparticles on porous materials. Methods and applications, Chem.Soc.Rev. 38 (2009) 481–494. https://doi.org/10.1039/b802654h

[5]   M.F.H. Jr, D.W. Mogk, J. Ranville, I.C. Allen, G.W. Luther, L.C. Marr, B.P. Mcgrail, M. Murayama, N.P. Qafoku, K.M. Rosso, N. Sahai, P.A. Schroeder, P. Vikesland, P. Westerhoff, Y. Yang, Natural, incidental, and engineered nanomaterials and their impacts on the Earth system, Science (80-.). 363 (2019) 6434. https://doi.org/10.1126/science.aau8299

[6]   M.F. Hochella, S.K. Lower, P.A. Maurice, R.L. Penn, N. Sahai, D.L. Sparks, B.S. Twining, Nanominerals, Mineral nanoparticles, and Earth Systems, Science. 1631 (2008) 1631–1635. https://doi.org/10.1126/science.1141134

[7]   M.K. Mcnutt, Convergence in the Geosciences, GeoHealth. 1 (2017) 2–3. https://doi.org/10.1002/2017GH000068

[8]   H. Shirakawa, The Discovery of Polyacetylene Film: The Dawning of an Era of Conducting Polymers (Nobel Lecture), Angew. Chem. Int. Ed. 40 (2000) 2574 - 2580.

[9]   J. Huang, L. Lin, D. Sun, H. Chen, D. Yang, Q. Li, Bio-inspired synthesis of metal nanomaterials and applications, Chem.Soc.Rev. 44 (2015) 6330–6374. https://doi.org/10.1039/c5cs00133a

[10]   C. Aur, T. Nesma, P. Juanes-velasco, A. Landeira-viñuela, H. Fidalgo-gomez, V. Acebes-fernandez, R. Gongora, A. Parra, R. Manzano-roman, M. Fuentes, Interactions of NPs and Biosystems : Microenvironment of NPs and Biomolecules in Nanomedicine, Nanomaterials. 9 (2019) 1365

[11]   C. Toumey, Reading Feynman Into Nanotechnology: A Text for a New Science, Res. Philos. Technol. 12 (2008) 133–168

Bioinspired Nanomaterials
Materials Research Foundations **111** (2021) 118-154

Materials Research Forum LLC
https://doi.org/10.21741/9781644901571-5

[12]   C.M. Niemeyer, Nanoparticles, Proteins, and Nucleic Acids: Biotechnology Meets Materials Science, Angew. Chem. Int. Ed. 40 (2001) 4128 ± 4158

[13]   H. Hosein, D.R. Strongin, T. Douglas, K. Rosso, A Bioengineering Approach to the Production of Metal and Metal Oxide NPs, 2005

[14]   A.R. Tao, S. Habas, P. Yang, Shape Control of Colloidal Metal Nanocrystals, Small 2008,. 4 (2008) 310–325. https://doi.org/10.1002/smll.200701295

[15]   E.D. Spoerke, A.K. Boal, G.D. Bachand, B.C. Bunker, Templated Nanocrystal Assembly on Biodynamic Artificial Microtubule Asters, ACS Nano. 7 (2019) 2012–2019

[16]   Z. Chen, C. Liu, F. Cao, J. Ren, X. Qu, DNA metallization: principles, methods, structures, and applications, Chem. Soc. Rev. 47 (2018) 4017–4072. https://doi.org/10.1039/C8CS00011E

[17]   Z. Zhang, Y. Wen, Y. Ma, J. Luo, L. Jiang, Y. Song, Mixed DNA-functionalized nanoparticle probes for surface-enhanced Raman scattering-based multiplex DNA detection, Chem. Commun. 47 (2011) 7407–7409. https://doi.org/10.1039/c1cc11062d

[18]   V. V Thacker, L.O. Herrmann, D.O. Sigle, T. Zhang, T. Liedl, J.J. Baumberg, U.F. Keyser, DNA origami based assembly of gold nanoparticle dimers for surface-enhanced Raman scattering, Nat. Commun. 5 (2014) 3448. https://doi.org/10.1038/ncomms4448

[19]   D. Janas, Towards monochiral carbon nanotubes: A review of progress in sorting of single-wall carbon nanotubes, Mater. Chem. Front. 2 (2017) 36–63. https://doi.org/10.1039/C7QM00427C

[20]   A. Mehmood, H. Ghafar, S. Yaqoob, U.F. Gohar, B. Ahmad, Mesoporous Silica NPs: A Review, J. Dev. Drugs. 6 (2017) 1–11. https://doi.org/10.4172/2329-6631.1000174

[21]   S.S. Sankar, K. Karthick, K. Sangeetha, R.S. Gill, S. Kundu, Annexation of Nickel vanadate ($Ni_3V_2O_8$) nanocubes on nanofibers: an excellent electrocatalyst for water oxidation, ACS Sustain. Chem. Eng. 8 (2020) 4572–4579. https://doi.org/10.1021/acssuschemeng.0c00352

[22]   P. Thiruvengetam, D.K. Chand, Oxidomolybdenum based catalysts for sulfoxidation reactions: A brief Review, J. Indian Chem. Soc. 95 (2018) 781–788

[23]   P. Thiruvengetam, R.D. Chakravarthy, D.K. Chand, A molybdenum based metallomicellar catalyst for controlled and chemoselective oxidation of activated

alcohols in aqueous medium, J. Catal. 376 (2019) 123–133.
https://doi.org/10.1016/j.jcat.2019.06.013

[24] Y. Cao, D. Li, F. Jiang, Y. Yang, Z. Huang, Engineering Metal Nanostructure for
SERS Application, J. Nanomater. 2013 (2013) 1–6

[25] D. Sharma, S. Kanchi, K. Bisetty, Biogenic synthesis of nanoparticles: A review,
Arab. J. Chem. 12 (2015) 3576–3600. https://doi.org/10.1016/j.arabjc.2015.11.002

[26] M. Razavi, Bio-based nanostructured materials, Elsevier Ltd., 2018.
https://doi.org/10.1016/B978-0-08-100716-7.00002-7

[27] R. Levy, N.T.K. Thanh, R.C. Doty, I. Hussain, R.J. Nichols, D.J. Schiffrin, M.
Brust, D.G. Fernig, Rational and Combinatorial Design of Peptide Capping Ligands
for Gold nanoparticles, J. Am. Chem. Soc. 126 (2004) 10076–10084

[28] M.S. Ekiz, G. Cinar, M.A. Khalily, M.O. Guler, Self-assembled peptide
nanostructures for functional materials, Nanotechnology. 27 (2016) 402002

[29] T. Maruyama, Y. Fujimoto, T. Maekawa, Synthesis of gold nanoparticles using
various amino acids, J. Colloid Interface Sci. 447 (2015) 447, 254–257.
https://doi.org/10.1016/j.jcis.2014.12.046

[30] S. Manivannan, I. Kang, Y. Seo, H. Jin, S. Lee, K. Kim, M13 Virus-Incorporated
Biotemplates on Electrode Surfaces to Nucleate Metal Nanostructures by
Electrodeposition M13 Virus-Incorporated Biotemplates on Electrode Surfaces to
Nucleate Metal Nanostructures by Electrodeposition, ACS Appl. Mater. Interfaces. 9
(2017) 32965–32976. https://doi.org/10.1021/acsami.7b06545

[31] S. Kumaravel, P. Thiruvengetam, S.R. Ede, K. Karthick, S. Anantharaj, S.S.
Sankar, S. Kundu, Cobalt tungsten oxide hydroxide hydrate (CTOHH) on DNA
scaffold: an excellent bi-functional catalyst for oxygen evolution reaction (OER) and
aromatic alcohol oxidation, Dalt. Trans. 48 (2019) 17117–17131.
https://doi.org/10.1039/c9dt03941d

[32] M.C. Bubalo, S. Vidovi, I. Radoj, S. Jokic, Green solvents for green technologies,
J Chem Technol Biotechnol. 90 (2015) 1631–1629. https://doi.org/10.1002/jctb.4668

[33] O.I. Wilner, I. Willner, Functionalized DNA Nanostructures, Chem. Rev. 112
(2012) 2528–2556. https://doi.org/10.1021/cr200104q

[34] N.C. Seeman, Nucleic Acid Junctions and Lattices, J. Theor. Biol. 99 (1982) 237–
247

[35]   J.D. Moroz, P. Nelson, Torsional directed walks, entropic elasticity, and DNA twist stiffness, Proc. Natl. Acad. Sci. 94 (1997) 14418–14422

[36]   E. Braun, Y. Eichen, U. Sivan, G. Ben-Yoseph, DNA-templated assembly and electrode attachment of a conducting silver wire, Nature. 391 (1998) 775-778

[37]   J.J. Storhoff, C.A. Mirkin, Programmed Materials Synthesis with DNA, Chem. Rev. 99 (1999) 1849–1862

[38]   C. Lin, Y. Liu, S. Rinker, H. Yan, DNA Tile Based Self-Assembly: Building Complex Nanoarchitectures, ChemPhysChem. 7 (2006) 1641–1647. https://doi.org/10.1002/cphc.200600260

[39]   J. Sharma, R. Chhabra, C.S. Andersen, K. V Gothelf, H. Yan, Y. Liu, Toward Reliable Gold Nanoparticle Patterning On Self-Assembled DNA Nanoscaffold, J. Am. Chem. Soc.. 130 (2008) 7820–7821

[40]   Y. Liu, C. Lin, H. Li, H. Yan, Aptamer-Directed Self-Assembly of Protein Arrays on a DNA Nanostructure, Angew. Chem. Int. Ed. 44 (2005) 4333–4338. https://doi.org/10.1002/anie.200501089

[41]   K. Sangeetha, S.S. Sankar, K. Karthick, S. Anantharaj, S.R. Ede, S. Wilson T., S. Kundu, Synthesis of ultra-small Rh nanoparticles congregated over DNA for catalysis and SERS applications, Colloids Surfaces B Biointerfaces. 173 (2019) 249–257. https://doi.org/10.1016/j.colsurfb.2018.09.052

[42]   S. Kumaravel, P. Thiruvengetam, K. Karthick, S.S. Sankar, S. Kundu, Detection of Lignin Motifs with RuO_2-DNA as an Active Catalyst via Surface-Enhanced Raman Scattering Studies, ACS Sustain. Chem. Eng. 7 (2019) 18463–18475. https://doi.org/10.1021/acssuschemeng.9b04414

[43]   S.R. Ede, S. Kundu, Microwave Synthesis of SnWO4 Nanoassemblies on DNA Scaffold: A Novel Material for High Performance Supercapacitor and as Catalyst for Butanol Oxidation, ACS Sustain. Chem. Eng. 3 (2015) 2321–2336. https://doi.org/10.1021/acssuschemeng.5b00627

[44]   S.R. Ede, A. Ramadoss, U. Nithiyanantham, S. Ananthara, and S. Kundu, Bio-molecule Assisted Aggregation of ZnWO4 nanoparticles (NPs) into Chain-like Assemblies: Material for High Performance Supercapacitor and as Catalyst for Benzyl Alcohol Oxidation, Inorg. Chem. 54 (2015) 3851–3863. https://doi.org/10.1021/acs.inorgchem.5b00018

[45]   K. Karthick, S. Anantharaj, S.N. Jagadeesan, P. Kumar, S.R. Ede, D.K. Pattanayak, S. Patchaiammal, S. Kundu, Advanced Cu3Sn and Selenized Cu3Sn@Cu

Foam as Electrocatalysts for Water Oxidation under Alkaline and Near-Neutral
Conditions, Inorg. Chem. 58 (2019) 9490–9499.
https://doi.org/10.1021/acs.inorgchem.9b01467

[46]   S. Anantharaj, K. Karthick, S. Kundu, $NiTe_2$ Nanowire Outperforms Pt/C in High-
Rate Hydrogen Evolution at Extreme pH Conditions, Inorg. Chem. 57 (2018) 3082–
3096. https://doi.org/10.1021/acs.inorgchem.7b02947

[47]   S. Anantharaj, T.S. Amarnath, E. Subhashini, S. Chatterjee, K.C.S. Swaathini, K.
Karthick, S. Kundu, Shrinking the Hydrogen Overpotential of Cu by 1 V and
Imparting Ultralow Charge Transfer Resistance for Enhanced H2 Evolution, ACS
Catal. 8 (2018) 5686–5697. https://doi.org/10.1021/acscatal.8b01172

[48]   I. Katayama, K. Shudo, J. Takeda, T. Shimada, A. Kubo, S. Hishita, D. Fujita, M.
Kitajima, Ultrafast Dynamics of Surface-Enhanced Raman Scattering Due to Au
Nanostructures, Nano Lett. 11 (2011) 2648–2654. https://doi.org/10.1021/nl200667t

[49]   S. Zong, C. Chen, Z. Wang, Y. Zhang, Y. Cui, Surface Enhanced Raman
Scattering Based in Situ Hybridization Strategy for Telomere Length Assessment,
ACS Nano. 10 (2016) 2950–2959. https://doi.org/10.1021/acsnano.6b00198

[50]   S. Kundu, M. Mandal, S.K. Ghosh, T. Pal, Photochemical deposition of SERS
active silver nanoparticles on silica gel, J. Photochem. Photobiol. A Chem. 162 (2004)
625–632. https://doi.org/10.1016/S1010-6030(03)00398-8

[51]   J.P. Camden, J.A. Dieringer, Y. Wang, D.J. Masiello, L.D. Marks, G.C. Schatz,
R.P. Van Duyne, Probing the Structure of Single-Molecule Surface-Enhanced Raman
Scattering Hot Spots, J. Am. Chem. Soc.. 130 (2008) 12616–12617

[52]   J. Shen, J. Su, J. Yan, B. Zhao, D. Wang, S. Wang, S. Mathur, K. Li, M. Liu, C.
Fan, Y. He, S. Song, Bimetallic nano-mushrooms with DNA-mediated interior
nanogaps for high-efficiency SERS signal amplification, Nano Res. 8 (2014) 731–742.
https://doi.org/10.1007/s12274-014-0556-2

[53]   C. Muehlethaler, C.R. Considine, V. Menon, W. Lin, Y. Lee, J.R. Lombardi,
Ultra-High Raman Enhancement on Monolayer MoS_2, ACS Photonics. 3 (2016) 1164–
1169. https://doi.org/10.1021/acsphotonics.6b00213

[54]   M. Fleischmann, P.J. Hendra, A.J. McQuillan, Raman Spectra of Pyridine
Adsorbed at a Silver Electrode, 1974

[55]   D. Mehn, C. Morasso, R. Vanna, M. Bedoni, D. Prosperi, F. Gramatica,
Vibrational Spectroscopy Immobilised gold nanostars in a paper-based test system for

surface-enhanced Raman spectroscopy, Vib. Spectrosc. 68 (2013) 45–50. https://doi.org/10.1016/j.vibspec.2013.05.010

[56]   Y. Yan, A.I. Radu, W. Rao, H. Wang, G. Chen, K. Weber, D. Wang, D. Cialla-May, J. Popp, P. Schaaf, Mesoscopically Bi-continuous Ag–Au Hybrid Nanosponges with Tunable Plasmon Resonances as Bottom-Up Substrates for Surface- Enhanced Raman Spectroscopy, Chem. Mater. 28 (2016) 7673–7682. https://doi.org/10.1021/acs.chemmater.6b02637

[57]   S. Chen, P. Xu, Y. Li, J. Xue, S. Han, W. Ou, L. Li, W. Ni, Rapid Seedless Synthesis of Gold Nanoplates with Microscaled Edge Length in a High Yield and Their Application in SERS, Nano-Micro Lett. 8 (2016) 328–335. https://doi.org/10.1007/s40820-016-0092-6

[58]   F. Pu, Y. Huang, Z. Yang, H. Qiu, J. Ren, Nucleotide-Based Assemblies for Green Synthesis of Silver nanoparticles with Controlled Localized Surface Plasmon Resonances and Their Applications, ACS Appl. Mater. Interfaces. 10 (2018) 9929–9937. https://doi.org/10.1021/acsami.7b18915

[59]   H. Wei, H. Xu, Hot spots in different metal nanostructures for plasmon- enhanced Raman spectroscopy, Nanoscale. 5 (2013) 10794–10805. https://doi.org/10.1039/c3nr02924g

[60]   S. Ding, J. Yi, J. Li, B. Ren, R. Panneerselvam, Z. Tian, Nanostructure-based plasmon- enhanced Raman spectroscopy for surface analysis of materials, Nat. Rev. Mater. 1 (2016) 1–16. https://doi.org/10.1038/natrevmats.2016.21

[61]   P.E. Noppadon Nuntawong, P. Eiamchai, S. Limwichean, M. Horprathum, V. Patthanasettakul, P. Chindaudom, Applications of surface-enhanced Raman scattering (SERS) substrate, in: Asian Conf. Def. Technol., 2015: pp. 92–95

[62]   J.M. Mclellan, A. Siekkinen, J. Chen, Y. Xia, Comparison of the surface-enhanced Raman scattering on sharp and truncated silver nanocubes, Chem. Phys. Lett. 427 (2006) 122–126. https://doi.org/10.1016/j.cplett.2006.05.111

[63]   J. Li, Y. Yanga, D. Qin, Hollow nanocubes made of Ag–Au alloys for SERS detection with sensitivity of 10-8 M for melamine, J. Mater. Chem. C. 2 (2014) 9934–9940. https://doi.org/10.1039/C4TC02004A

[64]   Y.W. Lee, M. Kim, S.W. Kang, S.W. Han, Polyhedral Bimetallic Alloy Nanocrystals Exclusively Bound by {110} Facets: Au–Pd Rhombic Dodecahedra, Angew. Chem. Int. Ed. 50 (2011) 3466–3470. https://doi.org/10.1002/anie.201007220

[65]   B. Ren, X. Lin, Z. Yang, G. Liu, R.F. Aroca, B. Mao, Z.-Q. Tian, Surface-Enhanced Raman Scattering in the Ultraviolet Spectral Region : UV-SERS on Rhodium and Ruthenium Electrodes, J. Am. Chem. Soc. 100 (2003) 9598–9599. https://doi.org/10.1021/ja035541d

[66]   C. Fasolato, F. Ripanti, A. Capocefalo, A. Sarra, F. Brasili, C. Petrillo, C. Fasolato, P. Postorino, DNA-functionalized gold nanoparticle assemblies for Surface Enhanced Raman Scattering, Colloids Surfaces A Physicochem. Eng. Asp. 589 (2019) 124399. https://doi.org/10.1016/j.colsurfa.2019.124399

[67]   L. Zhang, H. Ma, L. Yang, Design and fabrication of surface-enhanced Raman scattering substrate from DNA – gold NPs, RSC Adv. 4 (2014) 45207–45213. https://doi.org/10.1039/C4RA06947A

[68]   X. Shen, C. Song, J. Wang, D. Shi, Z. Wang, N. Liu, B. Ding, Rolling Up Gold Nanoparticle-Dressed DNA Origami into Three- Dimensional Plasmonic Chiral Nanostructures, J. Am. Chem. Soc.. 134 (2012) 146–149

[69]   J. Keum, M. Kim, J. Park, C. Yoo, N. Huh, S. Chul, DNA-directed self-assembly of three-dimensional plasmonic nanostructures for detection by surface-enhanced Raman scattering (SERS), Sens. Bio-Sensing Res. 1 (2014) 21–25. https://doi.org/10.1016/j.sbsr.2014.06.003

[70]   Y. Li, X. Qi, C. Lei, Q. Yue, S. Zhang, Simultaneous SERS detection and imaging of two biomarkers on the cancer cell surface by self-assembly of branched DNA–gold nanoaggregates, Chem. Commun. 50 (2014) 9907–9909. https://doi.org/10.1039/C4CC05226A

[71]   M. Liu, Z. Wang, S. Zong, R. Zhang, D. Zhu, S. Xu, C. Wang, SERS-based DNA detection in aqueous solutions using oligonucleotide-modified Ag nanoprisms and gold nanoparticles, Anal Bioanal Chem. 405 (2013) 6131–6136. https://doi.org/10.1007/s00216-013-7016-9

[72]   S. Anantharaj, U. Nithiyanantham, S.R. Ede, S. Kundu, Osmium organosol on DNA: Application in catalytic hydrogenation reaction and in SERS studies, Ind. Eng. Chem. Res. 53 (2014) 19228–19238. https://doi.org/10.1021/ie503667y

[73]   K. Sakthikumar, S. Anantharaj, S.R. Ede, K. Karthick, S. Kundu, A highly stable rhenium organosol on a DNA scaffold for catalytic and SERS applications, J. Mater. Chem. C. 4 (2016) 6309–6320. https://doi.org/10.1039/c6tc01250g

[74]   G. Wei, L. Wang, Z. Liu, Y. Song, L. Sun, T. Yang, Z. Li, DNA-Network-Templated Self-Assembly of Silver nanoparticles and Their Application in Surface-Enhanced Raman Scattering, J. Phys. Chem. B. 109 (2005) 23941–23947

[75]   C. Peng, Y. Song, G. Wei, W. Zhang, Z. Li, W. Dong, Self-assembly of λ-DNA networks/Ag nanoparticles: Hybrid architecture and active-SERS substrate, J. Colloid Interface Sci. 317 (2008) 183–190. https://doi.org/10.1016/j.jcis.2007.09.017

[76]   S. Kundu, M. Jayachandran, The self-assembling of DNA-templated Au nanoparticles into nanowires and their enhanced SERS and catalytic applications, RSC Adv. 3 (2013) 16486. https://doi.org/10.1039/c3ra42203h

[77]   L. Yang, G. Chen, J. Wang, T. Wang, M. Li, J. Liu, Sunlight-induced formation of silver-gold bimetallic nanostructures on DNA template for highly active surface enhanced Raman scattering substrates and application in TNT / tumor marker detection, J. Mater. Chem. 19 (2009) 6849–6856. https://doi.org/10.1039/b909600k

[78]   H. Ma, D. Lin, H. Liu, L. Yang, L. Zhang, J. Liu, Hot spots in photoreduced Au nanoparticles on DNA scaffolds potent for robust and high-sensitive surface-enhanced Raman scattering substrates, Mater. Chem. Phys. 138 (2013) 573–580. https://doi.org/10.1016/j.matchemphys.2012.12.021

[79]   S. Kundu, Y. Chen, W. Dai, L. Ma, A.M. Sinyukov, H. Liang, Enhanced Catalytic and SERS Activities of Size-selective Rh nanoparticles on DNA Scaffold, J. Mater. Chem. C. 5 (2017) 2577–2590. https://doi.org/10.1039/C6TC05529J

[80]   S. Kundu, S.-I. Yi, L. Ma, Y. Chen, W. Dai, A.M. Sinyukov, H. Liang, Morphology dependent catalysis and surface enhanced Raman scattering (SERS) studies using Pd nanostructures in DNA, CTAB and PVA scaffolds, Dalt. Trans. 46 (2017) 9678–9691. https://doi.org/10.1039/C7DT01474K

[81]   L. Sun, D. Zhao, Z. Zhang, B. Li, D. Shen, DNA-based fabrication of density-controlled vertically aligned ZnO nanorod arrays and their SERS applications, J. Mater. Chem. 21 (2011) 9674–9681. https://doi.org/10.1039/c1jm10830a

[82]   Z.Y. Jiang, X.X. Jiang, S. Su, X.P. Wei, S.T. Lee, Z.Y. Jiang, X.X. Jiang, S. Su, Y. He, Silicon-based reproducible and active surface-enhanced Raman scattering substrates for sensitive , specific , and multiplex DNA detection Silicon-based reproducible and active surface-enhanced Raman scattering substrates for sensitive , specific , and mul, Appl. Phys. Lett. 100 (2012) 203104. https://doi.org/10.1063/1.3701731

[83]   M. Kahraman, E.R. Mullen, A. Korkmaz, S. Wachsmann-hogiu, Fundamentals and applications of SERS-based bioanalytical sensing, Nanophotonics. 6 (2017) 831–852. https://doi.org/10.1515/nanoph-2016-0174

[84]   C.A.R. Auchinvole, P. Richardson, C. Mcguinnes, V. Mallikarjun, K. Donaldson, H. Mcnab, C.J. Campbell, Monitoring Intracellular Redox Potential Changes Using SERS Nanosensors, ACS Nano. 6 (2012) 888–896

[85]   Z.Q. Tian, B. Ren, D.Y. Wu, Surface-enhanced Raman scattering: From noble to transition metals and from rough surfaces to ordered nanostructures, J. Phys. Chem. B. 106 (2002) 9463–9483. https://doi.org/10.1021/jp0257449

[86]   J.D. Driskell, R.A. Tripp, Label-free SERS detection of microRNA based on affinity for an unmodified silver nanorod array substrate, Chem. Commun. 46 (2010) 3298–3300. https://doi.org/10.1039/c002059a

[87]   X. Yang, C. Gu, F. Qian, Y. Li, J.Z. Zhang, Highly Sensitive Detection of Proteins and Bacteria in Aqueous Solution Using Surface-Enhanced Raman Scattering and Optical Fibers, Anal. Chem. 83 (2011) 5888–5894

[88]   P. Negri, A. Kage, A. Nitsche, D. Naumann, R.A. Dluhy, Detection of viral nucleoprotein binding to anti-influenza aptamers via SERS, Chem. Commun. 47 (2011) 8635–8637. https://doi.org/10.1039/c0cc05433j

[89]   S. Efrima, B. V Bronk, Silver Colloids Impregnating or Coating Bacteria, J. Phys. Chem. B. 102 (1998) 5947–5950

[90]   I. Sayin, M. Kahraman, F. Sahin, D. Yurdakul, M. Culha, Characterization of Yeast Species Using Surface-Enhanced Raman Scattering, Appl. Spectrosc. 63 (2009) 1276–1282

[91]   M. Lee, S. Lee, J. Lee, H. Lim, G. Hun, E. Kyu, S. Chang, C. Hwan, J. Choo, Highly reproducible immunoassay of cancer markers on a gold-patterned microarray chip using surface-enhanced Raman scattering imaging, Biosens. Bioelectron. 26 (2011) 2135–2141. https://doi.org/10.1016/j.bios.2010.09.021

[92]   H. Jun, L. Liu, C. An, X. Zhang, M.Y. Lv, Y.M. Zhao, H.J. Xu, Study of surface-enhanced Raman scattering activity of DNA-directed self-assembled gold nanoparticle dimers, Appl. Phys. Lett. 107 (2015) 193106. https://doi.org/10.1063/1.4935543

[93]   S.S. Sankar, K. Sangeetha, K. Karthick, S. Anantharaj, S.R. Ede, S. Kundu, Pt nanoparticles Tethered DNA Assemblies for Enhanced Catalysis and SERS Applications, New J. Chem. 42 (2018) 15784-15792. https://doi.org/10.1039/C8NJ03940B

Bioinspired Nanomaterials   Materials Research Forum LLC
Materials Research Foundations **111** (2021) 118-154   https://doi.org/10.21741/9781644901571-5

[94]   D. Majumdar, A. Singha, P.K. Mondal, S. Kundu, DNA-mediated wirelike clusters of silver nanoparticles: An ultrasensitive SERS substrate, ACS Appl. Mater. Interfaces. 5 (2013) 7798–7807. https://doi.org/10.1021/am402448j

[95]   K. Sangeetha, K. Karthick, S. Sam Sankar, Arun Karmakar, S, Kundu, Prospects in interfaces of biomolecule DNA and nanomaterials as an effective way for improvising surface enhanced Raman scattering: A review, Advances in Colloid and Interface Science: 291 (2021) 102399

Materials Research Forum LLC
https://doi.org/10 21741/9781644901571-6

Chapter 6

Bio-Mediated Synthesis of Nanoparticles for Fluorescence Sensors

Somasundaram Anbu Anjugam Vandarkuzhali[1], Salman Ahmad Khan[3],
Subramanian Singaravadivel[2*], Gandhi Sivaraman[4*]

[1]National Centre for Catalysis Research, Indian Institute of Technology Madras, Chennai 600036, India

[2]Department of Chemistry, SSM Institute of Engineering and Technology, Dindigul 624002, India

[3]Department of Chemistry, Maulana Azad National Urdu University, Hyderabad, Telangana, India

[4]Department of Chemistry, Gandhigram rural Institue-Deemed to be university, Gandhigram - 624302, India

raman474@gmail.com

Abstract

The nature deeds alike a hefty "bio-laboratory" embracing of plants, algae, fungi, yeast etc. which are poised of biomolecules. These indeed befalling biomolecules must stood notorious to play an active role in the establishment of nanoparticles through diverse shapes and sizes thus acting as a driving force intended for the scheming of greener, safe and environmentally benign protocols for the synthesis of nanoparticles. The contemporary chapter targets the proportional biogenic synthesis and mechanisms of nanoparticles using biomolecules. The practice of biomolecules not only diminishes the price of synthesis but also curtails the need of using hazardous chemicals and arouses 'green synthesis'. It also emphases on aspects of binding of biomolecules to nanoparticles and certain of the applications of the biosynthesized nanoparticles as sensor for cations, anions and also biosensors.

Keywords

Chemosensors, Nanoparticles, Naked Eye Sensing, Fluorescence , Imaging

Contents

1.   Introduction

A number of enthralling progresses, technologies and prospects have materialized in recent years in the escalating field of nanotechnology.[1-3] As building blocks, nanomaterials (NMs) can play a vibrant part in nanotechnology owing to their bizarrely diverse properties as paralleled to their bulk counterparts.[4] MNMs have long existed in natural environments. Among several strategies, physical, chemical, and biological approaches are the most widely utilized methods for the MNMs synthesis. Indeed, physical and chemical methods have some major drawbacks such as the use of high-energy inputs, and costly and toxic chemicals in their synthetic roots.[5-7] Seminal reports on bioinspired synthesis date back to the late 90s once Ag and Au NPs were primed from Pseudomonas Stutzeri AG259[8] and alfalfa plant biomass,[9] respectively. Laterally with microorganisms and plants, viruses,[12] proteins[11] and DNA[12] have similarly become possibly expedient candidates for bio-inspired synthesis of MNMs. Herein, created on these biological candidates, bio-inspired synthesis embraces a collective application of biological perceptions, mechanisms and purposes for the proposal and development of pioneering bio-derived (nano) constituents with a number of applications.[13-16] Such collective application could be pertinent to synthesis or gathering of a wide range of inorganic NMs.[17-23]

The biological slant which embraces diverse types of microorganisms has remained used to synthesize diverse metallic NPs, which has a benefits over other chemical systems as this is greener, energy saving and cost effective. The coating of biological molecules on the outward of NPs makes them biocompatible in association to the NPs primed by chemical methods.[24-26] The biocompatibility of bioinspired NPs compromises very fascinating presentations in biomedicine and related fields.[27] These NPs contrived via biogenic enzymatic practice were superior to those synthesized via chemical methods as the use of affluent chemicals was inadequate and they obsessed advanced catalytic activity. Among diverse biological systems used for NP synthesis, innumerable forms of algae are currently being used as model schemes as these have remarkable facility of bioremediation of toxic metals thus adapting them into further flexible forms. Also, these are proficient in the fabrication of dissimilar metal and metal oxide NPs.[28] The biosynthesis of NPs using alage and waste constituents is an evolving and impending research.

Live microorganisms swarm a substantial array of metabolic responses essential for functions comprising nutrient dispensation, growth, and energy release.[29] The metabolic practice of live microorganisms valor be intricate in the bioreduction of metal ions to diminish the toxicity of metal ions. Through the transport system, microbes can intracellularly amass metals ions. Metal ions can be abridged by several falling species extant inside microbes[30] as well as extracellularly abridged by diverse metabolites.[31,32] In contrast to live microorganisms, dead objects are not reliant on metabolic practices. Metal ions are bound by microbial cells which then deliver privileged nucleation sites for MNM advance on their surface. Innumerable functional groups comprising thiol, hydroxyl, carboxyl, imidazole, amino, guanidine and imino groups have been revealed to have great affinity to bind metal ions.[33]

Alike to microorganisms, viruses frequently offer binding sites for metal ions and nucleation sites for MNM development in the incidence (and even absence) of additional reducing agents. Though there are various coating protein molecules in wild-type (WT) viruses, some active groups entrenched inside the viral matrices are not available to metal ions.[13] Definite vital processes should be consequently assumed to ailment WT viruses aforementioned to their use for the synthesis of MNMs (i.e. surface modification) to endorse the metallization of viruses by amassed their low affinity for metal precursors.[34]

The capsids of viruses can be adapted by innumerable approaches comprising charge and genetic amendments to offer uniform and accurately spaced binding sites for metal ions.[35,36]

Enthused by the bioremediation of heavy metals by plants,[37] live plants have developed as an substitute candidate in MNM biosynthesis.[38] Nevertheless, the general mechanism for the bioinspired synthesis of MNMs by live plants has not been abundantly assumed.[39,40] Water-soluble biomolecules from microbial or plant biomass normally play dual roles as reducing and protecting agents in bio-inspired syntheses.[40-42] Functional groups (e.g. hydroxyl groups) play reductive roles in the development of MNMs while the strong interaction amid biomolecules and MNMs hints to an exceptional stability of as-synthesized MNMs.[42]

2.   Synthesis of nanoparticles

2.1   Microorganism-mediated synthesis

Microorganisms can be principally categorized as prokaryotic and eukaryotic. Bacteria and fungi exemplify prokaryotic and eukaryotic microorganisms as foremost bio-inspired candidates for the synthesis of MNMs. Live microorganisms not only reveal intricate biological activity but also retain a convoluted hierarchical structure. A cumulative number of perceptions into metal resistance in the bioreduction practice arbitrated by microorganisms have been newly attained but have not been thriving addressed in presently existing reviews. Metal NPs can be instituting in the periplasmic space, on the cell wall and external the cells. It is assumed that innumerable enzymes take an active part in the bioreduction practice of transporting electrons from assured electron donors to metal electron acceptors for diverse microorganisms.[43,44]

Klaus et al.[8] engaged Pseudomonas stutzeri AG259 in the preparation of Ag NPs. Ag-resistance can encompass the development and amassing of Ag precipitates outside the cytoplasmic membrane, feasibly convoyed by metal efflux and metal binding.[8] Mukherjee et al.[45] later described the use of eukaryotic microorganisms Verticillium fungus (Verticillium sp.) in the intracellular synthesis of metal NPs. They exhibited that the toxic effects of Ag(I) could be curtailed by reducing Ag(I) ions to elemental Ag NPs within the cells, which were alive and could advance reproduce after reduction.[45] Ag NPs were also primed via AgNO3 reduction with Lactobacillus strains.[31] While the exact Ag-resistant mechanism was not explicated in this work, auxiliary studies correlated to the extracellular development mechanism of Ag NPs by Ag-resistant Morganella sp. Revealed by three homologous genes silE, silP and silS were closely allied with Ag-resistance.[32] Numerous Ag-specific proteins were institute to be secreted outside the cell throughout the progress and were advised as being accountable for the reduction of Ag(I) ions and formation of Ag NPs in the extracellular microenvironment.[32]

Lin et al.[46] also probed the biosynthesis of Ag NPs upon reduction of Ag(I) ions by the periplasmic nitrate reductase c-type cytochrome subunit NapC in a Ag-resistant Escherichia coli (E. coli) under anaerobic circumstances. Outcomes indicated that c-type cytochromes such as NapC located in the periplasm could reduce Ag(I) ions to Ag NPs (Fig. 1).

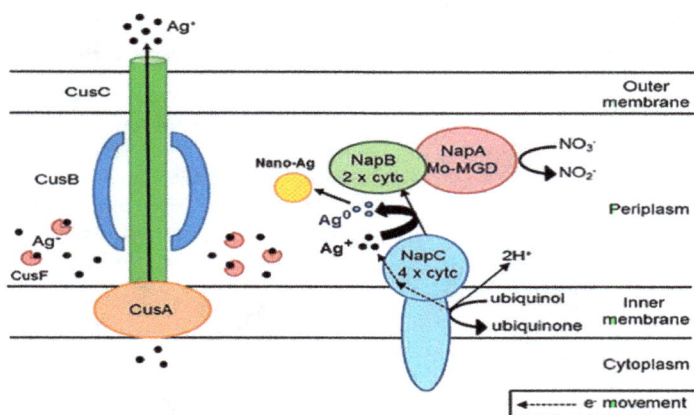

Fig. 1 A proposed model of biosynthesis of nano-Ag by periplasmic c-type cytochrome NapC in the Ag-resistant E. coli strain 116AR. Reprinted with permission from ref. 46.

Rösken et al.[47] observed the time-dependent growth of in vivo Au NPs in cyanobacteria Anabaena sp., when visible to Au(III) ions at 0.8 mM. All microorganisms appeared to be dead after 8 days due to the assimilation of Au NPs. Nevertheless, in some cases, Au(III) ions could not be utterly abridged by microorganisms. Instead, the comprehensive reduction of Au(III) ions essential the input of electron donors as revealed in the swift preparation of Au NPs by Shewanella algae (S. algae) ATCC 51181 using H_2 as electron donor (Au NPs formed in the periplasmic space).[48,49]

Ahmad et al.[50] consequently revealed the use of single-spore bacteria (Thermomonospora sp.) for the reduction of Au(III) ions to Au NPs at 50°C. The outcomes keen to proteins (molecular weight 80-10 000) playing an imperative role in the reduction of Au(III) ions at 50°C, also promising for the endurance of the thermophilic bacteria. The reduction of Pd(II) ions to produce Pd NPs with D. desulfuriacans ATCC 29577 was showed in the existence of sodium pyruvate, formate or H_2 as an electron donor.[50] Hydrogenase and cytochrome C3 were appealed to be possibly intricate in the reduction of Pd(II) ions.[51] Table 1 précises some instances of enzymes intricate in the microbial reduction of metal ions.

Table 1 *Examples of enzymes involved in the microbial reduction of metal ions.*

Microorganisms	MNMs	Size	Enzymes
Bacteria D. desulfuriacans ATCC 29577[52]	Pd NPs	-	Hydrogenase and cytochrome C3
M. psychrotolerans[53]	Ag nanoplates	100-150 nm	Ag reductase
S. maltophilia[54]	Ag NPs	93 nm	Chromium reductase
Shewanella oneidensis[55]	Ag NPs	24.4 ± 0.8 nm	c-Type cytochromes
E. coli[46]	Ag NPs	5-70 nm	Nitrate reductase
Fungi F. oxysporum[56]	Ag NPs	20-40 nm	NADH-dependent reductases
R. oryzae[57]	Au NPs	15 nm	Cytoplasmic proteins
L. edodes[58]	Au NPs	5-50 nm	Laccase, tyrosinase

2.2 Non-enzymatic reduction

The non-enzymatic reduction of metal ions to NPs based on dead cells has also acknowledged some consideration in recent years. This practice is independent on the metabolic method of microorganisms[59-62] and subsequently diverse from enzymatic reduction. The protocol principally necessitates a simple adsorption and reduction of metal ions on cell surfaces, which consequence in utterly extracellular metal NPs. Described instances embrace the formation of Au NPs using Shewanella oneidensis over a fast biosorption but gentle reduction process in the existence of an electron donor.[63] The reduction was appealed to be non-enzymatic as microorganism cells were killed at high Au precursor concentrations.

The bioreduction of metal ions has been broadly considered by Lin et al.[64-66] at Xiamen University, China. In the case of Pt(IV) ions bound by proteins on cell walls, polypeptide chains might amendment from b-folded to a-helical forms, with a-helical being possibly more expedient than b-folded to prevent Pt NPs from aggregation.[64] S. cerevisiae was revealed to have a significant affinity for Au(III) ions due to the existence of oxygen-containing functional groups (hydroxyl and carboxylate ion groups) on the cell wall.[65] Reduction of Au(III) ions to zero-valent Au was chiefly pretentious by the free aldehyde group of the reducing sugars.[65] Using infrared spectrometry revisions, polypeptides were

Bioinspired Nanomaterials   Materials Research Forum LLC
Materials Research Foundations **111** (2021) 155-184   https://doi.org/10.21741/9781644901571-6

anticipated to be hypothetically initiated by the intrusion of Au(III) ions via the molecular reconformation and intensely precious the course of Au(0) nucleation and crystal growth.[66]

Microbial reduction appeared as an innovative and sustainable substitute to chemical and physical procedures for synthesis of metal NPs in recent years. Normally, metal ions was adsorbed and reduced by the microbial surface, ensuing in very small NPs that progressively rose over the microorganisms. Nevertheless, the shape of MNMs cannot be efficiently controlled beneath microbial reduction. Yang et al.[67] extended the microorganism-mediated, surfactant-directed (MSD) method to synthesize hierarchically pronged AuNWs by using E. coli cells. Au nanohorns could be also synthesized using the MSD technique with E. coli cells and cetyl trimethyl ammonium chloride (CTAC).[68]

2.3   Microorganism-mediated surfactant-directed synthesis

Binary metal ions can be concurrently adsorbed by microorganisms, which may offer nuclei for the growth of bimetallic nanostructures.[69] Very recently, Chen et al.[70] designated a MSD proof of perception method to synthesize novel AuPd bimetallic nanoflowers (NFs) entailing of 1D pedicels and 3D horns in the incidence of CTAC (Fig. 2). The authors acceptable the improvement of the MSD methods to attain bimetallic nanostructures in one pot protocol at room temperature. Moreover, the delinquent of metal leaching that befalls in the galvanic standby reaction for synthesizing bimetallic nanostructures can be evaded. In accumulation, the bimetallic nanostructures could form application-oriented nano-composites with microorganisms which, for instance, could be straight used as active catalysts in the selective hydrogenation of 1,3-butadiene. Consequences also exhibited that all attained constituents were alloys with Pd-enriched surfaces. The diameters of horns amplified while those of pedicels decreased with the increase of Pd precursor feeding concentration.[70]

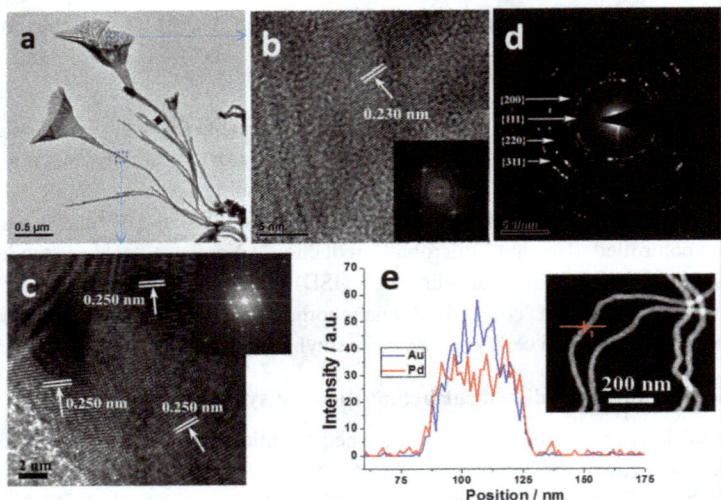

Fig. 2 (a) TEM image of AuPd NFs, (b) HRTEM image of blossom, (c) HRTEM image of pedicel, (d) SAED pattern of AuPd NFs, and (e) EDX line profiles of an individual AuPd nanowire (pedicel) corresponding to the framed part (dash) in (a). The insets indicate the corresponding fast Fourier transform (FFT) pattern. Reprinted with permission from ref. 70.

2.4 Virus-templated synthesis

A diversity of viruses has occurred as capable candidates in the bio-inspired synthesis of MNMs in the past decade. Some viruses have the benefits of unique dimensions and structures, well-spaced functionalities, high chemical stability and high yields, which stimulated research concern in the synthesis of virus-templated MNMs. Structurally, most viruses entail of two parts:[71] (i) the genetic substantial from either deoxyribonucleic acid (DNA) or ribonucleic acid (RNA); (ii) the capsid encompassing of a number of identically coated protein molecules. Commonly, the average size of viruses is about one-hundredth that of bacteria.[71] Mild reductants comprising dimethylamine borane complexes (DMAB),[72] sodium cyanoborohydride (NaBH$_3$CN),[35] hydrazine hydrate (N$_2$H$_5$OH),[10] hydroxylamine (NH$_2$OH),[73] hydroxylamine hydrochloride (NH$_2$OH.HCl),[74] sodium hypophosphite hydrate (NaH$_2$PO$_2$), and strong reductants such as NaBH$_4$[75] as well as UV irradiation[10] have been engaged to reduce metal precursors throughout virus-templated synthesis of MNMs.

Bioinspired Nanomaterials
Materials Research Foundations **111** (2021) 155-184

Materials Research Forum LLC
https://doi.org/10.21741/9781644901571-6

2.5   Metallization with wild-type viruses

Viruses habitually offer binding sites for metal ions and nucleation sites for MNMs formation with or without the essential of surplus reducing agents. Several approaches have been deliberated for WT viruses pretreatment to expand their metallization probable for the synthesis of MNMs. A suggested approach compacts with the instigation of viral surfaces for enriched nucleation. Knez et al.[76] first described the metallization of TMV plant viruses over a facile Ni deposition inside the TMV channels mediated by Pd(II).[10] Aljabali et al.[74] engaged WT icosahedral CPMV as a template for the electroless deposition of monodisperse metallic Co, Ni, Fe, Pt, Co-Pt and Ni-Fe NPs at room temperature with reduced Pd as nucleation sites.[74]

2.6   Assembly and size control of virus-templated MNMs

Virus-templated synthesis gives rise to diverse amassed structures and several MNMs. The assembled structures of MNMs on external surfaces or inside the channels of distinct virus have chiefly fascinated much consideration in recent years. The inclusive structures are contingent on geometrical shapes of viruses. Tunable Au-TMV NWs entailing of TMV and densely packed Au NPs over amendable addition-reduction cycles were also verified.[77] Ethanol addition subsequently the first addition-reduction cycle and enfolding of the Au-TMV nanohybrids with poly-L-lysine were assumed to confirm high homogeneity and stability in the nanowire suspension.[77] Khan et al.[78] also evidenced a simple, fast and high yield binding of citrate-coated Au NPs to deprotonated WT TMV. The deprotonation incredulous leading electrostatic repulsion so that attractive van der Waals forces subjugated.[78] In contrast to metal-virus NWs templated at the external surface of TMV, Ni and Co NWs could be synthesized inside TMV's internal channels in the absence of phosphate buffers reported by Knez and coworkers.[75]

2.7   Biomineralization using live plants.

The first demonstration of plant-mediated synthesis was stated in 1999 by Gardea-Torresdey et al.,[9] who designated the synthesis of Au NPs by bio-precipitation from Au(III) solutions with alfalfa biomass. Au and Ag NPs were fashioned using live alfalfa plants.[80] Some instances of alike biomineralization approaches in recent years are listed in Table 2. Spherical metal NPs (Au, Ag, Pd, Pt and AuAg) with a wide range of size distributions could be synthesized using live plants in most cases.

Table 2 Examples of biomineralization with live plants.

Plants	MNMs	Shape	Size (nm)
Sesbania[81]	Au NPs	Spherical	6-20
Brassica juncea[82]	Ag NPs	Spherical	2-35
Medicago sativa, Brassica juncea[83]	Pt NPs	Triangular, spherical	3-100
Brassica juncea[84]	Au, AuAg NPs	Spherical	< 50
Arabidopsis[42]	Pd NPs	Spherical	32
Populus deltoides[85]	Au NPs	Spherical	20-40
Arabidopsis thaliana L.[86]	Au NPs	Spherical, irregular	5-100

The contrivance of metal uptake and transport into the plant cells has been considered.[85] Sharma et al.[81] also confirmed that biomatrix-embedded Au NPs resulted from the reduction of Au(III) ions in the root cells and symplastic transportation of Au NPs to the aerial parts or shoots. The suggested arrangement of Au transformation in Arabidopsis thaliana L. ((1) uptake of the ionic Au monitored by (2) ensuing reduction of Au in planta to create NPs) was found to be a plausible route for Au NP formation.[86] The environment around Arabidopsis thaliana L. may affect the uptake or deposition of metallic ions.

2.8   Synthesis of MNMs using plant extracts.

Metal NPs embedded in plant environments are commonly challenging to harvest for auxiliary applications. Plant biomass and extracts can be a moderately better preference to whole plants. Nonetheless, the use of plant biomass stances the exertion in purification of as-synthesized metal NPs prior to their applications. Soluble biomolecules from plants or plant biomass have been subsequently desired for extracellular biosynthesis of metal NPs in the past decade.[87] Paralleled to microbial reduction, the biosynthesis of metal NPs with plant extracts retains higher reduction rates of metal precursors and facile shape control for metal NPs.[88-91] Plant-mediated syntheses can take benefit of existing plant resources and accomplish the extracellular synthesis of MNMs.

Plant-mediated synthesis of Au nanoplates has convert one of the major interests in the area of biosynthesis.[92-98] Interestingly, Zhan et al.[99] assumed Cacumen Platycladi (C. Platycladi) extracts as a weak reducing and capping agent to easily synthesize Au nanoplates in high yields beneath kinetic control deprived of other chemical reagents, as shown in Fig. 8. The reduction rate could be effortlessly controlled over modulating the investigational factors such as addition modes/rates of feeding solutions, temperature and pH based on a syringe-pump apparatus. A two-step size- and shape-separation of

Bioinspired Nanomaterials   Materials Research Forum LLC
Materials Research Foundations **111** (2021) 155-184   https://doi.org/10.21741/9781644901571-6

biosynthesized Au NPs by density gradient centrifugation and ensuing agarose gel electrophoresis, correspondingly, was also recently described.[100]

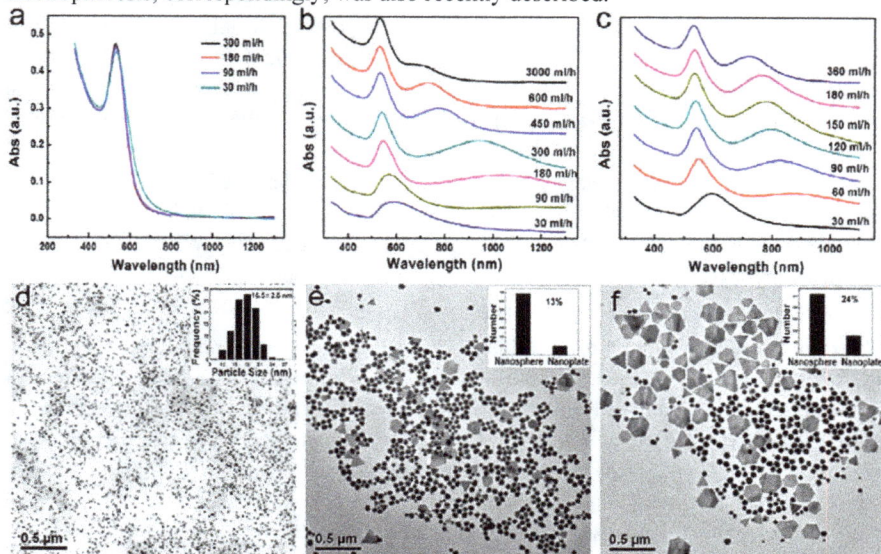

Fig. 3 (a-c) UV-vis-NIR spectra of Au sols prepared by addition modes (a) A, (b) B, and (c) C. (d-f) Representative TEM images of the Au NPs prepared by (d) mode A at an addition rate of 300 mL h⁻¹, (e) mode B at an addition rate of 450 mL h⁻¹, and (f) mode C at an addition rate of 90 mL h⁻¹. The reaction temperature and pH for all samples were 90 °C and 3.41, respectively. Three different mixing modes were adopted: mode A, involving the addition of Au precursor to C. Platycladi extract; mode B, involving an inverted addition of the extract to the Au precursor; and mode C, involving the simultaneous injection of the two feed solutions (C. Platycladi extract and Au precursor) with equal addition rates. Reprinted with permission from ref. 99.

3.   Interfaces of bio-metal systems as sensor

3.1   Colorimetric sensors

In recent years, colorimetric nanosensors obligate become prevalent because they offer numerous practical benefits, containing the ability for naked-eye visualization, cost efficiency, simplicity, portability and a low detection limit.[101] Of all the NMs, Au NPs are widely used for the colorimetric assays. The sensing mechanism is owing to the color

Bioinspired Nanomaterials   Materials Research Forum LLC
Materials Research Foundations **111** (2021) 155-184   https://doi.org/10.21741/9781644901571-6

variations of Au NPs, which are sensitive to the size, shape, capping agents, medium refractive index and the aggregation state of Au NPs. Li et al.[102] reported the BSA-stabilized Pt NPs with 57% Pt(0) and 43% Pt(II) alignment owning a high peroxidase-like activity, which can catalyze the substrates of TMB (3,30,5,50-tetramethylbenzidine) and H_2O_2. The colorimetric sensing mechanism of Hg(II) ions is based on the metallophilic interaction (Hg(II)‑Pt(0)), which down-regulate the peroxidase activity of Pt NPs. To a Pt nanoenzyme-TMB-H_2O_2 system, amassed the concentration of Hg(II) ions outcomes in a color advance from dark blue to colorless, enlightening the Hg(II) ions persuaded decrease of active Pt(0) species. The detection limit is 7.2 nM with a linear response range of 0‑120 nM. The as-prepared senor can be hypothetically pertinent for the quantitative fortitude of Hg(II) ions in a real water sample.

The revealing of glucose concentration in blood is an exclusively imperative part of observing the health of an individual. Nanoenzyme is becoming a novel tool for glucose detection due to its low cost, high chemical stability, easiness, as well as excellent biocompatibility. Established on the intrinsic peroxidase catalytic activity of apoferritin paired Au clusters (Au-Ft),[103] a colorimetric detection method has been advanced for glucose revealing. The linear range of glucose concentrations is from 2.0 mM to 10 mM, which is fit for the practical solicitation. In addition, other saccharides will not interfere with the detection of glucose, illuminating the good selectivity. Thus, the protein-Au NMs can be stared as a new forthcoming candidate for glucose sensor advance.

3.2   Fluorescent sensors

The notable photoluminescence properties and exceptional photo-stability of metal NCs have led to their extensive request in the analytical and biochemical fields. Protein- and DNA-templated metal NCs have been expansively discovered for the fluorescence revealing of biomolecules composites and metal ions. Protein template fluorescent noble metal NCs have been extensively used for the H_2O_2 detection. Zhang et al. employed horseradish peroxidase (HRP) enzyme to construct a dual functional Au NC retaining both catalytic and fluorescent properties for the detection of H_2O_2.[104] The HRP shell retains intrinsic catalytic activity, which permits catalytic reaction of HRP-Au itself, and H_2O_2, ensuing the appeasing of fluorescence properties of the core Au NC (Fig 5). Upon the addition of H_2O_2 beneath ideal conditions, the fluorescence strength was quenched linearly with high sensitivity (LOD = 30 nM).

Bioinspired Nanomaterials   Materials Research Forum LLC
Materials Research Foundations **111** (2021) 155-184   https://doi.org/10.21741/9781644901571-6

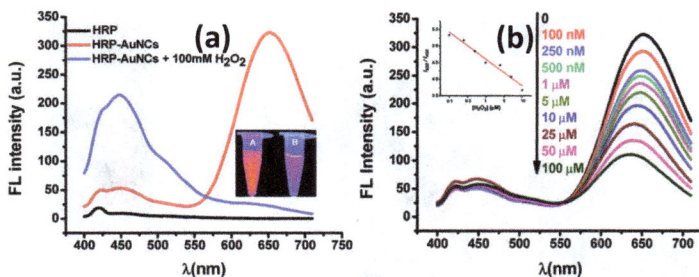

Fig 4. (a) Fluorescence spectra of HRP (black line), HRP−Au NCs in the absence (red line) and presence of 100 mM H_2O_2 (blue line). Inset: photograph displaying the fluorescence of HRP−Au NPs in the absence (A) and presence (B) of H_2O_2 upon excitation at 365 nm under a hand-held UV lamp. (b) Fluorescence responses of HRP−Au NCs after the addition of H_2O_2 (0-100 µM). Inset: Plot of the fluorescence ratio [I_{650}/I_{450}] of HRP−Au NCs versus the log concentration of H_2O_2.

Xie et al. employed BSA-templated Au NCs to intention a simple label-free process for the highly selective and ultrasensitive detection of Hg(II) ions.[105] The surface of as-prepared Au cluster is alleviated by a small amount of Au(I), which has a strong and precise interaction with Hg(II) ions. Based on the high-affinity metallophilic Hg(II)-Au(I) interactions, the red fluorescence of BSA-templated Au_{25} NCs could be efficiently quenched (Fig. 4). The fluorescence intensity of BSA-templated Au_{25} NCs decreased linearly over the Hg(II) concentration range of 1-20 nM. Furthermore, it had a remarkably high selectivity for Hg(II) over other metal ions and a low LOD value (0.5 nM), which could be advance advanced as a paper test strip for Hg(II) monitoring.

Lan et al.[106] institute that the fluorescence intensity of the DNA-templated Cu/Ag NCs was enriched when Cu(II) ions remained familiarized into a solution containing the DNA and Ag(I). Accumulative the Cu(II) concentration up to 250 nM would lead to constant fluorescence enhancement. Huang et al.[107] engaged a series of DNA templates to fabricate Ag NCs and then considered the effect of DNA length and base alignment on the fluorescent retorts of DNA-templated Ag NCs.

Materials Research Forum LLC

https://doi.org/10.21741/9781644901571-6

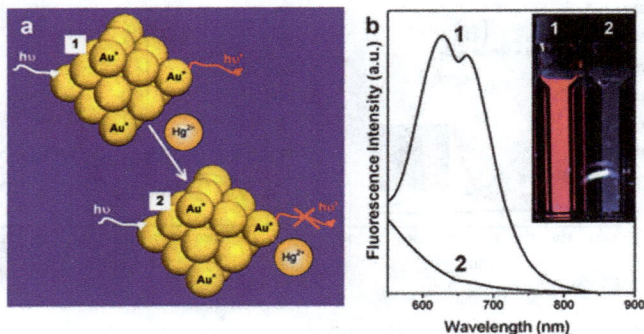

Fig. 5 (a) Schematic illustration of the Hg(II) sensing mechanism based on BSA-templated Au NCs. (b) Photoemission spectra and (inset) photographs of Au NCs under UV light in the absence and presence of Hg(II) ions. Reprinted with permission from ref. 105.

A novel, gold-nanocluster-based fluorescent sensor for cyanide in aqueous solution, based on the cyanide etching-induced fluorescence quenching of gold nanoclusters, is described by Lu et al.[108] In addition to posing high selectivity due to the distinctive Elsner reaction amid cyanide and the gold atoms of gold nanoclusters, this facile, ecologically friendly and cost-effective process runs high sensitivity. The lowermost concentration to enumerate cyanide ions could be down to 200×10^{-9} M , this fluorescent sensor displays exceptional recoveries (over 93%). This gold-nanocluster-based fluorescent sensor could find submissions in extremely sensitive and selective detection of cyanide in food, soil, water, and biological samples.

Chen et al. reports, a simple, eco-friendly and one-pot synthetic approach for the synthesis of fluorescent gold nanoclusters (GNCs) using a native polysaccharide (chondroitin sulfate) as the novel template. Furthermore, the chondroitin sulfate-capped gold nanoclusters (CS-GNCs) exhibited excitation and emission bands at 450 and 520 nm, respectively. The fluorescence quantum yield was 4.78%. Further they have used CS-stabilized gold nanoclusters selectively and sensitively for sensing Cu^{2+} in real samples via non-covalent d^{10}-d^{10} metallophilic bond. The linear range of the process was 0.01-1.8 µM with the limit of detection (LOD) was 8 nM (signal/noise = 3), and the recoveries of the spiked samples in lake water and tap water samples ranged from 90% to 108%, signifying that the projected technique could be used in the preparation for the detection of Cu^{2+}.[109]

Bioinspired Nanomaterials   Materials Research Forum LLC
Materials Research Foundations 111 (2021) 155-184   https://doi.org/10.21741/9781644901571-6

DNA-templated silver nanoclusters (DNA-AgNCs) have appeared as capable constituents for sensing, bio-labelling, and bio-imaging due to their fluorogenic properties. Using the 12 mer DNA with a cytosine-rich sequence, AgCNs have been designed effectively. Kim et al advanced two diverse types of DNA-AgNC systems for the detection of potassium ions (K^+) and nitric oxide (NO) by employing the structural change of DNA or DNA template transformation. In K^+ detection, a thrombin binding aptamer (TBA) was exploited as both the AgNC template and the K^+ binding probe. The fluorescence emission of the DNA1-AgNC probe was lessened correspondingly to the amount of K^+ detected, which alters the AgNC-adapted structure into G-quadruplex. On the other hand, NO detection has been proficient by NO-induced deamination of cytosine. The structural alteration of cytosine to deoxyuridine destabilizes the DNA2-AgNC structure, subsequent in quenching of fluorescence strength. Each system delivers a detection limit of 6.2 μM for K^+ and 0.1 μM for NO. We expect that these systems assist the opening of a new horizon for highly sensitive and selective detection systems.[110]

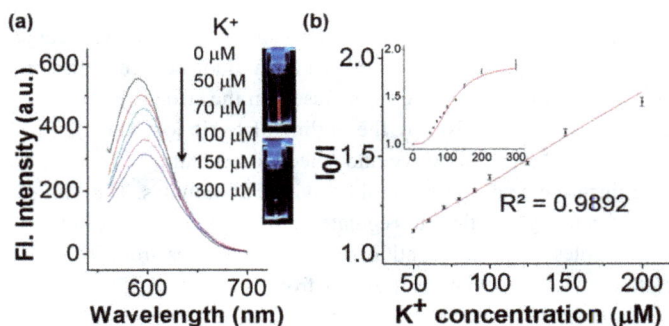

Fig. 6 Fluorescence emission spectra representing the quenching effect depending on K^+ concentrations on the fluorescent DNA1-AgNC probes. (a) Two photographs (right) of DNA1-AgNCs indicate the quenching effect under UV illumination. (b) Sensitivity of the fluorescent DNA1-AgNC probe toward K^+. The Stern-Volmer plot represents the quenching effect of K^+ on the fluorescence emission of DNA1-AgNCs at 590 nm.

Ren et al. advanced a new fluorescent technique for the label-free assay of thiol-containing amino acids and peptide by use of silver deposited DNA duplex and the intercalating dye. The sensing method is based on the precise interactions amid thiols and DNA-templated silver deposition via robust Ag-S bonds. In the incidence of thiols, the intercalating dye gives a intense surge in fluorescence as a consequence of the strong

Materials Research Forum LLC
https://doi.org/10.21741/9781644901571-6

interaction amid the intercalator and the released DNA from silver surfaces. The detection limit of this scheme is lower than or at least similar to preceding fluorescence-based methods, and the turn-on sensing mode offers additional benefit to proficiently reduce background noise. This technique also displays exceptional selectivity for these thiol-containing biomolecules over numerous other amino acids.[111]

Chattopadhyay et al. exhibit bio-mineralized synthesis of stable Au-Ag bimetallic NCs with tunable NIR fluorescence using bovine serum albumin (BSA) as a protein template. We also validate its presentation in the detection of toxic heavy metal ions Pb^{2+} *invitro* and inside cells. The tunability of fluorescence emission among 680 nm and 815 nm is accomplished by thoroughly variable the ratio of Au and Ag in the composite NCs. The bimetallic NCs when interacted with Pb^{2+} concentration existing a large surge in fluorescence intensity, which allowed sensitive detection of Pb^{2+}. We resolute a limit of detection (LOD) of 96 nM for the detection of Pb^{2+} under *invitro* condition, which is suggestively less than the safe level in drinking water. Its applicability has also been confirmed effectively in real water samples poised from local water bodies.[112]

Song et al. described molecule of vancomycin was engaged as reducer/stabilizer for facile one-pot synthesis of water revealed a bluish fluorescence emission at 410 nm within a short synthesis time about 50 min. Based on the strong fluorescence quenching due to electron transfer mechanism by the outline of ferric ions (Fe^{3+}), the Van-AuNCs were remarkably intended for sensitive and selective detecting Fe3+ with a limit of 1.4 $\mu mol\ L^{-1}$ in the linear range of 2-100 $\mu mol\ L^{-1}$ within 20 minutes. The Van-AuNCs based technique was effectively practical to regulate Fe^{3+} in tap water, lake water, river water and sea water samples with the quantitative spike recoveries from 97.50% to 111.14% with low relative standard deviations ranging from 0.49% to 1.87%, demonstrating the probable presentation of this Van-AuNCs based fluorescent sensor for environmental sample analysis.[113]

Cu^{2+} can synchronize with amine and carboxylic groups of proteins, protein-encapsulated AuNCs were familiarized for selective detection of Cu^{2+} in environmental water.[42,114-116] Fascinatingly, Luo et al. pronounced an ultrasensitive probe for sensing Cu^{2+} with the limit of detection (LOD) of 0.3 nM relying on the permutation of (i) Cu^{2+}-catalyzed reaction of persulfate and iodide and (ii) iodine-induced luminescence quenching of BSA-stabilized AuNCs.64 Moreover, GSH-capped AuNCs were shown to be well-suited for luminescence turn-off detection of Cu^{2+} through the complexation of negatively charged carboxylate groups of GSH with Cu^{2+}.66 Since a due-emission probe can offer precise quantification of the analyte with low signal deviation, Zhang et al. used GSH as a reducing agent to formulate riboflavin dwindling AuNCs with dual emission peaks at 530 and 840 nm.63 This dual-emission probe was used for ratiometric detection of Cu^{2+}

through the interplay of (i) fluorescence resonance energy transfer (FRET) from riboflavin to AuNCs and (ii) Cu^{2+}-triggered luminescence quenching of AuNCs. Besides, some of these heavy metals, even at a low concentration level, are highly poisonous to human health as a result of accumulation in organs.[42,105,114-131] Therefore, numerous NIR-emitting AuNCs have been employed as a platform to monitor whether the level of heavy metal ions surpasses the suggested limits (Table 3).

Table 3. AuNCs fluorescent probes for the detection of cations.

Nano cluster	Ex	Em	QY	Analyte	Linear range	LOD
BSA stabilized AuNCs[114]	500	680	6.0 %	Cu^{2+}	20-0.1 µM	10 nM
Riboflavin stabilized AuNCs[115]	375	530, 840	NA	Cu^{2+}	0-30 µM	0.9 µM
BSA stabilized AuNCs[42]	365	640	NA	Cu^{2+}	0.8-80 nM	0.3nM
Papain stabilized AuNCs[116]	470	660	4.3%	Cu^{2+}	10-250 nM	3.0 nM
GSH stabilized AuNCs[117]	420	730	NA	Cu^{2+}	0.1-6.2 µM	86.0 nM
Trypsin stabilized AuNCs[118]	520	665	7.30%	Cu2+	0.01-100 µM	5.2nM
				Co^{2+}	0.01100 µM	0.0078nM
				Hg^{2+}	0.0001-100 µM	0.005nM
Insulin stabilized Ag/AuNCs[119]	340	640		Cu^{2+}	20-500 nM	7.0 nM
				Hg^{2+}	20-2000 nM	5.0 nM
SMBD stabilized AuNCs[120]	360	630	9.7%	Hg^{2+}	4-20 µM	0.5nM
Keratin stabilized Ag-AuNCs[121]	400	725	6.5%	Hg^{2+}	2.44-2500 nM	2.31 nM
BSA stabilized Au-AgNCs[105]	470	640	6.0%	Hg^{2+}	1-20 nM	0.5 nM
Lysozyme stabilized AuNCs[122]	360	657	5.6%	Hg^{2+}	10-5000 nM	10 nM
Trypsin stabilized AuNCs[123]	365	645		Hg^{2+}	50-600 nM	50 nM
DHLA@AuNCs[124]	580	715	2.9%	Hg^{2+}	0-50 µM	0.5nM
LA capping Au/Ag NCs[125]	365	630	6.4%	Fe^{3+}	1-80 µM	0.5 µM
His-AuNCs stabilized BSA-Au/Ag NCs[126]	370	660	9.54%	Hg^{2+}	5 nM-5 µM	1.56 nM
				Fe^{3+}	5 µM-1 mM	1.1 µM
MUA stabilized AuNCs[127]	300	630	3.1%	Hg^{2+}	1.8-8 µM	0.04 µM
Zein stabilized AuNCs[128]	442	655	4.5%	Hg^{2+}	0.1-15 µM	25.0 nM
				Ag^+	1-15 µM	200 nM
GSH stabilized AuNCs[129]	410	710	1.5%	Cr(VI)	5-500 µg L^{-1}	0.5 µg L^{-1}
				Cr(III)	25-3800 µg L^{-1}	2.5 µg L^{-1}
BSA stabilized AuNCs[130]	365	620	NA	Pb^{2+}	10-1000 nM	1.0 nM
DTT-BSA stabilized AuNCs[131]	565	660	20%	Pb^{2+}	0.001-10 ppm	1.0 ppb

Inorganic anions, such as cyanide (CN^-), sulfide (S^{2-}), and (nitrite) NO^{2-}, retain the damaging influence on ecological systems as well as human health.[108,132-143] Among them, acquaintance to high levels of CN- leads to the compensations of organs and skin *via* the suppression of oxygen transportation.[108,132-136] Besides, hydrogen cyanide (HCN) is practical to be liberated from enzyme-catalyzed hydrolysis of cyanogenic glycosides that are commonly present in edible foods, like bamboo and cassava.[134,135] Given that CN^- triggers the dissolution of AuNCs *via* the formation of $Au(CN)^{2-}$, protein stabilized AuNCs were practical to be well-suited for the quantitative purpose of cyanide at alkaline pH conditions.[136] Wang et al. exploited BSA-stabilized Ce-Au NCs for ratiometric detection of CN^- in environmental water samples conferring to two consecutive processes: (i) cyanide persuaded the dissolution of the Au core of BSA-stabilized Ce-Au NCs, quenching the luminescence of AuNCs; (ii) the subsequent $Au(CN)^{2-}$ improved the luminescence of Ce^{4+} owing to the complexation reaction of BSA, $Au(CN)^{2-}$, and Ce^{4+}. Hu et al. also advanced a ratiometric membrane sensor for the fortitude of CN^- in water samples based on the electro spinning mediated assimilation of blue-emitting carbon dots and BSA stabilized AuNCs into a nanofibrous membrane.[144] The suggested membrane sensor delivered the LOD of cyanide corresponding to 0.15 µM, which is much lower than the MPB value of cyanide (1.9 µM) in drinking water set by World Health Organization.

Advance of the luminescent probe for sensing aminothiols has established notable credit in the recent past years outstanding to the vibrant role of aminothiols in physiological conduits such as detoxification and metabolism. Five kinds of low-molecular-weight aminothiols-cysteine, homocysteine, glutathione (GSH), cysteinyl glycine, and γ-glutamyl cysteine are contemporary in human fluids and their strange concentrations are concerned in a series of clinical illnesses, such as skin lesions, hair discoloration, edema, liver damage, sluggish development, cancer-related diseases, and Alzheimer's disease.[145-148] In assessment to the level of intracellular GSH ranging from 1 to 10 mM, the concentration of the utmost plentiful aminothiols, cysteine, in human plasma is observed to be 150-250 µM.[149-151] Fluorescent AuNCs were employed as an excellent probe for the sensing of anions and amino thiols in environmental, chemical, and biological samples were abridged in Table 4.

Table 4. AuNCs fluorescent probes for the detection of anions and aminothiols.

Nano cluster	Ex	Em	Q.Y	Analyte	Linear range	LOD
BSA stabilized Ce-AuNCs[132]	325	658	NA	CN⁻	0.1-15 μM	50 nM
Lysozyme stabilized AuNCs[133]	370	650	5.2%	CN⁻	5-120 μM	190 nM
BSA stabilized AuNCs[108]	365	640	N.A	CN⁻	1-9 μM	200 nM
BSA stabilized Ce/Au NCs[136]	325	410, 658	N.A	CN⁻	0.1-15 μM	50 nM
Papain stabilized AuNCs[137]	470	650	N.A	S^{2-}	0.5 to 80 μM	0.38 μM
DNA stabilized AuAgNCs[138]	460	630	4.5%	S^{2-}	0.01 to 9 μM	0.83 nM
BSA stabilized GSH @AuNCs[139]	330	650	-	NO^{2-}	1 to 20 μM	0.3 μM
BSA stabilized GSH @AuNCs[140]	375	660	6.0%	NO^{2-}-	100 nM to 100 μM	100 nM
BSA stabilized AuNCs[141]	350	670	NA	NO^{2-}-	0.2 nM to 50 μM	1.0 nM
BSA stabilized AuNCs[142]	502	622	NA	NO^{2-}-	0.1 to 50	30 nM
BSA stabilized AuNCs[143]	365	640	NA	NO^{2-}-	0 to 30.8 μM	80 nM
BSA stabilized AuNCs[145]	530	670	0.06	GSH	0.04-16 μM	7.0 nM
BSA stabilized AuNCs[146]	NA	644	NA	Cysteine	2-800 nmol mL⁻¹	1.2 nM
Transferrin stabilized AuNCs[147]	410	640	NA	GSH	0 to 150 μM	2.86 μM
Self-quenched AuNCs[148]	535	735	0.006	GSH	0.1 to 1.5 mM	0.004mM
AuAg NCs[149]	470	650	N.A	Cysteine	2.0-100 μM	1.1 μM
AuAg NCs[150]	370	650	N.A	Cysteine	0.02–80 mM	5.87 nM
				GSH	2.0-70 mM	1.01 μM
GSH stabilized Au/AgNCs[151]	380	618	0.072	Cysteine	0-5.0 μM	10.0 nM
				GSH	0-0.8 μM	9.0 nM
				Cysteine	0.05-50 μM	2.5 nM

Conclusions and future perspective

The practice of diverse sorts of algae in the synthesis of NPs have exhilarated the conniving of simple, green, cost and time operative methods thereby, curtailing the use of chemicals and solvents. The polysaccharides, proteins and lipids extant in the algal membranes act as capping agents and thus limit the use of non-biodegradable commercial surfactants which are challenging to eradicate after the synthesis of NMs. Recent progresses in the bio-inspired synthesis of metal NMs pointed out that a good number of bio-candidates (microorganisms, viruses, plants, proteins and DNA) could deliver an exclusive adaptability with a wide range of bio-templating systems and pioneering procedures intended to a careful control of critical limits in NP growth such as morphology, size and assembly of bio-MNMs and resultant super-structures. Significant benefits of these protocols embrace the use of mild synthetic conditions, environmentally friendly and biocompatible substrates, evading of any auxiliary capping agents and/or in convinced cases minor capacities of reducing agents, offering an attractive and progressively viable alternative to NP syntheses using pure physical/chemical synthesis.

The empathy of complex biomolecules, i.e. reductive and protecting agents, and their roles in shaping MNMs with diverse morphologies are crucial to restored apprehend the formation mechanism of MNMs. Additionally, a better indulgent of the roles of biological residues in processes (i.e. catalytic reaction) and interaction among metal NPs and support can enable the scheme of progressively effective bio-MNM materials in the future. Substantial efforts are dedicated on diverse capping agents such as biomolecules and polysaccaharides which can act as both chelating/reducing and capping agents for the synthesis of NMs. Consequently, the subsequent elements are dwindling from added reactions and aggregation, which increases their stability and longevity. Greener methods that have been used in NM synthesis are commonly single-pot reactions, deprived of the use of additional surfactants, capping agents and templates.

In spite of the experiments and current shortcomings of some bio-inpired protocols for the proposal and progress of MNMs, this influence clearly demonstrates the probable and acknowledged a number of investigation avenues for conducting added research to advance the field in the coming years which will definitely subsidize to situation some of the deliberated bio-synthetic protocols as seemly practical substitutes to current physico-chemical methods for MNM preparation for a more justifiable future. Research efforts must be redoubled to better appreciate the indispensable interactions, mechanisms and processes in the bio-design of MNMs which we hope can lead us to the next generation of bio-compatible NMs for progressive applications.

References

[1] C. Joachim, Nat. Mater. 4 (2005) 4, 107-109. https://doi.org/10.1038/nmat1319

[2] Y. Xia, Y. J. Xiong, B. Lim, S. E. Skrabalak, Angew. Chem. Int. Ed. 48 (2009) 60-103. https://doi.org/10.1002/anie.200802248

[3] Z. Y. Zhou, N. Tian, J. T. Li, I. Broadwell, S. G. Sun, Chem. Soc. Rev. 40 (2011) 4167-4185. https://doi.org/10.1039/c0cs00176g

[4] Y. N. Xia, P. D. Yang, Y. G. Sun, Y. Y. Wu, B. Mayers, B. Gates, Y. D. Yin, F. Kim, Y. Q. Yan, Adv. Mater. 15 (2003) 353-389. https://doi.org/10.1002/adma.200390087

[5] G. Chen, Y. Zhao, G. Fu, P. N. Duchesne, L. Gu, Y. Zheng, X. Weng, M. Chen, P. Zhang, C.-W. Pao, Science, 344 (2014) 495-499. https://doi.org/10.1126/science.1252553

[6] B. H. Wu, N. F. Zheng, Nano Today, 8 (2013) 168-197. https://doi.org/10.1016/j.nantod.2013.02.006

[7] L. Wei, J. Lu, H. Xu, A. Patel, Z.-S. Chen, G. Chen, Drug Discov. Today, 20 (2015) 595. https://doi.org/10.1016/j.drudis.2014.11.014

[8] T. Klaus, R. Joerger, E. Olsson, C. G. Granqvist, Proc. Natl. Acad. Sci. U. S. A. 96 (1999) 13611-13614. https://doi.org/10.1073/pnas.96.24.13611

[9] J. Gardea-Torresdey, K. Tiemann, G. Gamez, K. Dokken, S. Tehuacanero, M. Jose-Yacaman, J. Nanopart. Res. 1 (1999) 397-404. https://doi.org/10.1023/A:1010008915465

[10] E. Dujardin, C. Peet, G. Stubbs, J. N. Culver, S. Mann, Nano Lett. 3 (2003) 413-417. https://doi.org/10.1021/nl034004o

[11] J. Xie, Y. Zheng, J. Y. Ying, J. Am. Chem. Soc. 131 (2009) 888-889. https://doi.org/10.1021/ja806804u

[12] M. Mertig, L. C. Ciacchi, R. Seidel, W. Pome, A. De Vita, Nano Lett. 2 (2002) 841-844. https://doi.org/10.1021/nl025612r

[13] E. Dujardin, S. Mann, Adv. Mater, 14 (2002) 1-14. https://doi.org/10.1002/1521-4095(20020605)14:11<775::AID-ADMA775>3.0.CO;2-0

[14] A.-W. Xu, Y. Ma, H. Cölfen, J. Mater. Chem. 17 (2007) 415-449. https://doi.org/10.1039/B611918M

[15] F. C. Meldrum, H. Cölfen, Chem. Rev. 108 (2008) 4332-4432. https://doi.org/10.1021/cr8002856

[16] S. Mann, Biomineralization, Oxford University Press, Oxford, 2001.

[17] F. J. Eber, S. Eiben, H. Jeske, C. Wege, Angew. Chem. Int. Ed. 52 (2013) 7203-7207. https://doi.org/10.1002/anie.201300834

[18] B. Cao, Y. Zhu, L. Wang, C. Mao, Angew. Chem. Int. Ed. 52 (2013) 11750-11754. https://doi.org/10.1002/anie.201303854

[19] F. Wang, S. L. Nimmo, B. Cao, C. Mao, Chem. Sci. 3 (2012) 2639-2645. https://doi.org/10.1039/c2sc00583b

[20] C. Mao, F. Wang, B. Cao, Angew. Chem. Int. Ed. 51 (2012) 6411-6415. https://doi.org/10.1002/anie.201107824

[21] B. Cao, H. Xu, C. Mao, Angew. Chem. Int. Ed. 50 (2011) 6264-6268. https://doi.org/10.1002/anie.201102052

[22] F. Wang, B. Cao, C. Mao, Chem. Mater. 22 (2010) 3630-3636. https://doi.org/10.1021/cm902727s

[23] F. Wang, D. Li, C. Mao, Adv. Funct. Mater. 18 (2008) 4007-4013. https://doi.org/10.1002/adfm.200800889

[24] L. F. Hakim, J. L. Portman, M. D. Casper, A.W. Weimer, Powder Technology, 160 (2015) 149-160. https://doi.org/10.1016/j.powtec.2005.08.019

[25] P. Mukherjee, A. Ahmad, D. Mandal, S. Senapati, S.R. Sainkar, M.I. Khan, R. Parishcha, P.V. Ajaykumar, M. Alam, R. Kumar, M. Sastry, Nano Letters, 1 (2001) 515-519. https://doi.org/10.1021/nl0155274

[26] S.L. Tripp, S.V. Pusztay, A.E. Ribbe, A. Wei, J. Am. Chem. Soc, 124 (2004) 7914-7915. https://doi.org/10.1021/ja0263285

[27] J. Huang, L.Lin, D. Sun, H. Chen, D. Yang, Q. Li, Chem. Soc. Rev, 44 (2015) 6330-6374. https://doi.org/10.1039/C5CS00133A

[28] V. Patel, D. Berthold, P. Puranik, M. Gantar, Biotech. Rep. 5 (2015) 112-119. https://doi.org/10.1016/j.btre.2014.12.001

[29] K. P. Talaro, A. Talaro, G. Delisle, L. Tomalty,

[30] Foundations in microbiology, McGraw-Hill Higher Education, Burr Ridge, 1996.

[31] B. Nair, T. Pradeep, Cryst. Growth Des. 2 (2002) 293-298. https://doi.org/10.1021/cg0255164

[32] R. Y. Parikh, S. Singh, B. L. V. Prasad, M. S. Patole, M. Sastry, Y. S. Shouche, Chem Bio Chem, 9 (2009) 1415-1422. https://doi.org/10.1002/cbic.200700592

[33] D. Mandal, M. E. Bolander, D. Mukhopadhyay, G. Sarkar, P. Mukherjee, Appl. Microbiol. Biotechnol. 69 (2006) 485-492. https://doi.org/10.1007/s00253-005-0179-3

[34] L. Q. Lin, W. W. Wu, J. L. Huang, D. H. Sun, N. M. Waithera, Y. Zhou, H. T. Wang, Q. B. Li, Chem. Eng. J. 225 (2013) 857-864. https://doi.org/10.1016/j.cej.2013.04.003

[35] A. K. Manocchi, N. E. Horelik, B. Lee, H. Yi, Langmuir, 26 (2010) 3670-3677. https://doi.org/10.1021/la9031514

Materials Research Forum LLC
https://doi.org/10.21741/9781644901571-6

[36] C. Yang, C.-H. Choi, C.-S. Lee, H. Yi, ACS Nano, 7 (2013) 5032-5044. https://doi.org/10.1021/nn4005582

[37] W. Ernst, Appl. Geochem. 11 (1996) 163-167. https://doi.org/10.1016/0883-2927(95)00040-2

[38] J. Gardea-Torresdey, J. Parsons, E. Gomez, J. Peralta-Videa, H. Troiani, P. Santiago, M. J. Yacaman, Nano Lett. 2 (2002) 397-401. https://doi.org/10.1021/nl015673+

[39] H. L. Parker, E. L. Rylott, A. J. Hunt, J. R. Dodson, A. F. Taylor, N. C. Bruce, J. H. Clark, PLoS One. 9 (2014) e87192. https://doi.org/10.1371/journal.pone.0087192

[40] S. S. Shankar, A. Rai, B. Ankamwar, A. Singh, A. Ahmad, M. Sastry, Nat. Mater. 3 (2004) 482-488. https://doi.org/10.1038/nmat1152

[41] J. L. Huang, G. W. Zhan, B. Y. Zheng, D. H. Sun, F. F. Lu, Y. Lin, H. M. Chen, Z. D. Zheng, Y. M. Zheng, Q. B. Li, Ind. Eng. Chem. Res. 50 (2011) 9095-9106. https://doi.org/10.1021/ie200858y

[42] B. Zheng, T. Kong, X. Jing, T. Odoom-Wubah, X. Li, D. Sun, F. Lu, Y. Zheng, J. Huang, Q. Li, J. Colloid Interface Sci. 396 (2013) 138-145. https://doi.org/10.1016/j.jcis.2013.01.021

[43] H. R. Zhang, Q. B. Li, Y. H. Lu, D. H. Sun, X. P. Lin, X. Deng, N. He, S. Z. Zheng, J. Chem. Technol. Biotechnol. 80 (2005) 285-290. https://doi.org/10.1002/jctb.1191

[44] J. R. Lloyd, FEMS Microbiol. Rev. 27 (2003) 411-425. https://doi.org/10.1016/S0168-6445(03)00044-5

[45] P. Mukherjee, A. Ahmad, D. Mandal, S. Senapati, S. R. Sainkar, M. I. Khan, R. Parishcha, P. V. Ajaykumar, M. Alam, R. Kumar, M. Sastry, Nano Lett. 1 (2001) 515-519. https://doi.org/10.1021/nl0155274

[46] I. W.-S. Lin, C.-N. Lok, C.-M. Che, Chem. Sci. 5 (2014) 3144-3150. https://doi.org/10.1039/C4SC00138A

[47] L. M. Rösken, S. Körsten, C. B. Fischer, A. Schönleber, S. van Smaalen, S. Geimer, S. Wehner, J. Nanopart. Res. 16 (2014) 2370. https://doi.org/10.1007/s11051-014-2370-x

[48] Y. Konishi, T. Tsukiyama, T. Tachimi, N. Saitoh, T. Nomura, S. Nagamine, Electrochim. Acta, 53 (2007) 186-192. https://doi.org/10.1016/j.electacta.2007.02.073

[49] Y. Konishi, T. Tsukiyama, K. Ohno, N. Saitoh, T. Nomura, S. Nagamine, Hydrometallurgy, 81 (2006) 24-29. https://doi.org/10.1016/j.hydromet.2005.09.006

[50] A. Ahmad, S. Senapati, M. I. Khan, R. Kumar, M. Sastry, Langmuir, 19 (2003) 3550-3553. https://doi.org/10.1021/la0267721

[51] A. N. Mabbett, P. Yong, J. P. G. Farr, L. E. Macaskie, Biotechnol. Bioeng. 87 (2004) 104-109. https://doi.org/10.1002/bit.20105

[52] A. N. Mabbett, P. Yong, J. P. G. Farr, L. E. Macaskie, Biotechnol. Bioeng. 87 (2004) 104-109. https://doi.org/10.1002/bit.20105

[53] R. Ramanathan, A. P. O'Mullane, R. Y. Parikh, P. M. Smooker, S. K. Bhargava, V. Bansal, Langmuir, 27 (2011)714-719. https://doi.org/10.1021/la1036162

[54] M. Oves, M. S. Khan, A. Zaidi, A. S. Ahmed, F. Ahmed, E. Ahmad, A. Sherwani, M. Owais, A. Azam, PLoS One, 8 (2013) e59140. https://doi.org/10.1371/journal.pone.0059140

[55] C. K. Ng, K. Sivakumar, X. Liu, M. Madhaiyan, L. Ji, L. Yang, C. Tang, H. Song, S. Kjelleberg, B. Cao, Biotechnol. Bioeng. 110 (2013)1831-1837. https://doi.org/10.1002/bit.24856

[56] P. Mukherjee, S. Senapati, D. Mandal, A. Ahmad, M. I. Khan, R. Kumar, M. Sastry, Chem Bio Chem. 3 (2002) 461-463. https://doi.org/10.1002/1439-7633(20020503)3:5<461::AID-CBIC461>3.0.CO;2-X

[57] S. K. Das, J. Liang, M. Schmidt, F. Laffir, E. Marsili, ACS Nano. 6 (2012) 6165-6173. https://doi.org/10.1021/nn301502s

[58] E. P. Vetchinkina, E. A. Loshchinina, A. M. Burov, L. A. Dykman, V. E. Nikitina, J. Biotechnol. 182 (2014) 37-45. https://doi.org/10.1016/j.jbiotec.2014.04.018

[59] Y. Liu, J. Fu, H. Hu, D. Tang, Z. Ni, X. Yu, Chin. Sci. Bull. 46 (2001) 1709-1712. https://doi.org/10.1360/csb2001-46-20-1709

[60] J. K. Fu, Y. Y. Liu, P. Y. Gu, D. L. Tang, Z. Y. Lin, B. X. Yao, S. Z. Weng, Acta Phys.-Chim. Sin. 16 (2000) 779-782.

[61] Z. Lin, C. Zhou, J. Wu, H. Cheng, B. Liu, Z. Ni, J. Zhou, J. Fu, Chin. Sci. Bull. 47 (2002) 1262-1266.

[62] H. R. Zhang, Q. B. Li, H. X. Wang, D. H. Sun, Y. H. Lu, N. He, Appl. Biochem. Biotechnol. 143 (2007) 54-62. https://doi.org/10.1007/s12010-007-8006-1

[63] S. De Corte, T. Hennebel, S. Verschuere, C. Cuvelier, W. Verstraete, N. Boon, J. Chem. Technol. Biotechnol. 86 (2011) 547-553. https://doi.org/10.1002/jctb.2549

[64] Z. Lin, R. Xue, Y. Ye, J. Zheng, Z. Xu, BMC Biotechnol. 9 (2009) 62. https://doi.org/10.1186/1472-6750-9-62

Materials Research Forum LLC

https://doi.org/10.21741/9781644901571-6

[65] Z. Y. Lin, J. M. Wu, R. Xue, Y. Yang, Spectrochim. Acta, Part A, 61 (2005) 761-765. https://doi.org/10.1016/j.saa.2004.03.029

[66] Z. Lin, Y. Ye, Q. Li, Z. Xu, M. Wang, BMC Biotechnol. 11 (2011) 98. https://doi.org/10.1186/1472-6750-11-98

[67] H. X. Yang, M. M. Du, T. Odoom-Wubah, J. Wang, D. H. Sun, J. L. Huang, Q. B. Li, J. Chem. Technol. Biotechnol. 89 (2014) 1410-1418. https://doi.org/10.1002/jctb.4225

[68] X. L. Jing, D. P. Huang, H. M. Chen, T. Odoom-Wubah, D. H. Sun, J. L. Huang, Q. B. Li, J. Chem. Technol. Biotechnol. 90 (2015) 678-685. https://doi.org/10.1002/jctb.4353

[69] H. M. Chen, D. H. Sun, X. L. Jing, F. F. Lu, T. Odoom- Wubah, Y. M. Zheng, J. L. Huang, Q. B. Li, RSC Adv. 3 (2013) 15389-15395. https://doi.org/10.1039/c3ra41215f

[70] H. M. Chen, J. L. Huang, D. P. Huang, M. H. Shao, D. H. Sun, Q. B. Li, J. Mater. Chem. A, 3 (2015) 4846-4854. https://doi.org/10.1039/C4TA06226D

[71] W. X. Jia, Medical microbiology, People's Medical Publishing House, Beijing, (2005).

[72] C. Yang, J. H. Meldon, B. Lee, H. Yi, Catal. Today. 233 (2014) 108-116. https://doi.org/10.1016/j.cattod.2014.02.043

[73] M. Wnęk, M. Ł. Górzny, M. Ward, C. Wälti, A. Davies, R. Brydson, S. Evans, P. Stockley, Nanotechnology. 24, (2013) 025605. https://doi.org/10.1088/0957-4484/24/2/025605

[74] A. A. Aljabali, G. P. Lomonossoff, D. J. Evans, Biomacromolecules. 12 (2011) 2723-2728. https://doi.org/10.1021/bm200499v

[75] K. N. Avery, J. E. Schaak, R. E. Schaak, Chem. Mater. 21(2009) 2176-2178. https://doi.org/10.1021/cm900869u

[76] M. Knez, M. Sumser, A. Bittner, C. Wege, H. Jeske, S. Kooi, M. Burghard, K. Kern, J. Electroanal. Chem. 522 (2002) 70-74. https://doi.org/10.1016/S0022-0728(01)00728-8

[77] K. M. Bromley, A. J. Patil, A. W. Perriman, G. Stubbs, S. Mann, J. Mater. Chem. 18 (2008) 4796-4801. https://doi.org/10.1039/b809585j

[78] A. A. Khan, E. K. Fox, M. Ł. Górzny, E. Nikulina, D. F. Brougham, C. Wege, A. M. Bittner, Langmuir. 29 (2013) 2094-2098. https://doi.org/10.1021/la3044126

[79] M. Knez, M. Sumser, A. M. Bittner, C. Wege, H. Jeske, T. P. Martin, K. Kern, Adv. Funct. Mater. 14 (2004) 116-124. https://doi.org/10.1002/adfm.200304376

[80] J. L. Gardea-Torresdey, E. Gomez, J. R. Peralta-Videa, J. G. Parsons, H. Troiani, M. Jose-Yacaman, Langmuir, 19 (2003) 1357-1361. https://doi.org/10.1021/la020835i

[81] N. C. Sharma, S. V. Sahi, S. Nath, J. G. Parsons, J. L. Gardea- Torresdey, T. Pal, Environ. Sci. Technol. 41 (2007) 5137-5142. https://doi.org/10.1021/es062929a

[82] R. Haverkamp, A. Marshall, J. Nanopart. Res. 11 (2009) 1453-1463. https://doi.org/10.1007/s11051-008-9533-6

[83] R. Bali, R. Siegele, A. T. Harris, J. Nanopart. Res. 12 (2010) 3087-3095. https://doi.org/10.1007/s11051-010-9904-7

[84] C. W. Anderson, S. M. Bhatti, J. Gardea-Torresdey, J. Parsons, ACS Sustainable Chem. Eng. 1 (2013) 640-648. https://doi.org/10.1021/sc400011s

[85] G. Zhai, K. S. Walters, D. W. Peate, P. J. Alvarez, J. L. Schnoor, Environ. Sci. Technol. Lett. 1 (2014) 146-151. https://doi.org/10.1021/ez400202b

[86] A. F. Taylor, E. L. Rylott, C. W. Anderson, N. C. Bruce, PLoS One, 9 (2014) e93793. https://doi.org/10.1371/journal.pone.0093793

[87] Y. Zhou, W. Lin, J. Huang, W. Wang, Y. Gao, L. Lin, Q. Li, L. Lin, M. Du, Nanoscale Res. Lett. 5 (2015) 1351-1359. https://doi.org/10.1007/s11671-010-9652-8

[88] P. Mohanpuria, N. K. Rana, S. K. Yadav, J. Nanopart. Res. 10 (2008) 507-517. https://doi.org/10.1007/s11051-007-9275-x

[89] V. Kumar, S. K. Yadav, J. Chem. Technol. Biotechnol. 84 (2009) 151-157. https://doi.org/10.1002/jctb.2023

[90] D. Bhattacharya, R. K. Gupta, Crit. Rev. Biotechnol. 25 (2005) 199-204. https://doi.org/10.1080/07388550500361994

[91] J. Huang, Q. Li, D. Sun, Y. Lu, Y. Su, X. Yang, H. Wang, Y. Wang, W. Shao, N. He, Nanotechnology, 18 (2007) 105104. https://doi.org/10.1088/0957-4484/18/10/105104

[92] D. Philip, Spectrochim. Acta, Part A, 73 (2009) 374-381. https://doi.org/10.1016/j.saa.2009.02.037

[93] J. Kasthuri, S. Veerapandian, N. Rajendiran, Colloids Surf. B, 68 (2009) 55-60. https://doi.org/10.1016/j.colsurfb.2008.09.021

[94] J. Y. Song, B. S. Kim, Korean J. Chem. Eng. 25 (2008) 808-811.
https://doi.org/10.1007/s11814-008-0133-z

[95] K. B. Narayanan, N. Sakthivel, Mater. Lett. , 62 (2008) 4588-4590.
https://doi.org/10.1016/j.matlet.2008.08.044

[96] A. T. Harris, R. Bali, J. Nanopart. Res. 10 (2008) 691-695.
https://doi.org/10.1007/s11051-007-9288-5

[97] M. N. Nadagouda, R. S. Varma, Green Chem. 10 (2008) 859-862.
https://doi.org/10.1039/b804703k

[98] J. P. Xie, J. Y. Lee, D. I. C. Wang, Y. P. Ting, Small, 3 (2007) 672-682.
https://doi.org/10.1002/smll.200600612

[99] G. Zhan, L. Ke, Q. Li, J. Huang, D. Hua, A.-R. Ibrahim, D. Sun, Ind. Eng. Chem.
Res. 51 (2012) 1575315762. https://doi.org/10.1021/ie302483d

[100] W. W. Wu, J. L. Huang, L. F. Wu, D. H. Sun, L. Q. Lin, Y. Zhou, H. T. Wang, Q.
B. Li, Sep. Purif. Technol. 106 (2013) 117-122.
https://doi.org/10.1016/j.seppur.2013.01.005

[101] D. Liu, Z. Wang, X. Jiang, Nanoscale, 3 (2011) 1421-1433.
https://doi.org/10.1039/c0nr00887g

[102] W. Li, B. Chen, H. Zhang, Y. Sun, J. Wang, J. Zhang, Y. Fu, Biosens. Bioelectron.
66 (2015) 251-258. https://doi.org/10.1016/j.bios.2014.11.032

[103] X. Jiang, C. Sun, Y. Guo, G. Nie, L. Xu, Biosens. Bioelectron. 64 (2015) 165-170.
https://doi.org/10.1016/j.bios.2014.08.078

[104] F. Wen, Y. Dong, L. Feng, S. Wang, S. Zhang, X. Zhang, Anal. Chem. 83 (2011)
1193-1196. https://doi.org/10.1021/ac1031447

[105] J. Xie, Y. Zheng, J. Y. Ying, Chem. Commun. 46 (2010) 961-963.
https://doi.org/10.1039/B920748A

[106] G. Y. Lan, C. C. Huang, H. T. Chang, Chem. Commun. 46 (2010) 1257-1259.
https://doi.org/10.1039/b920783j

[107] Z. Huang, F. Pu, Y. Lin, J. Ren, X. Qu, Chem. Commun. 47 (2011) 3487-3489.
https://doi.org/10.1039/c0cc05651k

[108] Y. Liu, K. Ai, X. Cheng, L. Huo, L. Lu, Adv. Funct. Mater. 20 (2010) 951-956.
https://doi.org/10.1002/adfm.200902062

[109] R. Wen, H. Lia, B. Chen, L. Wang, Sens. Actu. B: Chem. 248 (2017) 63-70.

[110] J. Lee, J. Park, H. H. Lee, H. I. Kim, W. J. Kim, J. Mater. Chem. B, 2 (2014) 2616-2621. https://doi.org/10.1039/c3tb21446j

[111] Y. Lin, Y. Tao, J. Ren, F. Pu, X. Qu, Biosens. Bioelectron. 28 (2011) 339-343. https://doi.org/10.1016/j.bios.2011.07.040

[112] A. Sannigrahi, S. Chowdhury, I. Nandi, D. Sanyal, S. Chall, K. Chattopadhyay, Nanoscale. Adv. 1 (2019) 3660-3669. https://doi.org/10.1039/C9NA00459A

[113] M. Yu, Z. Zhu, H. Wang, L. Li, F. Fu, Y. Song, E. Song, Biosens. Bioelectron. 91 (2017) 143-148. https://doi.org/10.1016/j.bios.2016.11.052

[114] J. Wu, K. Jiang, X. Wang, C. Wang, C. Zhang, Microchim Acta. 184 (2017) 1315-1324. https://doi.org/10.1007/s00604-017-2111-9

[115] M. Zhang, H-N. Le, X-Q. Jiang, S-M. Guo, H-J. Yu, B-C. Ye, Talanta. 117 (2013)399-404. https://doi.org/10.1016/j.talanta.2013.09.034

[116] M. Luo, J. Di, L. Li, Y. Tu, J. Yan, Talanta. 187 (2018) 231-236. https://doi.org/10.1016/j.talanta.2018.05.047

[117] G-M. Zhang, Y-G. J. Li, C-H. Xu Zhang, S-M. Shuang, C. Dong, MM- F. Choi, Sens Actuators B. 183 (2013) 583-588. https://doi.org/10.1016/j.snb.2013.04.023

[118] W. Chen, X. Tu, X. Guo, Chem. Commun. 13 (2009) 1736-1738. https://doi.org/10.1039/b820145e

[119] E. Babaee, A. Barati, M-B. Gholivand, A. Taherpour, N. Zolfaghar, M. Shamsipur J. Hazard. Mater. 367 (2019) 437-446. https://doi.org/10.1016/j.jhazmat.2018.12.104

[120] Q. Niu, P-F. Gao, M-J. Yuan, G-M. Zhang, Y. Zhou, C. Dong, S-M. Shuang, Y. Zhang, Microchem. J. 146 (2019) 1140-1149. https://doi.org/10.1016/j.microc.2019.02.050

[121] J. Wang, S. Ma, J. Ren, J. Yang, Y. Qu, D. Ding, M. Zhang, G. Yang, Sens. Actuators B. 267 (2018)342-350. https://doi.org/10.1016/j.snb.2018.04.034

[122. H. Wei, Z. Wang, L. Yang, S. Tian, C. Hou, Y. Lu, Analyst, 135 (2010) 1406-1410. https://doi.org/10.1039/c0an00046a

[123] H. Kawasaki, K. Yoshimura, K. Hamaguchi, R. Arakawa, Anal. Sci Int. J. Jpn. Soc. Anal. Chem. 27 (2011) 591-596. https://doi.org/10.2116/analsci.27.591

[124] L. Shang, L. Yang, F. Stockmar, R. Popescu, V. Trouillet, M. Bruns, D. Gerthsen, G-U. Nienhaus, Nanoscale, 4 (2012) 4155-4160. https://doi.org/10.1039/c2nr30219e

Bioinspired Nanomaterials
Materials Research Foundations **111** (2021) 155-184

Materials Research Forum LLC
https://doi.org/10.21741/9781644901571-6

[125] H. Huang, H. Li, J. Feng, A-J. Wang, Sens. Actuators B. 223 (2016) 550-556. https://doi.org/10.1016/j.snb.2015.09.136

[126] J-J. Li, D. Qiao, J. Zhao, G-J. Weng J. Zhu, J-W. Zhao, Methods. Appl. Fluoresc. 7 (2019) 045001. https://doi.org/10.1088/2050-6120/ab34be

[127] Y. Yang, A. Han, R-X. Li, G-Z. Fang, J-F. Liu, S. Wang, Analyst, 142 (2017) 4486-4493. https://doi.org/10.1039/C7AN01348E

[128] B. Duan, M. Wang, Y. Li, S. Jiang, Y. Liu, Z-Z. Huang, New J. Chem. 43 (2019) 14678-14683. https://doi.org/10.1039/C9NJ03524A

[129] H-Y. Zhang, Q. Liu, T. Wang, Z-J, Yun, G-L, Li, J, Liu, G-B. Jiang, Anal. Chim Acta., 770 (2013) 140-146. https://doi.org/10.1016/j.aca.2013.01.042

[130] P. Li, J. Li, M-H. Bian, D. Huo, C. Hou, P. Yang, S. Zhang, C-H. Shen, M. A. Yang, Anal. Methods. 10 (2018) 3256-3262. https://doi.org/10.1039/C8AY00892B

[131] P. Nath, M. Chatterjee, N. Chanda, ACS. Appl. Nano. Mater. 1 (2018) 5108-5118. https://doi.org/10.1021/acsanm.8b01191

[132] C. Zong, L-R. Zheng, W. He, X. Ren, C. Jiang, L. Lu, Anal .Chem. 86 (2014) 1687-1692. https://doi.org/10.1021/ac403480q

[133] D. Lu, L. Liu, F. Li, S. Shuang, Y. Li, M. M. F. Choi et al. Spectrochim Acta A.121 (2014) 77-80. https://doi.org/10.1016/j.saa.2013.10.009

[134] C-Y. Liu, W-L. Tseng, Anal. Methods, 4 (2012) 2537-2542. https://doi.org/10.1039/c2ay25291k

[135] C-Y. Liu, W-L. Tseng, Chem.Commun. 47 (2011) 2550-2552. https://doi.org/10.1039/c0cc04591h

[136] C. Wang, Y. Chen, B-Y. Wu, C-K. Lee, Y-C. Chen, Y-H. Huang, H-T. Chang, Anal. Bioanal. Chem. 408 (2016) 287-294. https://doi.org/10.1007/s00216-015-9104-5

[137] L. Wang, G-Q. Chen, G-M. Zeng, J. Liang, H. Dong, M. Yan, Z. Li, Z. Guo, W, Tao, L. Peng, New. J. Chem. 39 (2015) 9306-9312. https://doi.org/10.1039/C5NJ01783A

[138] W-Y. Chen, G-Y. Lan, H-T. Chang, Anal. Chem. 83 (2011) 9450-9455. https://doi.org/10.1021/ac202162u

[139] B-Y. Wu, C-W. Wang, P-C. Chen, H-T. Chang, Sens. Actuators B.238 (2017) 1258-1265. https://doi.org/10.1016/j.snb.2016.09.071

[140] B. Unnikrishnan, S-C. Wei, W-J. Chiu, J-S. Cang, P-H. Hsu, C-C. Huang, Analyst, 139 (2014) 2221-2228. https://doi.org/10.1039/C3AN02291A

[141] H-Y. Liu, G-H. Yang, E-S. A. Halim,, J-J. Zhu, Talanta, 104 (2013) 135-139. https://doi.org/10.1016/j.talanta.2012.11.020

[142] Q-L.Yue, L-J. Sun, T-F. Shen, X-H. Gu, S-Q. Zhang, J-F. Liu, J. Fluoresc. 23 (2013) 1313-1318. https://doi.org/10.1007/s10895-013-1265-z

[143] J. Zhang, C-X. Chen, X-W. Xu, X-L. Wang, X-R. Yang, Chem. Commun. 49 (2013) 2691-2693. https://doi.org/10.1039/c3cc38298b

[144] Y. Hu, X-M. Lu, X-M. Jiang, P. Wu, J. Hazard Mater. 384 (2019) 121368. https://doi.org/10.1016/j.jhazmat.2019.121368

[145] D-H. Tian, Z-S. Qian, Y-S. Xia, C-Q. Zhu, Langmuir, 28 (2012) 3945-3951. https://doi.org/10.1021/la204380a

[146] M-L. Cui, J-M. Liu, X-X. Wang, L-P. Lin, L. Jiao, L-H. Zhang, Z-Y. Zheng, S-Q. Lin Analyst, 137 (2012) 5346-5351. https://doi.org/10.1039/c2an36284h

[147] H. Zhao, X-P. Wen, W-Y. Li, Y-Q. Li, C-X. Yin, J. Mater Chem B. 7 (2019) 2169-2176. https://doi.org/10.1039/C8TB03184C

[148] C. Dai, C-X. Yang, X-P. Yan, Nano. Res. 11 (2018) 2488-2497. https://doi.org/10.1007/s12274-017-1872-0

[149] T. Feng, Y. Chen, B-B. Feng, J. Yan, J. Di, Spectrochim. Acta. A. 206 (2019) 97-103. https://doi.org/10.1016/j.saa.2018.07.087

[150] Z-X. Wang, S-N. Ding, E-Y-J. Narjh, Anal. Lett. 48 (2015) 647-658. https://doi.org/10.1080/00032719.2014.956217

[151] J. Zhang, Y. Yuan, Y. Wang, F. Sun, G. Liang, Z. Jiang, S-H. Yu, Nano Res. 8 (2015) 2329-2339. https://doi.org/10.1007/s12274-015-0743-9

Bioinspired Nanomaterials
Materials Research Foundations **111** (2021) 185-223

Materials Research Forum LLC
https://doi.org/10.21741/9781644901571-7

Chapter 7

Bio-Mediated Synthesis of Quantum Dots for Fluorescent Biosensing and Bio-Imaging Applications

Selvaraj Devi[1] and Vairaperumal Tharmaraj[2]*

[1]Deparment of Chemistry, Cauvery College for Women (Autonomous), Tiruchirappalli-620018, Tamil Nadu, India

[2]State Key Laboratory of Chemo/ Bio-Sensing and Chemometrics, College of Chemistry and Chemical Engineering, Hunan University, Changsha, Hunan-410082, China

*tharmachem@gmail.com

Abstract

Quantum dots (QDs) have received great attention for development of novel fluorescent nanoprobe with tunable colors towards the near-infrared (NIR) region because of their unique optical and electronic properties such as luminescence characteristics, wide range, continuous absorption spectra and narrow emission spectra with high light stability. Quantum dots are promising materials for biosensing and single molecular bio-imaging application due to their excellent photophysical properties such as strong brightness and resistance to photobleaching. However, the use of quantum dots in biomedical applications is limited due to their toxicity. Recently, the development of novel and safe alternative method, the bio-mediated greener approach is one of the best aspects for synthesis of quantum dots. In this Chapter, bio-mediated synthesis of quantum dots by living organisms and biomimetic systems were highlighted. Quantum dots based fluorescent probes utilizing resonance energy transfer (RET), especially Förster resonance energy transfer (FRET), bioluminescence resonance energy transfer (BRET) and chemiluminescence resonance energy transfer (CRET) to probe biological phenomena were discussed. In addition, quantum dot nanocomposites are promising ultrasensitive bioimaging probe for in vivo multicolor, multimodal, multiplex and NIR deep tissue imaging. Finally, this chapter provides a conclusion with future perspectives of this field.

Keywords

Green Synthesis, Quantum Dots, Biosensor, Fluorescence Imaging, FRET, BRET, CRET, NIR

Contents

1.   Introduction

Quantum dots (QDs) are considered as a zero dimensional semiconductor nanocrystals with size in the range of 1-20 nm [1]. Quantum dots are usually synthesized by one-step or step-by-step formation of core shell with a protective layer [2-4]. Quantum dots have unique photophysical and optical properties that arises from the size and shape of the particles and composition [5]. The highly desirable properties of quantum dots such as size tunable emission and high fluorescent quantum yield allow it to pursue several applications such as biosensing and biomedical imaging [6-8].

Quantum confinement effect plays a major role in which the energy of the particle increases as the size of the box decreases. As particle size decreases, there will be an increase in the band gap resulting in shorten of emission wavelengths. This phenomenon in quantum dots results in tunable optical properties which are completely related to the size, shape and composition of quantum dots [9-13]. Cadmium selenide quantum dots

Materials Research Forum LLC
https://doi.org/10.21741/9781644901571-7

(CdSe QDs) is the typical example for the size-tunable fluorescence spectrum. The fluorescence wavelength increased with increase in the particles size as shown in figure 1 [14].

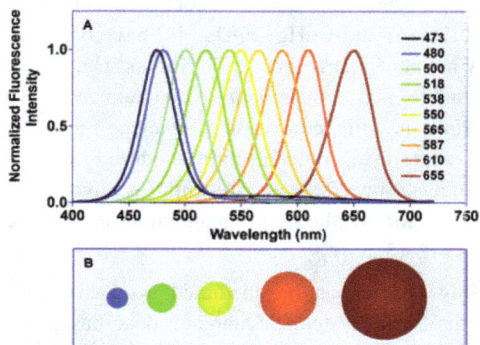

Figure 1. (A) *Fluorescence spectra of CdSe QDs depends on the size of CdSe QDsand* **(B)***pictures of CdSe QDs at different relative particle diameters such as 2.1 nm, 2.5 nm, 2.9 nm, 4.7 nm, and 7.5 nm. (Reproduced with the permission from Ref. No. 14).*

In the past few years, quantum dots are considered as an alternative to organic fluorophores for a range of applications such as biosensing and imaging [15-17]. Quantum dots based sensors have offered a new generation of nanomaterials to achieve a highly desirable, near-infrared region (NIR-II) fluorescence with lower autofluorescence background, deeper tissue penetration, and high signal-to-noise ratio [18]. Near-infrared quantum dots (NIR QDs) with water soluble and controllable particle size to nanometer scale leads a substantial change in the properties of materials that have a broad excitation spectrum, offering revolutionary fluorescence performance in novel applications such as single excitation multicolor fluorescence imaging, long term photostability for live cell imaging and cancer cell [19].

Usually, quantum dots can be prepared by physical, chemical and biological methods based on the application where it is used. Physical methods such as laser irradiation of larger particles [20-22], and physical vapor deposition [23] result in quantum dots with higher surface defect, lower quantum yield and poor solubility [24-25]. Quantum dots can be synthesized through chemical approaches at two different ways (i) organic-phase and (ii) water-phase. Both methods have good optical performance due to the monodipersity of quantum dots particles [26-28]. Eventhough, the chemical synthesis approach was used simple synthesis route [29,30] and controlled the size and shape of the quantum dots particles [31,32], the use of organic solvents for synthesis and modification process limits its uses in synthesizing quantum dots [33].

During the past decade, traditional quantum dots such as CdTe/CdSe core–shell [34-36], PbSe [37,38], PbS [39,40] and CdHgTeQDs [41] have been successfully developed but they contain heavy metal elements such as Pb, Cd and Hg. Therefore, the development of nontoxic, highly stable and biocompatible quantum dots with bright NIR-emitting fluorophore materials is still crucial and challenging. Alternatively, bio-mediated synthesis approaches are using mild reaction conditions with biocompatibility to avoid toxicity and fluorescence interferences. Recently, a novel and safe alternative method have been developed for synthesizing quantum dots in greener way from various biological materials [42-45].

This chapter mainly focuses on bio-mediated synthesis of quantum dots by living organisms and biomimetic systems. Quantum dots based targeted biosensors using Förster resonance energy transfer (FRET), bioluminescence resonance energy transfer (BRET) and chemiluminescence resonance energy transfer (CRET) processes were discussed. In addition, this chapter provided the application of quantum dots nanocomposite in vivo multicolor, multimodal, multiplex and NIR deep tissue imaging. Bio-mediated synthesized quantum dots display the tunable size of quantum dots with module of tunable broad spectral windows and their application in biosensors and bioimaging applications are shown in figure 2.

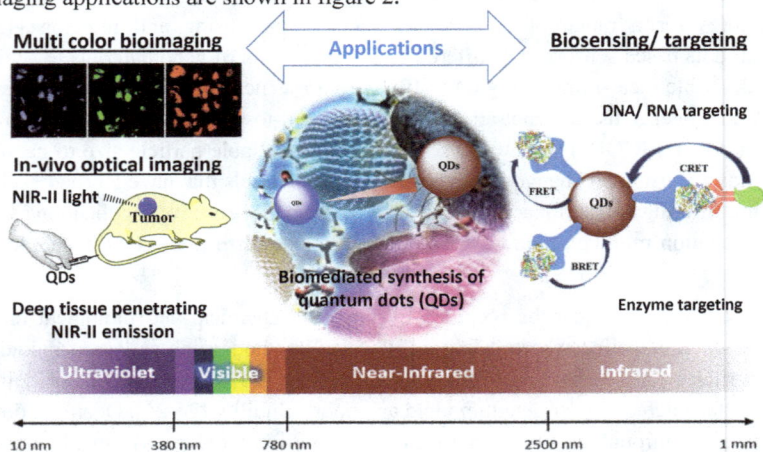

Figure 2. *Schematic representation of biomediated synthesis of quantum dots and their application in biosensing with energy transfer process and in-vivo imaging.*

Bioinspired Nanomaterials

Materials Research Foundations **111** (2021) 185-223

Materials Research Forum LLC

https://doi.org/10.21741/9781644901571-7

2. Bio-mediated synthesis of quantum dots

Bio-mediated synthetic method is a most desirable direction for synthesizing quantum dots when compared with conventional chemical or physical methods due to its environmentally friendly nature and mild reaction conditions. The bio-mediated synthesis of quantum dots does not require external stabilizing agents. Their microorganisms or extract itself act as stabilizing and capping agents. Another advantage of using biomaterials is that there is no contamination of the quantum dots surface by the toxic chemical compound and also microorganisms can control the sizes and shapes of the quantum dots. Therefore, bio-mediated synthesis method is more preferred for the biomedical applications of QDs due to its non-toxic, stable, environmentally friendly and cost effective.

In this section, we have described bio-mediated synthetic methods for synthesizing quantum dots via (i) living organisms, (ii) biomimetic systems and (iii) biomolecules functionalization with focusing on size control, modification strategies and optical properties.

2.1 Quantum dots synthesis from living organisms

Quantum dots have been successfully synthesized within various biological organisms such as bacteria, viruses and yeasts. Microorganisms play important roles in the growth of quantum dots where enzymatic reactions are able to reduce metal ions and offering an efficient approach to synthesis quantum dots [46]. For example, hydrogen sulfide was produced from many numbers of enzymatic reactions such as alkaline phosphatase and thiocholine from hydrolyzation of thiophosphate and acetylthiocholine respectively. In addition, the catalytic oxidation of 1-thio-β-D-glucose by glucose oxidase was used for producing CdS quantum dots [47-49]. Moreover, some enzymatic reactions produced a small biomolecule such as glutathione and phosphate which may help to stabilize quantum dots and acts as capping ligands.

Water-soluble cadmium telluride (CdTe) quantum dot was prepared from earthworms by metal detoxification pathway. Earthworms were exposed for 11 days in standard soil that spiked with $CdCl_2$ and Na_2TeO_3 precursors. M. Green et.al [50] proposed the mechanism for the formation of CdTe quantum dots from earthworms.

At first, tellurium precursor was formed by the reduction of tellurite *via* glutathione (GSH), glutathione reductase and nicotinamide adenine dinucleotide phosphate (NADPH). The reaction of tellurium precursor with readily available $CdCl_2$ forms CdTe quantum dots before isolation as seen in scheme 1. During the formation of quantum dots, the precursors can be transported *via* metallothioneins to chloragogenous tissues where the reaction occurred for producing quantum dots.

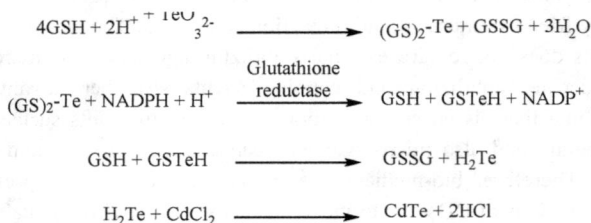

$$4GSH + 2H^+ + TeO_3^{2-} \longrightarrow (GS)_2\text{-}Te + GSSG + 3H_2O$$

$$(GS)_2\text{-}Te + NADPH + H^+ \xrightarrow{\text{Glutathione reductase}} GSH + GSTeH + NADP^+$$

$$GSH + GSTeH \longrightarrow GSSG + H_2Te$$

$$H_2Te + CdCl_2 \longrightarrow CdTe + 2HCl$$

Scheme 1. *The proposed mechanism for the formation of CdTe quantum dots from earthworms.*

Semiconductor quantum dots were successfully prepared in different living organisms and the basic properties of quantum dots such as size, emission wavelengths are summarized in Table 1.

Table 1. Basic properties of semiconductor quantum dots.

QDs	Living organisms	Size of the QDs (nm)	Emission wavelengths (nm)	References
CdSe	Living yeast cells	2.69	520, 560, 670	[51]
CdTe	E. coli	2.0-3.2	488-551	[52]
CdS	Tobacco mosaic virus	5	Not reported	[53]
CdS	Engineered E. coli	6	445-510	[54]
ZnSe andCdSe	Veillonellaatypica	3-6 (ZnSe) and 2-4 (CdSe)	368 (ZnSe) and 512 (CdSe)	[55]
ZnS	Sulfate-reducing bacteria	2-4	Not reported	[56]
ZnS	E. coli curli fibrils	5	490	[57]
CdTe	Gsh A-overexpressed E. coli	2-3	450	[58]
CdSe	Veillonellaatypica	2.3	510-531	[59]
CdSe	Staphylococcus aureus	1.8	520	[60]
CdS	bacterial cellulose nanofiber	8	417 and 437	[61]
CdS	engineered M13 bacteriophage	3-5	425	[62]

Biosynthesis of fluorescent protein-capped CdTe quantum dots by using yeastcells was achieved by Dongyuan Zhao et.al. [63]. The size of the extracellularly biogrown CdTe QDs was obtained around 2.0-3.6 nm with good crystallinity in aqueous solutions at relatively low temperatures (25-35 °C). The extracellular protein capped CdTe QDs produced from the yeast cells were dispersed throughout the cytoplasm and nucleus domains. The formation of protein capped CdTe QDs involved four steps. They are i) QDs was adhered with cell membrane through electrostatic interaction, ii) the deformation of membrane embedded QDs and endocytose into the cell, iii) the disruption of endosomal membrane released QDs from the endosome into the cytoplasm and iv) enter into the nucleus by a nuclear translocation.

Paknikar et al. reported the preparation of PbS quantum dots *via* feeding yeast Torulopsis [64] and Rose et al. reported the preparation of ZnS quantum dots *via* Saccharomyces cerevisiae [65] using lead sulfate and zinc sulfate as a metal precursor respectively. This is a very simple and easy method to produce metal sulfide quantum dots. Here, the metal ions stimulate the yeast cell to produce metal-chelating peptides with a primary structure of γ-(Glu-Cys)n-Glyfollowed by the formation of metal complexes. Then it reacted with intracellular sulfide to produce metal sulfide quantum dots. In addition, living yeast cells [66], Escherichia coli [67] and some funguses (e.g., Fusariumoxysporum, Phanerochaetechrysosporium, and Helminthosporumsolani) [68-71] and even the liver and kidneys of rats [72,73] can also be used for the synthesis of fluorescent quantum dots.

Near-infrared fluorescent silver sulfide quantum dots (NIR-Ag_2S QDs) in Hepatoma carcinoma (HepG2) cancer cells were successfully synthesized by Wan et al. [74]. The formation of Ag_2S quantum dot in cultured HepG2 cells was achieved by the reaction between the precursors such as silver nitrate and sodium sulfide *via* glutathione as a reducing agent. Figure (3i)a shows the intracellular synthesis of GSH-stabilized Ag_2S quantum dots in cultured cancer cells. The bright-field image of HepG2 cells after uptake of the precursors of silver nitrate and sodium sulfide is shown in figure (3i)b. The NIR image of precursors contained cultured HepG2 cells in the beginning of the aging process shows no observed fluorescence signals (Fig. (3i)c). The NIR fluorescence emission observed 20 h after uptake of silver nitrate and sodium sulfide indicated that the intracellular formation of Ag_2S quantum dots as seen in figure (3i)d. The morphology of synthesized NIR-Ag_2S quantum dots was observed by transmission electron microscopy image (figure (3i)e) that shows the average particles size of QDs as 5.45nm.

(i) Synthesis and characterization of Ag₂S quantum dot   (ii) Spectroscopy and imaging analysis of Ag₂S quantum dot

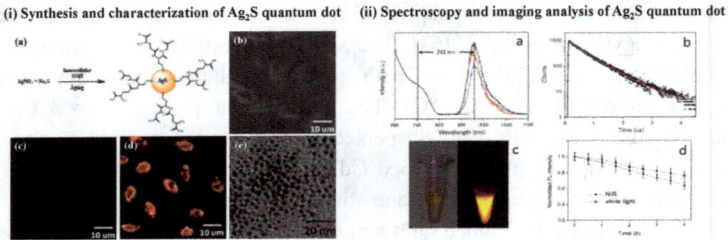

Figure3. (i) *(a) Synthetic approach of Ag₂S QDs in cultured HepG2 cancer cells. (b) Bright-field image of HepG2 cells immediately after uptake of precursors. NIR image of HepG2 cells (c) immediately and (d) after 20 h uptake of precursors. (e) TEM image of formed Ag₂S QDs in cultured HepG2 cancer cells.* ***(ii)****(a) Absorption (left; 16 h) and NIR emission spectra (right; 16, 20, and 12 h, from top to bottom) of isolated Ag₂S quantum dots in cells. (b) Fluorescence lifetime biexponential fitting of isolated Ag₂S quantum dots which corresponded to $\tau 1 = 204.7 \pm 8.52$ ns, $\tau 2 = 655.6 \pm 15.1$ ns. (c) Visual and fluorescence images of Ag₂S quantum dots in aqueous solution. (d) Photostability diagram of Ag₂S quantum dots under irradiation of laser (758 nm) and white light (100 W). (Reproduced with the permission from Ref. No. 74).*

The absorption spectra of Ag₂S quantum dots at 16 h after uptake of starting precursors was shown in (Fig.(3ii) a(left)) that showed an absorption peak at 744 nm. The emission spectra of isolated Ag₂S quantum dots at 16, 20, and 12 h (from top to bottom) after uptake of precursors showed the emission peak at wavelength of 945, 941 and 943 nm respectively (Figure (3ii) a(right)). The isolated Ag₂S quantum dots fluorescence lifetime measurements showed two decay components as seen in figure (3ii) b. The longer component 655.6 ns indicated the presence of Ag₂S quantum dots and the shorter component 204.7 ns was attributed to exciton scattering from the nonradiative recombination centers of quantum dot surface. The visual and NIR fluorescence images of isolated Ag₂S quantum dots was shown in figure(3ii) c. The photostability of isolated Ag₂S quantum dots was tested from continuous irradiation of NIR laser (758 nm) and white light (100W) for 4 h. The result showed that 75.9 and 63.6% of the original fluorescence intensity was retained for NIR laser and white light respectively after 4 h, indicating an excellent photostability of cultured HepG2 cancer cells isolated Ag₂S quantum dots.

Bioinspired Nanomaterials   Materials Research Forum LLC
Materials Research Foundations 111 (2021) 185-223   https://doi.org/10.21741/9781644901571-7

2.2   Quantum dots synthesis from biomimetic systems

Quantum dots can be directly synthesized from natural biomolecules such as nucleic acids, peptides, proteins, and enzymes and also from artificial cellular structures like biomimetic membranes. The direct approach possesses some advantages such as the biophysical and biochemical properties of biomolecules, the three-dimensional structure act as stabilizing agent and higher solubility in aqueous solution. The nucleic acids have metal binding functional groups such as phosphate, hydroxyl, and nitrogen atoms which may chelate with metal ion like template to grow the quantum particles [75,76]. DNA mediated quantum dots growth mechanisms was studied and reported [77-79].

DNA-templated controlled growth of CdS quantum dots based on novel gene delivery strategy was reported by Li Gao and Nan Ma [80]. DNA-templated CdS quantum dots involved two steps as follows: (i) The negatively charged DNA backbones bound with cadmium ions in supersaturated microenvironment facilitate the heterogenous nucleation and (ii) The sulfide ions were introduced for the formation of CdS nuclei and its subsequent growth into nanocrystals. The DNA template could be released after the formation of quantum dots by the intercellular ligand exchange between DNA and thiol moieties in which the thiol group was strongly bound with Cd^{2+} ion, interrupting the interaction of DNA phosphate backbone with cadmium ion. Glutathione (GSH), a thiol containing molecule, plays an important role in DNA release within the living cells. These steps were involved to achieve DNA plasmid-template CdS quantum dots and the growth strategy of DNA plasmid-template CdS quantum dots with controlling the packing and unpacking of DNA plasmids are shown in figure 4(i).

Quantum dots surface modified with DNA provided an excellent solubility and stability of quantum dots in water [81,82]. Yan Liu et al. have prepared single stranded DNA (ssDNA) functionalized IR emitting $Cd_xPb_{1-x}Te$ alloyed quantum dots *via* one pot synthesis [83].The staple DNA strands was extended with some complementary sequences of binding domain at specific locations to bind the quantum dots and the assembly of DNA origami process occurred in two steps as seen in figure 4(ii).The first step involved the assembling of origami with the required staple strand, capture strands and DNA for overnight annealing from 90 to 4 °C. The excess staple strands and capture strands was removed by washing and filtration of annealed samples. The second step involved the mixing of ssDNA functionalized quantum dots with pre-assembled DNA origami structures and annealed from 40 to 4 °C for 24 hours. The formed DNA origami was confirmed by atomic force microscopy (AFM) images with height profiles and transmission electron microscopy (TEM), indicating the formation of ~50% yield of $Cd_xPb_{1-x}Te$ alloyed quantum dots origami with a diameter of ~10.5 nm (Fig. 4(ii)B (i - vi)).

(i) CdS QDs in DNA-template

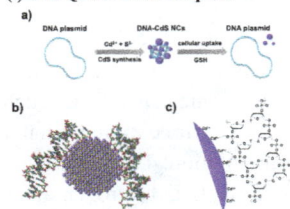

(ii) Cd$_x$Pb$_{1-x}$Te alloyed quantum dots in DNA origami

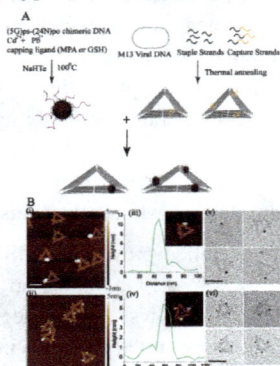

(iii) PbS QDs in luciferase enzymatic template

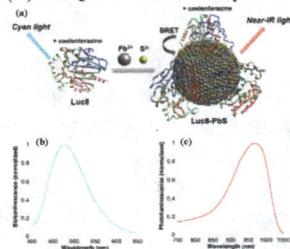

Figure 4 (i) (a)*CdS quantum dots growth of DNA plasmid induced DNA packing and GSH-mediated DNA unpacking. **(b)** Hybrid nanostructured image of double stranded DNA-CdS quantum dots. **(c)** Image of electrostatic interaction between phosphate backbone and surface Cd^{2+} ions.(Reproduced with the permission from Ref. No. 80).**(ii) (A)** Schematic diagram indicate the synthesis and self-assembly of DNA functionalized Cd$_x$Pb$_{1-x}$Te alloyed quantum dots with DNA origami. **(B)** AFM images with height profiles and TEM images of self-assembled Cd$_x$Pb$_{1-x}$Te quantum dots of triangular origami (i, iii & v) with GSH capped and (ii, iv & vi) MPA capped respectively. The scale bar is 100 nm in all images. (Reproduced with the permission from Ref. No. 83). **(iii). (a)** Schematic illustration of luciferase enzymatic template assist synthesis of Luc-PbS quantum dots hybrid nanostructure. **(b)** Bioluminescence spectrum of Luc8 in the presence of coelenterazine and **(c)** Fluorescence emission spectrum of Luc8-PbS quantum dots (excitation: 480 nm).(Reproduced with the permission from Ref. No. 86).*

Luciferase is one of the bioluminescent proteins that emits the light by luciferin oxidization. Luciferases emits in the visible region that can be tuned further by the functionalization of quantum dots with living organisms or biomimetic systems *via* a bioluminescence resonance energy transfer (BRET) [84,85]. For example, Jianghong Rao et al. reported the synthesis and growth of PbS quantum dots using luciferase enzyme as a template [86]. Luc-PbS hybrid quantum dots was synthesized by incubating luciferase withPb^{2+}ion at ambient conditions followed by the injection of S^{2-}intoLuc8-Pb^{2+}

Bioinspired Nanomaterials
Materials Research Foundations **111** (2021) 185-223

Materials Research Forum LLC
https://doi.org/10 21741/9781644901571-7

complex. The formed PbS quantum dots exhibited an emission peak in the range of 800-1050 nm in NIR luminescence as seen in figure 4(iii). In additior., the Luc8 enzyme remains active within the Luc8-PbS complex and emit the NIR light due to the bioluminescence resonance energy transfer between the Luc8 and PbS quantum dots.

The semiconductor quantum dots synthesized from various biomimetic systems and the basic properties such assize and emission wavelengths are summarized in Table 2.

Table 2. The properties of QDs synthesized from biomimetic systems.

QDs	Biomimetic systems	Size of the QDs (nm)	Emission wavelengths (nm)	References
Ag_2S	Bovine serum albumin	1.6-6.8	1050-1294	[87]
Ag_2S	Ribonuclease-A	5.6	980	[88]
CdTe	DNA	not reported	513-584	[89]
CdTe	Dithiol-peptide, trypsin	2.5-5.5	400-650	[90]
CdTe	Glutathione	3-6	500-600	[91]
PbS	Achymotrypsin	3	550	[92]
PbS	Thrombin-binding aptamer	3-6	1050	[93]

3.   Quantum dots for targeted biosensors

Quantum dots have unique photophysical properties when compared with traditional fluorescent probes such as organic dyes and fluorescent proteins. Quantum dots have unique optical and electronic properties such as narrow and symmetic emission spectra, broad absorption spectra that enable the simultaneous excitation of multiple fluorescent colors and size-tunable light emissions. Quantum dots are also remarkably brighter and more resistant to photobleaching than that of other materials [94-97]. Quantum dots based fluorescence biosensors systems usually depend on the functional biomolecules and its affinity towards bioanalyte interactions such as transport protein-receptor protein, antigen-antibody, biotin-streptavidin and carbohydrate–lectin. The detection method depends on the fluorescence sensing mechanisms such as aggregation of quantum dots, fluorescence resonance energy transfer (FRET), bioluminescence resonance energy transfer (BRET) and chemiluminescence resonance energy transfer (CRET) that have been widely used approaches in the field of quantum dots biosensing [98-100].

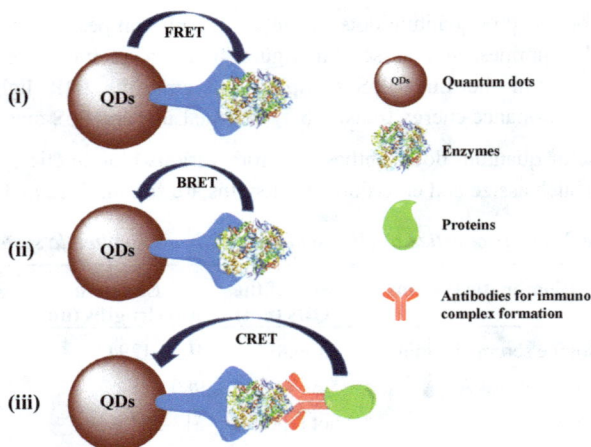

Figure 5. *Schematic representation of fluorescence biosensing mechanisms for three different approaches:* ***(i)*** *Quantum dots based fluorescence resonance energy transfer (FRET).* ***(ii)*** *Quantum dots based bioluminescence resonance energy transfer (BRET) and* ***(iii)*** *Quantum dots based chemiluminescence resonance energy transfer (CRET).*

The quantum dots based fluorescence biosensors were fabricated by three different approaches: The first approach is fluorescence resonance energy transfer (FRET) where a non-radiative energy transfer occur between an energy donor quantum dots and an acceptor biomolecules. Here quantum dots used as fluorescent labels to detect targeted biomolecules (Fig. 5(i)). The second approach is bioluminescence resonance energy transfer (BRET) in which quantum dots used as an energy acceptor from the target biomolecules (Fig. 5(ii)). The third approach is chemiluminescence resonance energy transfer (CRET) which involves a nonradiative energy transfer from a chemiluminescent donor to an energy acceptor of quantum dots (Fig. 5(iii)).

3.1 Fluorescence resonance energy transfer based biosensors

The quantum dots based fluorescence resonance energy transfer biosensors involved quantum dots as a donor and fluorescent dye as the energy acceptor for the detection of various biomolecules such as DNA, proteins, saccharides and dopamine [101-103]. CdSe/ZnS core/shell quantum dots as a simultaneous donors and acceptors in a time-gated FRET relay was used for the multiplexed detection of protease activity [104]. The fluorescent substrate of CdSe/ZnS core/shell quantum dots was assembled with multiple

Bioinspired Nanomaterials   Materials Research Forum LLC
Materials Research Foundations **111** (2021) 185-223   https://doi.org/10.21741/9781644901571-7

luminescent Tb^{3+}complexes (Tb) and Alexa Fluor 647 (A647) fluorescent dyes as shown in figure 6(i). This fluorescent substrate has two different functions: the energy was transferred from Tb to quantum dots ($FRET_1$) and then from quantum dots to A647 ($FRET_2$). A multiplexed FRET configuration with a single color of CdSe/ZnS core/shell quantum dots can offer to track the coupled biological processes in real time. In addition, the quantitative analysis of time-gated FRET sensing of TRP, ChT, or TRP +ChT, evaluating the inhibition in a multiplexed format and tracking the activation of pro-ChT to ChT.

The low cost, quantitative and ratio metric detection platform of green-emitting CdSeS/ZnS core/shell quantum dots (gQDs) as donors with Cy3 as an acceptor in FRET assays on paper substrates was demonstrated for the detection of single nucleotide polymorphism (SNP) [105]. Hybridization paper based substrate was prepared in two different method (i) direct and (ii) sandwich format. The paper surface was modified with imidazole group and assembled by oligonucleotide conjugate QDs-probe as shown in figure 6 (ii)a. The Cy3 dye was in close proximity to the surface of gQDs in the presence of nucleic acid that can induced FRET and appeared an analytical emission signal where the green-emitting quantum dots act as donors and Cy3 as an acceptor. The photoluminescence (PL) intensities of gQDs and Cy3 were converted into green (G) and red (R) color imaging channels from iPad mini camera as shown in figure 6 (ii)b.

The normalized photoluminescence spectral response of direct assay format with increasing the concentration (0 to 45 pmol) of fully complementary target as SMN1 FC TGT (3'-TAA AAC AGA CTT TGG GAC ATT CCT TTTATT TCC T-5') with increase in FRET sensitized Cy3 emission were shown in figure 6 (ii) c. Figure 6 (ii) d showed the ratio metric response for the ratio of FRET and R/ G (from digital image) in sandwich assay format in target SMN1 FC TGT with the concentration of 0 to 45 pmol. This enhancement was helped for the detection of subpicomole quantities of oligonucleotide targets and the detection limit as low as 450 fmol and 30 fmolin digital camera imaging method and epifluorescence microscope detection method respectively.

Green-emitting QDs based paper substrates in dry format showed higher analytical sensitivity and the digital image, R (Cy3) and G (gQDs) channel images of SMN1 FC TGT in a sandwich assay format was shown in figure 6. (ii) e. The PL color response of spots was changed from green to yellow on increasing the concentration of SMN1 FCTGT. Also, the increase in color intensity of R channel with parallel decrease in color intensity of G channel was occurred due to the fluorescence resonance energy transfer from gQDs to Cy3.

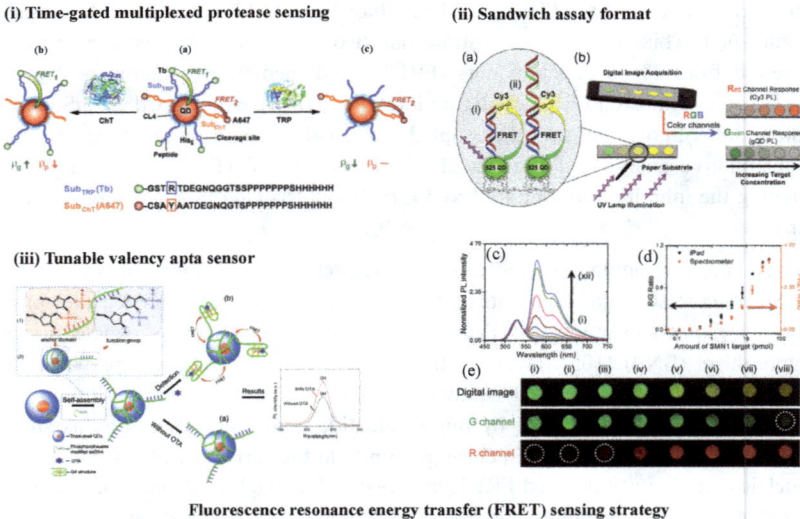

Fluorescence resonance energy transfer (FRET) sensing strategy

Figure 6. (i)*Schematic illustration of time-gated FRET relay for multiplexed protease sensing. **(a)** The substrate of CdSe/ZnS quantum dots coated with zwitterionic ligands (CL4) and assembled with polyhistidine (His6)-appended peptide. **(b)**FRET$_1$of the peptides labeled with Tb assisted Sub$_{ChT}$ for ChT. **(c)**FRET$_2$of the peptides labeled with A647 assist Sub$_{TRP}$ for TRP and FRET gated ratios of ρ_p and ρ_g used for analytical signals. (Reproduced with the permission from Ref. No. 104). **(ii)(a)** Design of green-emitting CdSeS/ZnS core/shell quantum dots (donors) with acceptor (Cy3) hybrid paper-based substrate for (i) direct and (ii) sandwich format. **(b)** R and G color channels indicates an increasing the target concentration, a significant decreasing QD photoluminescence with increase in Cy3 photoluminescence. **(c)** Normalized PL spectra of SMN1 FC TGT in a sandwich assay format at (i) 0 pmol, (ii) 0.057 pmol, (iii) 0.12 pmol, (iv) 0.23 pmol, (v) 0.47 pmol, (vi) 0.94 pmol, (vii) 1.9 pmol, (viii) 3.8 pmol, (ix) 7.5 pmol, (x) 15 pmol, (xi) 30 pmol, and (xii) 45 pmol. **(d)** Normalized PL spectra with calibration curves shows the response of R/G ratio (black) and the FRET ratio (red). **(e)** Digital image and R (Cy3) and G (gQDs) channel images of SMN1 FC TGT in a sandwich assay format at (i) 0 pmol, (ii) 0.94 pmol, (iii) 1.9 pmol, (iv) 3.8pmol, (v) 7.5 pmol, (vi) 15 pmol, (vii) 30 pmol, and (viii) 45 pmol. The white-dashed circles indicated the clear visibility of spots locations. (Reproduced with the permission from Ref. No. 105). **(iii)** Schematic illustration of self-assembling tunable valency aptasensor through incubation of CdSe/CdS QDs with phosphorothioate modified ssDNA (1 & 2) and the design for monitoring OTA in the system (a & b). (Reproduced with the permission from Ref. No. 106).*

CdSe/CdS QDs coated in ten layer CdS shells was developed for monitoring ochratoxin A (OTA), a poisonous contaminant widespread in foodstuffs [106]. The FRET sensing strategy of CdSe/CdS QDs thick-shell quantum dot was shown in figure 6(iii) in which QDs and OTA acts as an acceptor and donor molecules respectively. ssDNA sequence have two domains: the anchor domain preferentially bound with metal ions because its containing phosphodiester backbone with sulphur; the functional domain was composed of G-rich oligonucleotide sequences that undergo a conformation transition to form G-quadruplexes in the presence of OTA. It is due to the affinity of OTA with aptamer and pull the OTA molecules towards the CdSe/CdS QDs, indicating the enhancement of fluorescence signals. In addition, the increase in fluorescence signals were also due to the formation of G-quadruplexes on increasing the valence of apta sensors DNA from 1 to 3 that able to operate the FRET distance in OTA binding and achieved better sensitivity at the concentration of OTA ranging from 1 ng ml^{-1} to 30 ng ml^{-1}.

3.2 Bioluminescence resonance energy transfer (BRET) based biosensors

Bioluminescence resonance energy transfer is a naturally occurring phenomenon in which the energy was transferred from a donor luminescence enzyme into a complementary acceptor fluorophore. The excitation of fluorophore occurred only if the donor molecules were in close proximity for allowing BRET that can be used to measure protein-protein interactions in live cells [107]. BRET based bioassays were used to measure the activation of several G protein coupled receptors (GPCR) signaling pathways by expression of specially designed BRET sensors [108,109]. Quantum dots using BRET based biosensor was employed in luciferase bioluminescence instead of physical light as the excitation source which allows it to work in any complicated biological media.

Carboxyl-coated QDs-625 as a selective and sensitive nucleic acid sensing system based on BRET was developed by Sapna K. Deo et al. [110]. Carboxyl-coated QDs-625 based nucleic acid target sensing system have adjacent binding of oligonucleotide probes labeled with Renilla luciferase (Rluc) and quantum dot (QD). Here, Rluc, a bioluminescent protein, act as an energy donor molecule and the quantum dot accepted the energy. In the presence of target nucleic acid, the two small antisense oligonucleotide sequences were covalently attached to Rluc and QD which were modified with 1,3-diamino-2-hydroxypropane (DAP) and the QD-labeled probes (QD-D-P) bound adjacent to each other in a head-to-head fashion as shown in figure 7(i).

QD655-Luc8 conjugates sandwich type of sensing system was consisted of bioluminescent proteins as BRET donor and quantum dots as BRET acceptor for the detection of metalloproteinase in complex biological samples [111].The design of

QD655-Luc8 BRET based sensors for the detection of protease activity used a peptide as a linker between the bioluminescent protein and the quantum dots in which BRET process occurred between QDs and the protein. The protease will cleave the linker between QDs and protein, resulting in the disruption of BRET process with decreasing the emission as shown in figure 7(ii).

Renillaluciferase-immobilized quantum dots-655 (QD-RLuc8) as BRET mediated self-illuminated photodynamic therapy (PDT) was developed by Ping-Shan Lai et al. [112]. The method was used bioluminescent quantum dots as an internal light source for meta-tetra-hydroxyphenyl-chlorin (mTHPC, Foscan) for PDT as shown in figure 7 (iii). This was due to the wavelength tunable properties of QD-RLuc8 that help to activate the specific photosensitizers and the internal emitted light in the body may overcome the limitation of light penetration for PDT.

(i) Oligonucleotide probes labeled QDs

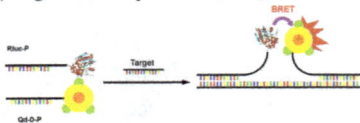

(ii) QD655-Luc8 conjugates sandwich type

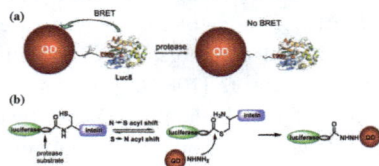

(iii) Renilla luciferase-immobilized quantum dots-65

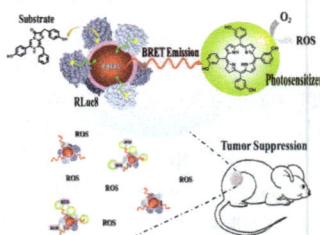

Bioluminescence resonance energy transfer (BRET) sensing strategy

Figure 7. (i) *Schematic illustration of BRET based sensing system. The Rluc-P and QDs-D-P hybridized to the target in a head-to-head fashion permitting BRET between Rluc and QDs. (Reproduced with permission from Ref. No. 110).* **(ii)** *Schematic illustration of QD-BRET based nanosensors for the detection of MMP. (a) Schematic representation of BRET process from luciferase proteins (Luc8) to QDs in the presence and absence of MMP peptide substrate. (b) intein-mediated site-specific conjugation of Luc8 fusion proteins with QDs. (Reproduced with permission from Ref. No. 111).* **(iii)** *Schematic representation of bioluminescence resonance energy transfer mediated self-illuminated photodynamic therapy (PDT) used in RLuc8-immobilized QDs. (Reproduced with permission from Ref. No. 112).*

3.3 Chemiluminescence resonance energy transfer (CRET) based biosensors

Chemiluminescence resonance energy transfer can occur by the oxidation of donor that emits chemiluminescence photons to excite the acceptor molecules [113, 114]. When compared with fluorescence resonance energy transfer, CRET doesn't need any external excitation light source. Thus, the sensitivity of CRET based biosensors sensing system avoids the autofluorescence and photo bleaching which make CRET become an attractive resonance energy transfer process (RET) for sensing mechanism of biomolecules.

Quantum dots (QDs) were used as an acceptor of CRET process due to their unique properties such as wide excitation, narrow emission, high quantum yield, size-dependent character and resistance to chemical and photo degradation [115-117]. At first time, Huang et al. observed the CRET phenomenon between luminol as donor and CdTe QDs as an acceptor [118]. Various QDs such as CdTe [119, 120], CdSe-ZnS composite [121, 122] and CdS [123] were used for the detection of biomolecules *via* CRET phenomenon.

Hairpin nucleic acid functionalized with CdSe/ZnS QDs showed CRET sensing system for the detection of aptamer or DNA [124]. The hemin/G-quadruplex acts as an internal light source to transfer the chemiluminescence resonance energy to CdSe/ZnS QDs. Three different sized CdSe/ZnS QDs(λ_{em}- 490 nm, λ_{em}- 560 nm, λ_{em}- 620 nm) were modified with hairpin structures namely (5′GGGTTGGGCGGGAT-GGGTTACCTCAGTGCTTATTCGAAACCCAAA(CH_2)$_6$-SH-3′), (5′-GGGTAG-GGCGGGTTGGGCTATCATCTTGGCAATTTATTAGCCC(CH_2)$_6$-SH-3′), and (5′-GGGTAGGGCGGGTTGGGTTCCTGGGGGAGTATTGATAGAACCC(CH_2)$_6$-SH-3′) to develop three different sensing platform for analysis of three different DNAs as shown in figure 8(i) a-c.

The luminescence spectrum for the mixture of three different sized CdSe/ZnSQDs modified with hairpins interact with three different DNA analytes of (5′-TCGAATAAGCACTGAGGT-3′), (5′-ATAAATTGCCAAGATGAT-3′) and (5′-TATCAATACTCCCCCAGG-3′) in the presence of hemin and H_2O_2/luminol was given in Fig. 17. The 5′-TCGAATAAGCACTGAGGT-3′ DNA analyte selectively triggered the luminescence of 620 nm CdSe/ZnS QDs functionalized 5′-GGGTAGGGCGGGTTGGGTTCCTGGGGGAGTATTGATAGAACCC (CH_2)$_6$-SH-3′ during their interactions, resulting in the opening of the hairpin and formed hemin/G-quadruplex. Likewise, (5′-TCGAATAAGCACTGAGGT-3′) and (5′-ATAAATTGCCAAGATGAT-3′) DNA analytes selectively opened the hairpin and formed hemin/G-quadruplex in hairpin nucleic acid functionalized CdSe/ZnSQDs by triggering the luminescence of 560 and 490 nm respectively. CdSe/ZnSQDs mixture with all three DNA analytes showed the CRET emission of all sized QDs at 490, 560, and 620 nm as seen in figure 8(i) d.

As similar, Itamar Willner et al. have reported the detection of thrombin and ATP by using nucleic acid-modified CdSe/ZnS QDs, operating *via* the CRET process [125]. The detection of thrombin and ATP from the chemiluminescence of CdSe/ZnSQDs which stimulated the formation of hemin/G-quadruplex/thrombin complex and the hemin/G-quadruplex/ATP nanostructure, in the presence of luminol/H_2O_2. The chemiluminescence operated by hemin/G-quadruplex/thrombin complex at the wavelength of 420 nm was suitable for exciting CdSe/ZnS QDs during the detection of thrombin. The close proximity between the light sources of hemin/G-quadruplex/thrombin complex and CdSe/ZnS QDs lead to the process of chemiluminescence resonance energy-transfer to CdSe/ZnS QDs as shown in figure 8(ii) a. The luminescence spectra of hemin/G-quadruplex aptamers assembled CdSe/ZnSQDs in the absence and presence of different concentration of thrombin was given in figure 8(ii) b. The lower chemiluminescence resonance energy-transfer signal was observed in the absence of thrombin due to the diffusion of hemin. The higher chemiluminescence signal was observed for CdSe/ZnS QDs at λ= 615 nmin the presence of thrombin. Furthermore, the chemiluminescence intensity was also increased with increasing the concentration of thrombin due to the generation of larger number of DNAzyme units. In the case of ATP detection, the chemiluminescence operated by hemin/G-quadruplex/ATP complex which included the aptamer sequence for ATP, tethered to the QDs with an oligo-T spacer as shown in figure 8(ii) c. The luminescence spectra of hemin/G-quadruplex aptamers assembled CdSe/ZnS QDs in the presence of different concentration of ATP was shown in 8(ii) d. As the concentration of ATP increased, the chemiluminescence intensity was also increased due to the generation of larger number of DNAzyme units.

Materials Research Forum LLC
https://doi.org/10.21741/9781644901571-7

(i) Hairpin nucleic acid functionalized with CdSe/ZnS QDs

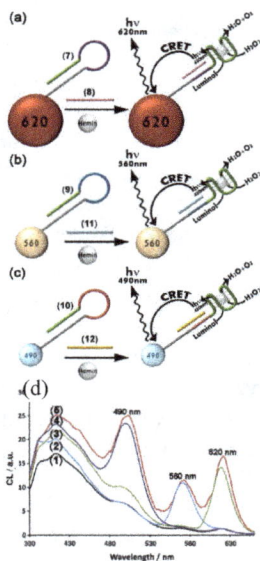

(ii) Hemin/G-quadruplex/t hrombin complex assembled CdSe/ ZnS QDs

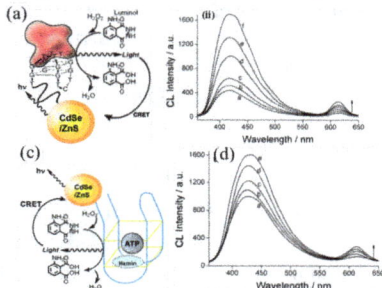

(iii) Hemin/G-quadruplex-modified GOx functionalized CdSe/ZnS QDs

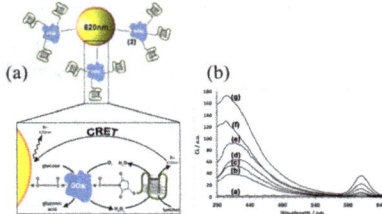

Chemiluminescence resonance energy transfer (CRET) sensing strategy

Figure 8. (i) Three different sized CdSe/ZnS QDs emittedchemiluminescence at (a) 620 nm, (b) 560 nm and (c) 490 nm for sensing three different target DNAs. (d)The luminescence spectrum of (1)CdSe/ZnS QDs mixture. (2)CdSe/ZnS QDs mixture with 5′-ATAAATTGCCAAGATGAT-3′. (3)CdSe/ZnS QDs mixture with 5′-TCGAATAAGCACTGAGGT-3′.(4)CdSe/ZnS QDs mixture with 5′-TATCAATACTCCCCCAGG-3′. (5)CdSe/ZnS QDs mixture with allthe DNA analytes.(Reproduced with the permission from Ref. No. 124). (ii) (a) Schematic diagram for detection of thrombin under chemiluminescence resonance energy-transfer process from luminol to hemin/G-quadruplexaptamer assembled CdSe/ZnS QDs. (b) Luminescence spectrum of (a) hemin/G-quadruplexaptamer assembled CdSe/ZnS QDs in the absence and presence of thrombin at different concentrations (b) 1.43, (c) 14.3, (d) 28.6, (e) 57 and (f) 86 nM. (c) Schematic diagram for detection of ATP under chemiluminescence resonance energy-transfer process from luminol to hemin/G-quadruplex ATP assembled CdSe/ZnS QDs. (d) Luminescence spectrum of (a) hemin/G-quadruplex ATP assembled CdSe/ZnS QDs in the absence and presence of thrombin at different concentrations (b) 0.05, (c) 0.1, (d) 0.15 and (e) 0.2 mM.(Reproduced with the permission from Ref. No. 125). (iii) (a) Schematic illustration ofhemin/G-quadruplex-

Bioinspired Nanomaterials   Materials Research Forum LLC
Materials Research Foundations **111** (2021) 185-223   https://doi.org/10.21741/9781644901571-7

modified GOx functionalized CdSe/ZnS QDs for the detection of glucose via a chemiluminescence resonance energy-transfer. **(b)** *Luminescence spectrum of (a) hemin/G-quadruplex-modified GOx functionalized CdSe/ZnS QDs in the absence and presence of glucose at different concentrations, (b) 5 mM, (c) 10 mM, (d) 25 mM, (e) 37 mM, (f) 50 mM and (g) 75 mM.(Reproduced with the permission from Ref. No. 126).*

Glucose oxidase-functionalized CdSe/ZnS QDs with hemin/G-quadruplex generated CRET for the detection of glucose [126]. The CdSe/ZnS QDs (λ_{em} = 620 nm) was functionalized with hemin/G-quadruplex-modified GOx. Then, nucleic acid 5'-HS-(CH2)6-TTTTTTGGGTAGGGCGGGTTGGG-3' was attached covalently with functionalized quantum dots. In the presence of hemin, luminol and glucose, the HRP-mimicking DNAzyme catalyzed the oxidation of luminol by GOx-generated H_2O_2, resulting in the occurrence of chemiluminescence resonance energy-transfer from luminal to functionalized CdSe/ZnS QDs as shown in figure 8(iii) a. The luminescence spectra of hemin/G-quadruplex-modified GOx functionalized CdSe/ZnS QDs in the presence of different concentrations of glucose was given in figure 8(iii) b. As the concentration of glucose increases, chemiluminescence intensity was also increased due to the consistent with a higher content of GOx-generated H_2O_2.

4.   Quantum dots for targeted bioimaging

Recently, quantum dots become a popular imaging agent, competing with other fluorescent nanomaterials and organic fluorescent dyes [127-131]. Quantum dots have unique optical features including narrow light emission, wideband excitation, photostability and multi-color fluorescence imaging [132]. In addition, the NIR-quantum dots for in vivo deep-tissue imaging are desirable because of their autofluorescence in the NIR range of 750-940 nm and 1200-1700 nm. Most importantly, the polymer surface was functionalized to get the water-soluble NIR quantum dots materials which is more suitable for biomedical-targeted in vivo imaging applications [133-136].

4.1   In vivo multicolor, multimodal and multiplex imaging

Folic acid-conjugated CuInS$_2$/ZnS quantum dots were successfully developed and utilized for in vivo tumor target labeling and imaging [137]. In vivo luminescence images of tumor-bearing mice were obtained *via* injecting bio-conjugated and non-bioconjugated CuInS$_2$/ZnS quantum dots intravenously. In the bio-conjugated CuInS$_2$/ZnS quantum dots injected mouse one clearly sees the intense red quantum dots signals that was easily differentiated from the auto-fluorescence background (green). Micelle functionalized CuInS$_2$/ ZnS quantum dots were used for the whole body multiplex in vivo tumor imaging. Two-different color motive CuInS$_2$ quantum dots were injected in a mouse at different spots by subcutaneous injection. The different intensity luminescence spots on the treated mouse can be easily identified from one another. This evidences proved that the micelle-encapsulated CuInS$_2$ quantum dots were used for multicolored optical contrast agent for in vivo multiplex imaging.

Highly luminescent NIR-emitting CdTe/CdSe quantum dots were prepared and successfully applied for the ultrasensitive, multicolor, and multiplex in vivo imaging in living animals [138]. The prepared CdTe/CdSe quantum dots emits in three different wave lengths of 700, 750, and 800 nm that has capability for in vivo multiplex imaging in living animals. 2 mg of CdTe/CdSe quantum dots diluted in 200 mL of PBS and subcutaneously injected into various locations on the back side of the athymic nude mouse, resulting in a stacked image that showed three clearly distinguished spots as seen in figure 9i (a). The observed fluorescent intensity indicated that the ability of CdTe/CdSe quantum dots produced different light levels at the same excitation wavelength and longer emission wavelength look brighter than shorter wavelength as shown in figure 9i (b). The signal-to-noise ratio (S/N) for living images at 700 nm, 750 nm and 800 nmof CdTe/CdSe quantum dots are 65 ± 3,148 ± 15 and 1130 ± 60 (Fig. 9i (c)) respectively. These results indicated that CdTe/CdSe quantum dots fluorescent probe was excellent for in vivo multiplex imaging and QDs have high sensitivity with enhanced signal-to-noise ratio in longer emission wavelengths.

Chung et al. also developed perfluorocarbon (PFC)/CdSe/ZnS quantum dots which emits light at three different wavelengths of 525, 596 and 596 nm that was used for multicolor optical imaging in three different immune cells (macrophage cells, T cells, and dendritic cells) [139]. The visible and fluorescence images at three different wavelengths after injecting PFC/CdSe/ZnS quantum dots in three different types of immune cells in mice was shown in Fig. 22. The visible images of (A) macrophage cells for PFC/CdSe/ZnS QDs at 525 nm, (B) T cells for PFC/CdSe/ZnS QDs at 596 nm and (C) dendritic cells for PFC/CdSe/ZnS QDs at 596 nm was shown in figure 9ii (a). The fluorescence images indicated that the PFC/CdSe/ZnS QDs fluorescence imaging probes provided an exact detection and localization of labeled immune cells in vivo condition. An optical filter was used for selectively detect the spot A for green (525WB20) and B, C for red (600WB20) as seen in figure 9ii (b&c).

(i) CdTe/CdSe quantum dots   **(ii) Perfluorocarbon (PFC)/CdSe/ZnS quantum dots**

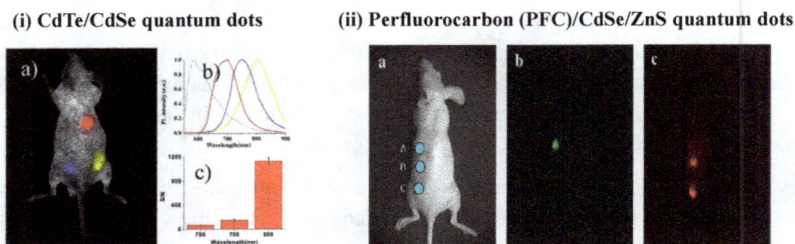

Figure 9. (i)(a) Multiplex in vivo imaging for three different wavelengthsof 700, 750, and 800 nm of CdTe/CdSe quantum dots in live animals. (b) Photoluminescence spectra of CdTe/CdSe quantum dots treated mouse in three different spots. (c) The signal-to-noise ratio (S/N) and error bars (five measurements). (Reproduced with the permission from Ref. No. 138). (ii)(a) Visible image of CdSe/ZnSquantum dots injected mouse labeled with three different types of immune cells [525 nm in macrophage cells (A), 596 nm in T cells (B), and 596 nm in dendritic cells (C)]. Fluorescent images of (b) green (c) red filter. (Reproduced with the permission from Ref. No. 139).

4.2   Near-Infrared (NIR) quantum dots for deep tissue imaging

Multi-functional ZnAgInSe/ZnS core/shell QDs for dual-targeting imaging strategy was fabricated for in vitro and in vivo of cancer cell [140]. The NIR fluorescence images of ZnAgInSe/ZnS core/shell QDs injected in nude mice *via* a vein at different times and after 5 min and the bright fluorescence arose in the whole body of the mice as shown in figure 24. Post-injection at 1 h, fluorescence was observed in U87MG tumor mice and maximum fluorescence contrast ratio reached at 4-8 h as seen in figure 10 (i) A. However, the post-injection of MCF-7 tumor mice identified fluorescence up to 8 h due to the EPR effect and suitable particle size of ZnAgInSe/ZnS core/shell QDs as shown in figure 10 (i) B.

A novel albumin mediated Gd^{3+} of paramagnetic NIR Ag_2S QDs ($Ag_2S@BSA-DTPA^{Gd}p$ QDs) was fabricated and successfully applied for bimodal imaging contrast agent for tiny tumor diagnosis [141]. $Ag_2S@BSA-DTPA^{Gd}p$ QDs are capable for in vivo imaging for both MRI and fluorescence imaging due to their paramagnetic properties. MRI and fluorescence bimodal in vivo imaging of $Ag_2S@BSA-DTPA^{Gd}p$ QDs in tiny tumor model before and after injection of quantum dots at different time points were shown in figure 10 (ii)a. The post-injection in point of 1.5 h have clearly seen the enhanced contrast of tiny tumors. In addition, $Ag_2S@BSA-DTPA^{Gd}p$ QDs become weak at 3 h post-injection and disappeared at 12 h post-injection that indicated the MRI pattern is time-dependent. The bar diagram indicated the corresponding quantified signal-to-noise

Bioinspired Nanomaterials
Materials Research Foundations **111** (2021) 185-223

Materials Research Forum LLC
https://doi.org/10.21741/9781644901571-7

ratio (SNR) of intensity (post)-to-intensity (pre) contrast in tumor as seen in figure 10 (ii)b. Tumor model of $Ag_2S@BSA$-$DTPA^{Gd}p$ QDs treated mice of in vivo NIR-fluorescence imaging was shown in figure10 (ii) c.

(i) NIR fluorescence imaging of ZnAgInSe/ZnS QDs

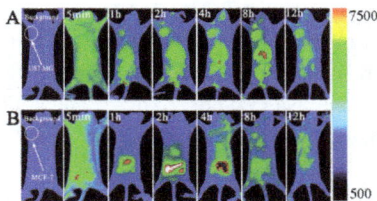

(ii) Tumor model of $Ag_2S@BSA$-$DTPA^{Gd}$ pQDs

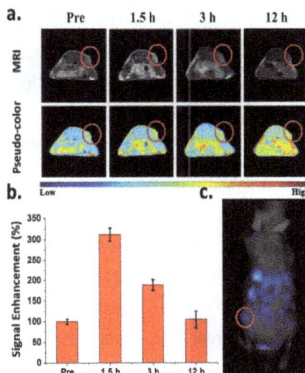

(iii) Upconverting $NaYF_4$:Yb^{3+}, Tm^{3+} NPs mixture with CdTeSe QDs multiplexing

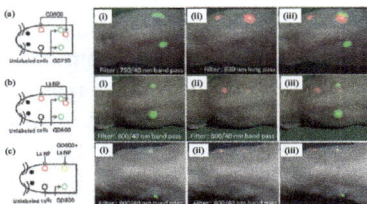

Figure 10. (i) NIR fluorescence imaging of ZnAgInSe/ZnS QDs injected nude mice in *(A)* $\alpha_v\beta_3$-positive U87MG tumor and *(B)* $\alpha_v\beta_3$-negative MCF-7 tumor. (λ_{ex} = 660 nm). *(Reproduced with permission from Ref. No. 140). (ii) (a)* In vivo MRI images of $Ag_2S@BSA$-$DTPA^{Gd}pQDs$ pre- and post-injected mice at different time. *(b)* Their corresponding quantified signal-to-noise ratio (SNR) of intensity (post)-to-intensity (pre) contrast in tumor. *(c)* Fluorescent image of $Ag_2S@BSA$-$DTPA^{Gd}pQDs$ intravenous injected tiny tumor-bearing nude mice after 2 h. (Reproduced with permission from Ref. No. 141). *(iii) (a)*Schematic illustration of CdTeSeQDs750 (green) and CdTeSeQD900 (red) mixture with HeLa cell injected mouse and NIR fluorescent images of (i) CdTeSeQDs750 (ii) CdTeSeQDs900 and (iii) Merged image of i & ii and illuminated with 660 nm laser. *(b)* Schematic illustration of CdTeSeQD800 (green) and La NP (red) mixture with HeLa cell injected mouse and NIR fluorescent images of (i) illuminated using 660 nm laser, (ii) excited using 980 nm diode laser and (iii) Merged image of i& ii. *(c)* Schematic illustration of CdTeSeQD800 (green), La NP (red) and CdTeSeQD800 + La NP (yellow)mixture with HeLa cell injected mouse and NIR fluorescent images of (i) illuminated using a halogen lamp with 660 nm laser, (ii) excited using 980 nm diode laser and (iii) Merged image of i& ii.(Reproduced with permission from Ref. No. 142).

A novel, multiplexed NIR in vivo imaging agent was developed by using CdTeSe quantum dotes and upconverting NaYF$_4$:Yb^{3+}, Tm^{3+} nanoparticles. The 'temporal' multiplexing was also demonstrated by alternating the excitation wavelengths with unmixing the emissions of different probes [142]. In vivo multiplexing was demonstrated by using CdTeSe quantum dotes and upconverting NaYF$_4$:Yb^{3+}, Tm^{3+} nanoparticles mixture with HeLa cell was injected to hairless mice and the signals of green and red false colors visualized simultaneously as shown in figure 10 (iii). These images were clearly indicated the compatibility of upconverting NaYF$_4$:Yb^{3+}, Tm^{3+} nanoparticles mixture with CdTeSe quantum dotes multiplexing.

Conclusions and future perspectives

Biomediated synthesis of quantum dots have rapid development in the field of biological sensing and in vivo imaging. The new approach and new strategy for synthesis and functionalization of QDs with targeting biomolecules such as nucleic acids, peptides, antibodies, and small molecules have been developed. Many of the bio-conjugated QDs based nanosensors probe have achieved with high selective and sensitive fluorescence signals and showed a low limit of detection (LOD) with wide linear range of concentration. Biosynthesized QDs are highly fluorescent, photostable and biocompatibility that are more suitable for in vivo studies.

Currently, many of the researcher have reported significant improvement in the quality of biosynthesized quantum dots. Water-soluble CdTe QDs were prepared from earthworms by metal detoxification pathway [50]. Extracellularly biogrown CdTe QDs by using yeast cells were obtained in the size range of 2.0-3.6 nm with good crystallinity in aqueous medium [63]. Ag$_2$S QDs were prepared from cultured HepG2 cells with high uniformity of QDs [74]. The controlled growth of CdS QDs was prepared by using DNA-template with high uniformity [80]. "One-pot synthesis" of NIR emitting Cd$_x$Pb$_{1-x}$Te alloyed QDs was prepared by using functionalized single stranded DNA (ssDNA) [83]. NIR luminescence Luc-PbS hybrid QDs were prepared from Luc8 enzyme and the emission occur in range at 800-1050 nm [86].

The functionalization of quantum dots with various biomolecules such as nucleic acids, antibodies, peptides, and aptamers were employed in biosensing DNA, proteins, nucleic acid, aptamers, ATP, thrombin, nucleotide, ochratoxin A and saccharides with suitable operating signaling mechanisms of resonance energy transfer process such as FRET,BRET and CRET. Further, the quantum dots were used as a fluorescent imaging agent for targeting in vivo multicolor, multimodal and multiplex imaging in living animals. Also, near-infrared quantum dots were used for deep tissue imaging in cancer cells.

Bioinspired Nanomaterials   Materials Research Forum LLC
Materials Research Foundations **111** (2021) 185-223   https://doi.org/10.21741/9781644901571-7

The biomediated synthesis of heavy metal free quantum dots may provide a promising opportunity to overcome the toxicity problems caused by cadmium based quantum dots. However, a new approach for synthesis of biocompatible quantum dots in vivo, especially in the human body is still developing. In future, we will move biocompatible quantum dots for real-life clinical applications.

References

[1] I. L. Medintz, H. T. Uyeda, E. R. Goldman and H. Mattoussi, Quantum dot bioconjugates for imaging, labelling and sensing, Nature Mater., 4 (2005) 435. https://doi.org/10.1038/nmat1390

[2] X. He, L. Gao and N. Ma, One-step instant synthesis of protein-conjugated quantum dots at room temperature, Sci. Rep., 3, (2013)2825. https://doi.org/10.1038/srep02825

[3] J. Wu, J. Dai, Y. Shao and Y. Sun, One-step synthesis of fluorescent silicon quantum dots (Si-QDs) and their application for cell imaging, RSC Adv., 5 (2015) 83581-83587. https://doi.org/10.1039/C5RA13119G

[4] R. Gui and X. An, Layer-by-layer aqueous synthesis, characterization and fluorescence properties of type-II CdTe/CdS core/shell quantum dots with near-infrared emission, RSC Adv., 3 (2013) 20959-20969. https://doi.org/10.1039/C3RA43120G

[5] H. Han, G. D. Francesco and M. M. Maye, Size control and photophysical properties of quantum dots prepared via a novel tunable hydrothermal route, J. Phys. Chem. C,45(2010) 19270-19277. https://doi.org/10.1021/jp107702b

[6] Q. Ma and X. Su, Recent advances and applications in QDs-based sensors, Analyst, 136 (2011) 4883-4893. https://doi.org/10.1039/c1an15741h

[7] J. Li and J-J. Zhu, Quantum dots for fluorescent biosensing and bio-imaging applications, Analyst, 138 (2013) 2506-2515. https://doi.org/10.1039/c3an36705c

[8] I. L.Medintz, H.Mattoussi and A. R. Clapp, Potential clinical applications of quantum dots, Int J Nanomedicine. 3(2) (2008) 151-167. https://doi.org/10.2147/IJN.S614

[9] X. H. Zhong, Y. Y. Feng, W. Knoll and M. Y. Han, Alloyed $Zn_xCd_{1-x}S$ nanocrystals with highly narrow luminescence spectral width. J. Am. Chem. Soc., 125 (2003) 13559-13563. https://doi.org/10.1021/ja036683a

[10] X. H. Zhong, M. Y. Han, Z. Dong, T. J. White and W. Knoll, Composition-tunable $Zn_xCd_{1-x}Se$ nanocrystals with high luminescence and stability, J. Am. Chem. Soc., 125 (2003) 8589-8594. https://doi.org/10.1021/ja035096m

[11] R. E. Bailey and S. M. Nie, Alloyed semiconductor quantum dots: tuning the optical properties without changing the particle size, J. Am. Chem. Soc., 125 (2003) 7100-7106. https://doi.org/10.1021/ja035000o

[12] S. Kim, B. Fisher, H. J. Eisler and M. Bawendi, Type-II quantum dots: CdTe/CdSe(core/ shell) and CdSe/ZnTe(Core/Shell) heterostructures, J. Am. Chem. Soc., 125 (2003) 11466-11467. https://doi.org/10.1021/ja0361749

[13] B. L. Wehrenberg, C. J. Wang and P. Guyot-Sionnest, Interband and intraband optical studies of PbSe colloidal quantum dots, J. Phys. Chem. B, 106 (2002) 10634-10640. https://doi.org/10.1021/jp021187e

[14] A. M. Smith and S. Nie, Chemical analysis and cellular imaging with quantum dots, Analyst, 129 (2004) 672–677. https://doi.org/10.1039/b404498n

[15] W. R. Algar, A. Tavares and U. J. Krull, Beyond labels: A review of the application of quantum dots as integrated components of assays, bioprobes, and biosensors utilizing optical transduction, Anal. Chim. Acta, 673 (2010) 1-25. https://doi.org/10.1016/j.aca.2010.05.026

[16] M. F. Frasco and N. Chaniotakis, Semiconductor quantum dots in chemical sensors and biosensors, Sensors, 9 (2009) 7266-7286. https://doi.org/10.3390/s90907266

[17] J. Costa-Fernandez, R. Pereiro and A. Sanz-Medel, The use of luminescent quantum dots for optical sensing, TrAC, Trends Anal. Chem., 25 (2006) 207-218. https://doi.org/10.1016/j.trac.2005.07.008

[18] R. Gui, J. Sun, D. Liu, Y. Wang, H. Jin, A facile cation exchange-based aqueous synthesis of highly stable and biocompatible Ag_2S quantum dots emitting in the second near-infrared biological window, Dalton Trans, 43 (2014) 16690−16697. https://doi.org/10.1039/C4DT00699B

[19] S. Chinnathambi and N. Shirahata, Recent advances on fluorescent biomarkers of near-infrared quantum dots for in vitro and in vivo imaging, Sci. Technol. Adv. Mater. 20 (2019) 337-355. https://doi.org/10.1080/14686996.2019.1590731

[20] W. T.Wu, H. Liu, C. Dong, W. J. Zheng, L. L. Han, L. Li, S. Z.Qiao, J. Yang and X. W. Du, High-quality colloidal quantum dots directly from natural minerals. Langmuir.31 (2015) 2251−2255. https://doi.org/10.1021/la5044415

[21] J. Yang, T. Ling, W-T. Wu, H. Liu, M-R.Gao, C. Ling, L. Li, X-W. Du, A top-down strategy towards monodisperse colloidal lead sulphide quantum dots. Nat. Commun., 4 (2013) 1695. https://doi.org/10.1038/ncomms2637

[22] H. B.Zeng, S. K. Yang, W. P.Cai, Reshaping formation and luminescence evolution of Zno quantum dots by laser-induced fragmentation in liquid. J. Phys. Chem. C.,115 (2011) 5038-5043. https://doi.org/10.1021/jp109010c

[23] K. Jagannadham, J. Howe andL. F. Allard, laser physical vapor deposition of nanocrystalline dots using nanopore filters. Appl. Phys. A: Mater. Sci. Process.,98 (2010) 285-292. https://doi.org/10.1007/s00339-009-5432-7

[24] H. B. Zeng, X. W. Du, S. C. Singh, S. A.Kulinich, S. K. Yang, J. P. He and W. P.Cai, Nanomaterials via laser ablation/irradiation in liquid: A Review. Adv. Funct. Mater., 22 (2012) 1333-1353. https://doi.org/10.1002/adfm.201102295

[25] S. C. Singh, S. K. Mishra, R. K. Srivastava and R. Gopal, Optical properties of selenium quantum dots produced with laser irradiation of water suspended Se nanoparticles. J. Phys. Chem. C,114 (2010) 17374-17384. https://doi.org/10.1021/jp105037w

[26] C. B. Murray, D. J. Norris and M. G.Bawendi, synthesis and characterization of nearly monodisperse CdE(E = S, Se, Te) semiconductor nanocrystallites. J. Am. Chem. Soc.,115 (1993) 8706-8715. https://doi.org/10.1021/ja00072a025

[27] N. Gaponik, D. V. Talapin, A. L.Rogach, K. Hoppe, E. V. Shevchenko, A.Kornowski, A.Eychmuller, and H. Weller, Thiolcapping of CdTe nanocrystals: an alternative to organometallic synthetic routes. J. Phys. Chem. B,106 (2002) 7177-7185. https://doi.org/10.1021/jp025541k

[28] T. Rajh, O. I. Micic and A. J.Nozik, Synthesis and characterization of surface-modified colloidal cdte quantum dots. J. Phys.Chem., 97 (1993) 11999-12003. https://doi.org/10.1021/j100148a026

[29] D. Zhou, M. Lin, Z. Chen, H. Sun, H. Zhang, H. Sun and B.Yang, Simple synthesis of highly luminescent water-soluble CdTe quantum dots with controllable surface functionality. Chem. Mater., 23 (2011) 4857-4862. https://doi.org/10.1021/cm202368w

[30]   L. H.Qu, Z. A. Peng and X. G. Peng, Alternative routes toward high quality CdSenanocrystals. Nano Lett., 1 (2001) 333-337. https://doi.org/10.1021/nl0155532

[31]   Q. Wang, T. Fang, P. Liu, B. Deng, X. Min, X. Li, Direct synthesis of high-quality water-soluble CdTe:Zn^{2+} quantum dots. Inorg. Chem..51 (2012) 9208-9213. https://doi.org/10.1021/ic300473u

[32]   L. H. Qu and X. G. Peng, Control of photoluminescence properties of CdSe nanocrystals in growth. J. Am. Chem. Soc. 124 (2002) 2049-2055. https://doi.org/10.1021/ja017002j

[33]   Z. A. Peng, X. G. Peng, Formation of high-quality CdTe, CdSe, and CdSnanocrystals using CdO as precursor. J. Am. Chem. Soc..123 (2001) 183-184. https://doi.org/10.1021/ja003633m

[34]   K. Yu, B. Zaman, S. Romanova, D-S. Wang and J. A. Ripmeester, Sequential synthesis of type II colloidal CdTe/ CdSe core-shell nanocrystals, Small, 1 (2005) 332-338. https://doi.org/10.1002/smll.200400069

[35]   S. Kim, B. Fisher, H.-J. Eisler and M. Bawendi, Type-II quantum dots: CdTe/ CdSe (Core/ Shell) and CdSe/ ZnTe (Core/ Shell) heterostructures, J. Am. Chem. Soc., 125 (2003) 11466-11467. https://doi.org/10.1021/ja0361749

[36]   S. Kim, W. Shim, H. Seo, J. H. Bae, J. Sung, S. H. Choi, W. K. Moon, G. Lee, B. Lee and S.-W. Kim, Bandgap engineered reverse type-I CdTe/InP/ZnS core–shell nanocrystals for the near-infrared, Chem. Commun., (2009) 1267-1269. https://doi.org/10.1039/b820864f

[37]   J. M. Pietryga, D. J. Werder, D. J. Williams, J. L. Casson, R. D. Schaller, V. I. Klimov and J. A. Hollingsworth, Utilizing the lability of lead selenide to produce heterostructured nanocrystals with bright, stable infrared emission, J. Am. Chem. Soc., 130 (2008) 4879-4885. https://doi.org/10.1021/ja710437r

[38]   R. Cui, Y.-P. Gu, Z.-L. Zhang, Z.-X. Xie, Z.-Q. Tian and D.-W. Pang, Controllable synthesis of PbSe nanocubes in aqueous phase using a quasi-biosystem, J. Mater. Chem., 22 (2012) 3713-3716. https://doi.org/10.1039/c2jm15691a

[39]   M. A. Hines and G. D. Scholes, Colloidal PbS nanocrystals with size-tunable near-infrared emission: observation of post-synthesis self-narrowing of the particle size distribution, Adv. Mater., 15 (2003) 1844-1849. https://doi.org/10.1002/adma.200305395

[40]   T. Rauch, M. Bo "berl, S. F. Tedde, J. Fu"rst, M. V. Kovalenko, G. Hesser, U. Lemmer, W. Heiss and O. Hayden, Near-infrared imaging with quantum-dot-sensitized organic photodiodes, Nat. Photonics, 3 (2009) 332-336. https://doi.org/10.1038/nphoton.2009.72

[41]   N. Gaponik, I. L. Radtchenko, M. R. Gerstenberger, Y. A. Fedutik, G. B. Sukhorukov and A. L. Rogach, Labeling of biocompatible polymer microcapsules with near-infrared emitting nanocrystals, Nano Lett., 3 (2003) 369-372. https://doi.org/10.1021/nl0259333

[42]   H. Korbekandi, S. Iravani andS. Abbasi, Production of nanoparticles using organisms. Crit. Rev. Biotechnol. 29 (2009) 279-306. https://doi.org/10.3109/07388550903062462

[43]   K. B. Narayanan andN. Sakthivel, biological synthesis of metal nanoparticles by microbes. Adv. Colloid Interface Sci. 156 (2010) 1-13. https://doi.org/10.1016/j.cis.2010.02.001

[44]   J. R. Lloyd, J. M. Byrne andV. S. Coker, Biotechnological synthesis of functional nanomaterials. Curr. Opin. Biotechnol. 22 (2011) 509-515. https://doi.org/10.1016/j.copbio.2011.06.008

[45]   G. S. Dhillon, S. K.Brar, S. Kaur and M. Verma, Green approach for nanoparticle biosynthesis by fungi: current trends and applications. Crit. Rev. Biotechnol. 32 (2012) 49-73. https://doi.org/10.3109/07388551.2010.550568

[46]   V. Pavlov, Enzymatic growth of metal and semiconductor nanoparticles in bioanalysis. Particle & Particle Systems Characterization, 31 (2014) 36-45. https://doi.org/10.1002/ppsc.201300295

[47]   L. Saa, A.Virel, J. Sanchez-Lopez andV. Pavlov, analytical applications of enzymatic growth of quantum dots. Chem. - Eur. J. 16 (2010) 6187-6192. https://doi.org/10.1002/chem.200903373

[48]   L. Saa and V. Pavlov, Enzymatic growth of quantum dots: applications to probe glucose oxidase and horseradish peroxidase and sense glucose. Small, 8 (2012) 3449-3455. https://doi.org/10.1002/smll.201201364

[49]   L. Saa, J.M. Mato and V. Pavlov, Assays for methionine gammalyase and s-adenosyl-l-homocysteine hydrolase based on enzymatic formation of CdSquantum dots in situ. Anal. Chem. 84 (2012) 8961-8965. https://doi.org/10.1021/ac302770q

Materials Research Forum LLC
https://doi.org/10.21741/9781644901571-7

[50]   S. R. Stuerzenbaum, M.Hoeckner, A. Panneerselvam, J. Levitt, J. S. Bouillard, S. Taniguchi, L. A. Dailey, R. A. Khanbeigi, E. V. Rosca and M.Thanou, Biosynthesis of luminescent quantum dots in an earthworm. Nat. Nanotechnol. 8 (2013) 57-60. https://doi.org/10.1038/nnano.2012.232

[51]   R. Cui, H-H. Liu, H-Y. Xie, Z-L. Zhang, Y-R. Yang, D-W. Pang, Z-X.Xie, B-B. Chen, B. Hu andP. Shen, Living yeast cells as a controllable biosynthesizer for fluorescent quantum dots. Adv. Funct. Mater. 19 (2009) 2359-2364. https://doi.org/10.1002/adfm.200801492

[52]   H. Bao, Z. Lu, X. Cui, Y.Qiao, J.Guo, J. M. Anderson and C. M. Li, Extracellular microbial synthesis of biocompatible CdTequantum dots. Acta. Biomater. 6 (2010) 3534-3541. https://doi.org/10.1016/j.actbio.2010.03.030

[53]   W. Shenton, T. Douglas, M. Young, G. Stubbs andS. Mann, inorganic-organic nanotube composites from template mineralization of tobacco mosaic virus. Adv. Mater. 11 (1999) 253-256. https://doi.org/10.1002/(SICI)1521-4095(199903)11:3<253::AID-ADMA253>3.0.CO;2-7

[54]   C. Mi, Y. Wang, J. Zhang, H. Huang, L. Xu, S. Wang, X. Fang, J. Fang, C. Mao andS. Xu, Biosynthesis and characterization of CdSquantum dots in genetically engineered escherichia coli. J. Biotechnol.,153 (2011) 125-132. https://doi.org/10.1016/j.jbiotec.2011.03.014

[55]   C. I. Pearce, V. S. Coker, J. M. Charnock, R. A. D.Pattrick, J. F. W.Mosselmans, N. Law, T. J. Beveridge and J. R. Lloyd, Microbial manufacture of chalcogenide-based nanoparticles via the reduction of selenite using veillonellaaatypica: an in situ EXAFS study. Nanotechnology, 19 (2008) 155603. https://doi.org/10.1088/0957-4484/19/15/155603

[56]   M. Labrenz, G. K. Druschel, T. Thomsen-Ebert, B. Gilbert, S. A. Welch, K. M.Kemner, G. A. Logan, R. E. Summons, G. De Stasio, and P. L. Bond, Formation of sphalerite (ZnS) deposits in natural biofilms of sulfate-reducing bacteria. Science, 290 (2000) 1744-1747. https://doi.org/10.1126/science.290.5497.1744

[57]   A. Y. Chen, Z. Deng, A. N. Billings, U. O. S.Seker, M. Y. Lu, R. J.Citorik, B. Zakeri andT. K. Lu, synthesis and patterning of tunable multiscale materials with engineered cells. Nat. Mater. 13 (2014) 515-523. https://doi.org/10.1038/nmat3912

[58]   J. P. Monras, V. Diaz, D. Bravo, R. A. Montes, T. G.Chasteen, I.O. Osorio-Roman, C. C. Vasquez andJ. M. Perez-Donoso, Enhanced glutathione content allows

214

the in vivo synthesis of fluorescent CdTe nanoparticles by escherichia coli. PLoS One 7 (2012) e48657. https://doi.org/10.1371/journal.pone.0048657

[59] J. W. Fellowes, R. A. D.Pattrick, J. R. Lloyd, J. M. Charnock, V. S. Coker, J. F. W.Mosselmans, T. C.Weng, and C. I. Pearce, Ex situ formation of metal selenide quantum dots using bacterially derived selenide precursors. Nanotechnology, 24 (2013) 145603. https://doi.org/10.1088/0957-4484/24/14/145603

[60] L-H. Xiong, R. Cui, Z-L. Zhang, X. Yu, Z. Xie, Y-B. Shi andD-W. Pang, Uniform fluorescent nanobioprobes for pathogen detection. ACS Nano, 8 (2014) 5116-5124. https://doi.org/10.1021/nn501174g

[61] X. Li, S. Chen, W. Hu, S. Shi, W. Shen, X. Zhang and H. Wang, In situ synthesis of CdS nanoparticles on bacterial cellulose nanofibers. Carbohydr. Polym., 76 (2009) 509-512. https://doi.org/10.1016/j.carbpol.2008.11.014

[62] C. B. Mao, C. E. Flynn, A.Hayhurst, R. Sweeney, J. F. Qi, G. Georgiou, B. Iverson and A. M. Belcher, Viral assembly of oriented quantum dot nanowires. Proc. Natl. Acad. Sci. U. S. A. 100 (2003) 6946-6951. https://doi.org/10.1073/pnas.0832310100

[63] H. Bao, N.Hao, Y. Yang and D. Zhao, Biosynthesis of biocompatible cadmium telluride quantum dots using yeast cells. Nano Res. 3 (2010) 481-489. https://doi.org/10.1007/s12274-010-0008-6

[64] M. Kowshik, W. Vogel, J. Urban, S. K. Kulkarni andK. M. Paknikar,Microbial synthesis of semiconductor PbSnanocrystallites. Adv. Mater., 14(2002) 815-818. https://doi.org/10.1002/1521-4095(20020605)14:11<815::AID-ADMA815>3.0.CO;2-K

[65] J. G. S. Mala and C. Rose, Facile production of ZnS quantum dot nanoparticles by saccharomyces cerevisiae MTCC 2918. J. Biotechnol., 170 (2014) 73-78. https://doi.org/10.1016/j.jbiotec.2013.11.017

[66] C. T. Dameron, R. N. Reese, R. K.Mehra, A. R.Kortan, P. J. Carroll, M. L.Steigerwald, L. E. Brus and Winge, Biosynthesis of cadmium sulphide quantum semiconductor crystallites. Nature,338 (1989) 596-597. https://doi.org/10.1038/338596a0

[67] R. Y. Sweeney, C. B. Mao, X. X. Gao, J. L. Burt, A. M. Belcher, G. Georgiou and B. L. Iverson, Bacterial biosynthesis of cadmium sulfide nanocrystals. Chem. Biol.,11 (2004) 1553-1559. https://doi.org/10.1016/j.chembiol.2004.08.022

[68] A. K. Suresh,Extracellular bio-production and characterization of small monodispersed CdSe quantum dot nanocrystallites. Spectrochim. Acta, Part A, 130 (2014) 344-349. https://doi.org/10.1016/j.saa.2014.04.021

[69] G. Chen, B. Yi, G. Zeng, Q.Niu, M. Yan, A. Chen, J. Du, J. Huang andQ. Zhang, Facile green extracellular biosynthesis of CdS quantum dots by white rot fungus phanerochaetechrysosporium. Colloids Surf., B, 117 (2014) 199-205. https://doi.org/10.1016/j.colsurfb.2014.02.027

[70] A. Syed andA. Ahmad, Extracellular biosynthesis of CdTe quantum dots by the fungus fusariumoxysporum and their anti-bacterial activity. Spectrochim. Acta, Part A, 106 (2013) 41-47. https://doi.org/10.1016/j.saa.2013.01.002

[71] A. Ahmad, P. Mukherjee, D. Mandal, S.Senapati, M. I. Khan, R. Kumar and M.Sastry, Enzyme mediated extracellular synthesis of CdS nanoparticles by the fungus, fusariumoxysporum. J. Am. Chem. Soc.,124 (2002) 12108-12109. https://doi.org/10.1021/ja0272960

[72] H. Trabelsi, I. Azzouz, M. Sakly andH.Abdelmelek, Subacute toxicity of cadmium on hepatocytes and nephrocytes in the rat could be considered as a green biosynthesis of nanoparticles. Int. J. Nanomed.,8 (2013) 1121-1128. https://doi.org/10.2147/IJN.S39426

[73] H. Trabelsi, I. Azzouz, S.Ferchichi, O.Tebourbi, M. Sakly andH.Abdelmelek, Nanotoxicological evaluation of oxidative responses in rat nephrocytes induced by cadmium. Int. J. Nanomed.,8 (2013) 3447-3453. https://doi.org/10.2147/IJN.S49323

[74] L. Tan, A. Wan andH. Li, Synthesis of near-infrared quantum dots in cultured cancer cells. ACS Appl. Mater. Interfaces,6 (2014) 18-23. https://doi.org/10.1021/am404534v

[75] A. Kumar andV. Kumar, Biotemplated inorganic nanostructures: supramolecular directed nanosystems of semiconductor(s)/metal(s) mediated by nucleic acids and their properties. Chem. Rev.,114 (2014) 7044-7078. https://doi.org/10.1021/cr4007285

[76] N. Ma, G. Tikhomirov and S. O. G.; Kelley, Nucleic acid-passivated semiconductor nanocrystals: biomoleculartemplating of form and function. Acc. Chem. Res., 43 (2010) 173-180. https://doi.org/10.1021/ar900046n

[77] S. Hinds, B. J. Taft, L.Levina, V.Sukhovatkin, C. J. Dooley, M. D. Roy, D. D.MacNeil, E. H. Sargent and S. O. Kelley, Nucleotidedirected growth of

Materials Research Forum LLC
https://doi.org/10.21741/9781644901571-7

semiconductor nanocrystals. J. Am. Chem. Soc., 128 (2006) 64-65. https://doi.org/10.1021/ja057002+

[78] L. Berti andG. Burley, Nucleic acid and nucleotide-mediated synthesis of inorganic nanoparticles. Nat. Nanotechnol.,3 (2008) 81-87. https://doi.org/10.1038/nnano.2007.460

[79] G. Tikhomirov, S. Hoogland, P.E. Lee, A. Fischer, E. H. Sargent andS. O. Kelley, DNA-based programming of quantum dot valency, self-assembly and luminescence. Nat. Nanotechnol.,6 (2011) 485-490. https://doi.org/10.1038/nnano.2011.100

[80] L. Gao and N. Ma, DNA-templated semiconductor nanocrystal growth for controlled DNA packing and gene delivery. ACS Nano,6 (2012) 689-695. https://doi.org/10.1021/nn204162y

[81] Q. B. Wang, Y. Liu, Y. G. Ke and H. Yan, Quantum dot bioconjugation during core-shell synthesis, Angew. Chem., Int. Ed., 47 (2008) 316-319. https://doi.org/10.1002/anie.200703648

[82] Z. Deng, A. Samanta, J. Nangreave, H. Yan and Y. Liu, Robust DNA-functionalized core/shell quantum dots with fluorescent emission spanning from UV–vis to near-IR and compatible with DNA-directed self-assembly, J. Am. Chem. Soc., 134 (2012) 17424-17427. https://doi.org/10.1021/ja3081023

[83] A. Samanta, Z. Deng and Y. Liu, Infrared emitting quantum dots: DNA conjugation and DNA origami directed self-assembly. Nanoscale,6 (2014) 4486-4490. https://doi.org/10.1039/C3NR06578B

[84] Y. Xu, D. W. Piston and C. H. Johnson, A bioluminescence resonance energy transfer (BRET) system: application to interacting circadian clock proteins, Proc. Natl. Acad. Sci. U.S.A. 96 (1999) 151-156. https://doi.org/10.1073/pnas.96.1.151

[85] M-K. So, C. Xu, A. M.Loening, S. S. Gambhir and J. Rao, Self-illuminating quantum dot conjugates for in vivo imaging, Nat. Biotechnol. 24 (2006) 339-343. https://doi.org/10.1038/nbt1188

[86] N. Ma, A. F. Marshall and J. H. Rao, Near-infrared light emitting luciferase via biomineralization. J. Am. Chem. Soc.,132 (2010) 6884-6885. https://doi.org/10.1021/ja101378g

[87] H-Y. Yang, Y-W. Zhao, Z-Y. Zhang, H-M. Xiong and S-N. Yu, One-pot synthesis of water-dispersible Ag_2S quantum dots with bright fluorescent emission in the

second near-infrared window. Nanotechnology,24 (2013) 055706. https://doi.org/10.1088/0957-4484/24/5/055706

[88] J. Chen, T. Zhang, L. Feng, M. Zhang, X. Zhang, H. Su and D. Cui, Synthesis of Ribonuclease-A conjugated Ag_2S quantum dots clusters via biomimetic route. Mater. Lett.,96 (2013) 224-227. https://doi.org/10.1016/j.matlet.2012.11.067

[89] C. Zhang, J. Xu, S. Zhang, X. Ji and Z. He, One-pot synthesized DNA-CdTe quantum dots applied in a biosensor for the detection of sequence-specific oligonucleotides. Chem. - Eur. J.,18 (20120 8296-8300. https://doi.org/10.1002/chem.201200107

[90] X. He andN. Ma, Biomimetic synthesis of fluorogenic quantum dots for ultrasensitive label-free detection of protease activities. Small,9 (2013) 2527-2531. https://doi.org/10.1002/smll.201202570

[91] J. M. Perez-Donoso, J. P. Monras, D. Bravo, A. Aguirre, A. F. Quest, I. O. Osorio-Roman, R. F.Aroca, T. G. Chasteen andC. C. Vasquez, biomimetic, mild chemical synthesis of CdTe-GSH quantum dots with improved biocompatibility. PLoS One 7 (2012) e30741. https://doi.org/10.1371/journal.pone.0030741

[92] S. S. Narayanan, R. Sarkar and S. K. Pal, Structural and functional characterization of enzyme-quantum dot conjugates: covalent attachment of CdS nanocrystal to alpha-chymotrypsin. J. Phys. Chem. C,111 (2007) 11539-11543. https://doi.org/10.1021/jp072636j

[93] J. H. Choi, K. H. Chen andM. S. Strano, Aptamer-capped nanocrystal quantum dots: a new method for label-free protein detection. J. Am. Chem. Soc.,128 (2006) 15584-15585. https://doi.org/10.1021/ja066506k

[94] T. Jamieson, R. Bakhshi, D. Petrova, R. Pocock, M. Imani, A. M. Seifalian, Biological applications of quantum dots, Biomaterials, 28 (2007) 4717-4732. https://doi.org/10.1016/j.biomaterials.2007.07.014

[95] B. A. Kairdolf, A. S. Smith, T. H. Stokes, M. D. Wang, A. N. Young, S. Nie, Semiconductor quantum dots for bioimaging and biodiagnostic applications, Annu Rev Anal Chem (Palo Alto Calif)., 6 (2013) 143-162. https://doi.org/10.1146/annurev-anchem-060908-155136

[96] H. Kuang, Y. Zhao, W. Ma, L. Xua, L. Wang, C. Xu, Recent developments in analytical applications of quantum dots, Trends Analyt Chem., 30 (2011) 1620-1636. https://doi.org/10.1016/j.trac.2011.04.022

[97] K. D. Wegner, N. Hildebrandt, Quantum dots: bright and versatile in vitro and in vivo fluorescence imaging biosensors, ChemSoc Rev., 21 (2015) 4792-834. https://doi.org/10.1039/C4CS00532E

[98] H. C. Ishikawa- Ankerhold, R. Ankerhold, G. P. Drummen, Advanced fluorescence microscopy techniques-FRAP, FLIP, FLAP, FRET and FLIM,Molecules., 17 (2012) 4047-4132. https://doi.org/10.3390/molecules17044047

[99] U. Resch-Genger, M. Grabolle, S. Cavaliere-Jaricot, R. Nitschke, T. Nann, Quantum dots versus organic dyes as fluorescent labels, Nat Methods., 5 (2008) 763-775. https://doi.org/10.1038/nmeth.1248

[100] A. C. Vinayaka, M. S. Thakur, Photoabsorption and resonance energy transfer phenomenon in CdTe-protein bioconjugates: an insight into QD-biomolecular interactions, Bioconjug Chem., 22 (2011) 968-975. https://doi.org/10.1021/bc200034a

[101] R. Gill, I. Willner, I. Shweky and U. Banin, Fluorescence Resonance Energy Transfer in CdSe/ZnS-DNA conjugates: probing hybridization and DNA Cleavage. J. Phys. Chem. B.109 (2005) 23715-23719. https://doi.org/10.1021/jp054874p

[102] M. O. Noor, A. Shahmuradyan andU. J.Krull, Paper-based solid-phase nucleic acid hybridization assay using immobilized quantum dots as donors in fluorescence resonance energy transfer. Anal. Chem.,85 (2013) 1860-1867. https://doi.org/10.1021/ac3032383

[103] R. Freeman, R. Gill, I.Shweky, M. Kotler, U. Banin andI. Willner, Biosensing and probing of intracellular metabolic pathways by NADH-sensitive quantum dots. Angew. Chem., Int. Ed..48 (2009) 309-313. https://doi.org/10.1002/anie.200803421

[104] W. R. Algar, A. P. Malanoski, K. Susumu, M. H. Stewart, N. Hildebrandt and I. L. Medintz, Multiplexed tracking of protease activity using a single color of quantum dot vector and a time-Gated Förster Resonance Energy Transfer relay, Anal. Chem., 84 (2012) 10136-10146. https://doi.org/10.1021/ac3028068

[105] M. O. Noor and U. J. Krull, Paper-based solid-phase multiplexed nucleic acid hybridization assay with tunable dynamic range using immobilized quantum dots as donors in fluorescence resonance energy transfer. Anal. Chem.,85 (2013) 7502-7511. https://doi.org/10.1021/ac401471n

[106] X. Chu, X. Dou, R. Liang, M. Li, W. Kong, X. Yang, J. Luo, M. Yang and M. Zhao, A self-assembly aptasensor based on thick-shell quantum dots for sensing of ochratoxin A, Nanoscale, 8 (2016) 4127. https://doi.org/10.1039/C5NR08284F

[107] N. Boute, The use of resonance energy transfer in high-throughput screening: BRET versus FRET. Trends PharmacolSci, 23 (2002) 351-4. https://doi.org/10.1016/S0165-6147(02)02062-X

[108] K. Matthiesenand J. Nielsen, Cyclic AMP control measured in two compartments in HEK293 cells: phosphodiesterase K(M) is more important than phosphodiesterase localization. PLoS One, 6 (2011) e24392. https://doi.org/10.1371/journal.pone.0024392

[109] Y. Percherancier, Bioluminescence resonance energy transfer reveals ligand-induced conformational changes in CXCR4 homo- and heterodimers. J BiolChem, 280 (2005) 9895903. https://doi.org/10.1074/jbc.M411151200

[110] M. Kumar, D. Zhang, D. Broyles and S. K. Deo, A rapid, sensitive, and selective bioluminescence resonance energy transfer (BRET)-based nucleic acid sensing system, Biosens. Bioelectron., 30 (2011) 133-139. https://doi.org/10.1016/j.bios.2011.08.043

[111] Z. Xia, Y. Xing, M-K.So, A. L.Koh, R. Sinclair and J. Rao, multiplex detection of protease activity with quantum dot nanosensors prepared by intein-mediated specific bioconjugation, Anal. Chem.,80 (2008) 8649-8655. https://doi.org/10.1021/ac801562f

[112] C-Y. Hsu, C-W. Chen, H-P. Yu, Y-F. Lin and P-S. Lai, Bioluminescence resonance energy transfer using luciferase-immobilized quantum dots for self-illuminated photodynamic therapy, Biomaterials 34 (2013) 1204-1212. https://doi.org/10.1016/j.biomaterials.2012.08.044

[113] X. Y Huang andJ. C. Re, Nanomaterial-based chemiluminescence resonance energy transfer: a strategy to develop new analytical methods. Trac-Trends Anal Chem, 40 (2012) 77-89. https://doi.org/10.1016/j.trac.2012.07.014

[114] M. Amjadi, J. L. Manzoori, T. Hallaj, M. H. Sorouraddin, Strong enhancement of the chemiluminescence of the cerium(IV)-thiosulfate reaction by carbon dots, and its application to the sensitive determination of dopamine. MicrochimActa, 181(2014)671-677. https://doi.org/10.1007/s00604-014-1172-2

[115] S. Q. Han, J. B. Wang and S. Z. Ji, Determination of formaldehyde based on the enhancement of the chemiluminescence produced by CdTe quantum dots and hydrogen peroxide. MicrochimActa, 181(2014)147-153. https://doi.org/10.1007/s00604-013-1083-7

Bioinspired Nanomaterials
Materials Research Foundations **111** (2021) 185-223

Materials Research Forum LLC
https://doi.org/10.21741/9781644901571-7

[116] L. Chen and H. Y. Han, Recent advances in the use of near-infrared quantum dots as optical probes for bioanalytical, imaging and solar cell application. MicrochimActa, 181(2014)1485-1495. https://doi.org/10.1007/s00504-014-1204-y

[117] F. Liu, W. P. Deng, Y. Zhang, S. G. Ge, J. H. Yu, M. Yan, Highly sensitive hybridization assay using the electrochemiluminescence of an ITO electrode, CdTe quantum dots functionalized with hierarchical nanoporousPtFe nanoparticles, and magnetic graphene nanosheets. MicrochimActa, 181(2014)213-222. https://doi.org/10.1007/s00604-013-1102-8

[118] X. Y. Huang,L. Li, H. F. Qian, C. Q. Dong and J. C. Ren, A resonance energy transfer between chemiluminescent donors and luminescent quantum-dots as acceptors (CRET). AngewChemInt Ed, 45(2006) 5140-5143. https://doi.org/10.1002/anie.200601196

[119] Z. Li, Y. X. W, G. X. Zhang, W. B. Xu and Y. J. Han, Chemiluminescence resonance energy transfer in the luminol-CdTe quantum dots conjugates. J Lumin, 130(2010) 995-999. https://doi.org/10.1016/j.jlumin.2010.01.013

[120] Z. M. Zhou, Y. Yu, Y. D. Zhao, A new strategy for the detection of adenosine triphosphate by aptamer/quantum dot biosensor based on chemiluminescence resonance energy transfer. Analyst, 137(2012)4262-4266. https://doi.org/10.1039/c2an35520e

[121] X. Q. Liu, R. Freeman, E. Golu and I. Willner, Chemiluminescence and chemiluminescence resonance energy transfer (CRET) aptamer sensors using catalytic hemin/G-quadruplexes. ACS Nano, 5(2011)7648-7655. https://doi.org/10.1021/nn202799d

[122] H. Q. Wang, Y. Q. Li, J. H. Wang, Q. Xu, X. Q. Li and Y. D. Zhao, Influence of quantum dot's quantum yield to chemiluminescent resonance energy transfer. Anal ChimActa, 610(2008) 68-73. https://doi.org/10.1016/j.aca.2008.01.018

[123] E. Golub, A. Niazov, R. Freeman, M. Zatsepin and I. Willner, Photoelectrochemical biosensors without external irradiation: probing enzyme activities and DNA sensing using hemin/G-quadruplex-stimulated chemiluminescence resonance energy transfer (CRET) generation of photocurrents. J PhysChem C, 116(2012)13827-13834. https://doi.org/10.1021/jp303741x

[124] R. Freeman, X. Liu and I.Willner, Chemiluminescent and chemiluminescence resonance energy transfer (CRET) detection of DNA, metal ions, and aptamers

substrate complexes using hemin/G-quadruplexes and CdSe/ZnSquantum dots, J. Am. Chem. Soc. 133 (2011) 11597-11604. https://doi.org/10.1021/ja202639m

[125] X. Liu, R. Freeman, E. Golub and I.Willner, chemiluminescence and chemiluminescence resonance energy transfer (CRET) aptamer sensors using catalytic hemin/G-quadruplexes, ACS Nano, 5 (2011) 7648-7655. https://doi.org/10.1021/nn202799d

[126] A.Niazov, R. Freeman, J.Girsh and I.Willner, Following glucose oxidase activity by chemiluminescence and chemiluminescence resonance energy transfer (CRET) processes involving enzyme-DNAzymeconjugates, Sensors, 11 (2011) 10388-10397; https://doi.org/10.3390/s111110388

[127] X. Michalet, F. F. Pinaud, L. A. Bentolila, Quantum dots for live cells, in vivo imaging, and diagnostics, Science., 307 (2005) 538-544. https://doi.org/10.1126/science.1104274

[128] S. W. Lee, C. Mao, C. E. Flynn, Ordering of quantum dots using genetically engineered viruses, Science., 296 (2002) 892-895. https://doi.org/10.1126/science.1068054

[129] J. Li, J. J. Zhu, Quantum dots for fluorescent biosensing and bio-imaging applications, Analyst., 2013, 138, 2506–2515. https://doi.org/10.1039/c3an36705c

[130] Y. Park, S. Jeonga, S. Kim, Medically translatable quantum dots for biosensing and imaging, J PhotochemPhotobiol C., 30 (2017) 51-70. https://doi.org/10.1016/j.jphotochemrev.2017.01.002

[131] I. Martinic, S. V. Eliseeva, S. Petoud, Near-infrared emitting probes for biological imaging: Organic fluorophores, quantum dots, fluorescent proteins, lanthanide(III) complexes and nanomaterials, J Lumin., 189 (2017) 19-43. https://doi.org/10.1016/j.jlumin.2016.09.058

[132] A. P. Alivisatos, semiconductor clusters, nanocrystals, and quantum dots, Science., 271 (1996) 933-937. https://doi.org/10.1126/science.271.5251.933

[133] H. S. Choi, W. Liu, P. Misra, Renal clearance of quantum dots, Nat Biotechnol. 25 (2007) 1165-1170. https://doi.org/10.1038/nbt1340

[134] R. Wang, F. Zhang, NIR luminescent nanomaterials for biomedical imaging,J Mater Chem B., 2 (2014) 2422-2443. https://doi.org/10.1039/c3tb21447h

[135] E. Hemmer, A. Benayas, F. Legare, Exploiting the biological windows: current perspectives on fluorescent bioprobes emitting above 1000 nm,Nanoscale Horiz., 1 (2016) 168-184. https://doi.org/10.1039/C5NH00073D

[136] J. Zhao, D. Zhong, S. Zhou, NIR-I-to-NIR-II fluorescent nanomaterials for biomedical imaging and cancer therapy,J Mater Chem B., 6 (2018) 349-365. https://doi.org/10.1039/C7TB02573D

[137] K-T. Yong, I. Roy, R. Hu, H. Ding, H.Cai, J. Zhu, X. Zhang, E. J. Bergeya and P. N. Prasada, Synthesis of ternary CuInS2/ZnS quantum dot bioconjugates and their applications for targeted cancer bioimaging, ntegr. Biol., 2 (2010) 121-129. https://doi.org/10.1039/b916663g

[138] D. Hu, P. Zhang, P. Gong, S.Lian, Y. Lu, D. Gao and L.Cai, A fast synthesis of near-infrared emitting CdTe/CdSe quantum dots with small hydrodynamic diameter for in vivo imaging probes, Nanoscale, 3 (2011) 4724-4732. https://doi.org/10.1039/c1nr10933b

[139] Y. T. Lim, Y-W. Noh, J-H. Cho, J. H. Han, B. S. Choi, J. Kwon, K. S. Hong, A.Gokarna, Y-H. Cho and B. H. Chung, Multiplexed imaging of therapeutic cells with multispectrally encoded magnetofluorescent nanocomposite emulsions, J. Am. Chem. Soc. 131 (2009) 17145-17154. https://doi.org/10.1021/ja904472z

[140] T. Deng, Y. Peng, R. Zhang, J. Wang, J. Zhang, Y.Gu, D. Huang and D. Deng, Water-solubilizing hydrophobic ZnAgInSe/ZnS QDs with tumortargetedcRGD-sulfobetaine-PIMA-histamine ligands via a self-assembly strategy for bioimaging, ACS Appl. Mater. Interfaces,9 (2017) 11405-11414. https://doi.org/10.1021/acsami.6b16639

[141] J. Zhang, G.Hao, C. Yao, J.i Yu, J. Wang, W. Yang, C. Hu and B. Zhang, Albumin-mediated biomineralization of paramagnetic NIR Ag2S QDs for tiny tumor bimodal targeted imaging in vivo, ACS Appl. Mater. Interfaces,8 (2016) 16612-16621. https://doi.org/10.1021/acsami.6b04738

[142] S.Jeong, N. Won, J. Lee, J. Bang, J.Yoo, S. G. Kim, J. Ah Chang, J. Kim and S. Kim, Multiplexed near-infrared in vivo imaging complementarily using quantum dots and upconverting NaYF4:Yb^{3+},Tm^{3+} nanoparticles, Chem. Commun., 47 (2011) 8022-8024. https://doi.org/10.1039/c1cc12746b

Bioinspired Nanomaterials
Materials Research Foundations **111** (2021) 224-263

Materials Research Forum LLC
https://doi.org/10.21741/9781644901571-8

Chapter 8

Bio-Mediated Synthesis of Nanomaterials for Electrochemical Sensor Applications

Ponnaiah Sathish Kumar[1], Vellaichamy Balakumar[1,2,*] and Ramalingam Manivannan[3]

[1]Department of Chemistry&National Centre of Excellence (MHRD), Thiagarajar College, Madurai-625 009, Tamilnadu, India

[2]Department of Earth Resources Engineering, Faculty of Engineering, Kyushu University, 744 Motooka, Nishiku, Fukuoka 819-0395, Japan

[3]Department of Advanced Organic Materials Engineering, Chungnam National University, 220 Gung-dong, Yuseong-gu, Daejeon, 305-764, South Korea

*chembalakumar@gmail.com, kumar@mine.kyushu-u.ac.jp

Abstract

The bio-mediated nanomaterials have expected growing responsiveness due to an increasing requirement to develop naturally nonthreatening technologies in nanomaterial synthesis. Biotic ways to prepare nanomaterials through extracts from the plant (includes stems, leaves, flowers, and roots) and microorganisms were recommended as likely replacements for physical and chemical routes due to their solvent medium and environment eco-friendliness and nontoxicity. This chapter focuses on electrocatalyst prepared by various bio-mediated synthetic ways and used as a green and eco-friendly electrocatalyst to recognize extensive chemical and biologically essential molecules with improved selectivity and sensitivity with low detection limit. The bio-mediated nanocomposite formation processes and their unique properties surface functionalization and electron transfer mechanism discussed in connection with the design and fabrication of sensors. As a final point, the encounters and prospects in developing bio-mediated nanomaterials-based electrochemical sensing technology was outlined.

Keywords

Bio-Mediated Synthesis, Nanomaterials, Electrochemical Sensor, Electron Transfer Mechanism, Chemical and Biological Molecules

Contents

1.   Introduction

In the earlier years, an investigation into nanotechnologies, nanoscience has developed exclusively and encouraged a greater number of the scientific community and technological fields, which is about one or more dimensions of the order of 1 to 100 nm [1, 2]. For the first time on December 29, 1959, the physicist Richard Feynman described in detail in his talk about the perception of nanotechnology and molecular machines constructed through atomic accuracy in his speech titled "There is Plenty of Room at the Bottom" at the California Institute of Technology [3, 4, 5]. The term nanotechnology indicates depicting, gathering, controlling, and exploiting structures to control shape and size in nanoscale [6]. Nanoparticles have excellent dissimilar kinds of stuff to that of analogous material in bulk and extended consideration due to their exclusive morphology as well as physiochemical stuff such as tiny size, shape (needles, prisms, disks, leaves, cubes, spheres, sheets, flowers, rods, wires, belts, and tubes), and proper distribution of size [7, 8, 9]. There are several chemical and physical approaches for the preparation of objects from nano-particles such as chemical auto-combustion [10], sol-gel [11], conventional ceramic process [12], RF-sputtering [13], chemical reduction [14], micro-emulsion [15], reverse micelles [16], electrochemical reduction [17], Langmuir–Blodgett [18], microwave [19], UV irradiation [20], pyrolysis [21], lithography [22], and laser ablation [23]. However, there is a thoughtful necessity to swap the present methods with an uncontaminated, no-toxicity, and naturally tolerable green chemistry approach [24]. A biologically benevolent solvent and sustainable reducing and capping agents are the three dynamic essentials for a widespread green synthesis approach [25].

Nowadays, a growing need to develop sustained, facile, economical, and eco-friendly bio mediated ways for the nanomaterials synthesis and produced great importance due to the promptly growing demand for ecological fortification [26, 27, 28]. In this regard, the development of a facile and green protocol to fabricate bio-mediated synthesis of nanoparticles (bacteria, algae, fungi, yeasts, actinomycetes, and plant (flower, peel, root, leaf) extracts or specific biomolecules from plants (including the mixture of terpenoids, lipids, polyphenols, flavonoids, saponins, essential oils, starch, amines, alkaloids, proteins, and carbohydrates) is highly needed [29, 30, 31]. These nanomaterials display their inherent electronic property and extraordinary surface-to-volume proportions, explored for several applications like environmental remediation, optics, catalysis, energy science, water treatment, bioengineering, metal-based consumer products, space industry, chemical industries, cancer therapeutics, single-electron transistors, electronics, light emitters, nonlinear optical devices, labels for cells and biomolecules especially for sensors [32, 33, 34, 35]. Bio-mediated nanomaterials growth has increased attention for numerous applications, comprising the essential biotic investigation, checking of healthiness, clinical diagnostics, pharmacological examination, and food safety [36, 37, 38, 39, 40]. The most important assembly of toxic substances is heavy metals, hazardous pesticides, anions mostly inorganic, substituted phenol complexes, pharmaceutical analysis, and chemical warfare reagents [41, 42]. Most of all, they have the toxicity that can posture health hazards to a living organism, which including cancer, heart disease, affects the central nervous system, high cholesterol, high blood pressure, neurogenic infection, and organ dysfunction, so that the accurate assessment detection of them becomes crucial in the present way of life [43, 44]. Hence, an appropriate method, it is

Bioinspired Nanomaterials   Materials Research Forum LLC
Materials Research Foundations **111** (2021) 224-263   https://doi.org/10.21741/9781644901571-8

essential to pursue real-time tools to perceive even small quantities of toxic chemicals [45]. Up to now, different techniques, including absorption spectrophotometry [46, 47], fluorescence spectroscopy [48, 49], and electrochemical method [50, 51] have been established for the detection of toxic chemicals. However, the analysis includes an electrochemical technique, an effective strategy for detection, since accuracy, small sample volumes, credibility, high sensitivity and selectivity, fast response time, portable, stability, instrumental simplicity, and low cost [52, 53, 54]. Different scientific papers and reviews describe nanoscale preparation of electrodes as crucial in presenting electrochemical detecting display place to identify target analytes [55]. Here, numerous approaches have been advanced to make electrode modifications, including drop-casting, electrospray, electrodeposition, and electrospun [56]. Very notably, the developed nanoscale electrode material is established to be a superior candidate through the extreme surface area showing to guest modules environmental sensor applications [57]. Still, some typical disadvantages are noticed, counting mass, electron transport that is deleteriously obstructed with functional and stabilization molecules, poor stability during sensing process, capable of endorsing the kinetics of electron transmission responses, and the recurrent manifestation of dissolution and nanomaterials aggregation [58, 59]. Moreover, the electrode surfaces direct growth of the nanocomposite-taking place has a capable purpose for the above-discussed problems [60].

This chapter mainly concentrates on synthesizing bio-mediated nanocomposites that have been fabricated and applied for the electrochemical determination of toxic chemicals. The current decision is to deliver an impeccable and transitory visualization of research, things of bio-mediated nanocomposites electrochemical performances towards toxic chemicals including, heavy metals, hazardous pesticides, anions (inorganic), phenolic moieties, pharmaceutical analysis, and chemical warfare reagents. Furthermore, we briefly highlight the recent development of bio-mediated synthesis nanocomposite used to remediate the improved electrochemical sensors applications, which could stimulate more comprehensive benefits in several parts.

2.   Why bio-mediated synthesis is more important

Bio-mediated nanocomposite synthesis is considered a key role in redefining the present environmental remediation and science field.

Wolfgang Pauli famous quote from that expressed long ago:

"God made the bulk; the surface was invented by the Devil".

By this decision, in materials nano-sized, the Devil's monarchy significantly prolonged. One significant outcome of the excellent surface area/unit mass is the reactivity of the nano-size materials. If for illustration, the nanoparticles surface not secure over surface passivation, relations readily happen between the particles [61]. The green-synthesis have several advantages mainly it is a replacement for toxic and expensive substance. This green approach (bacteria, algae, fungi, yeasts, actinomycetes, and plant (flower, peel, root, leaf, fruit, seed) a consistent, harmless, unsoiled, biocompatible, and environmentally friendly process [62]. This is likely as this bio mediated contains bioactive components. Typically, flavonoids, terpenoids, phenolic acids, alkaloids, alcohol, polyphenols, vitamins, antioxidants by plants, enzymes, proteins, NADPH, peptides, NADH, nitrogenous biomacromolecules by fungi and cellulose, chitosan, lignin, alginate, polypeptides, protein by biopolymers that act as supporting/reducing/capping agents, while still tolerating regulator is the fundamental properties at lab scale for the nanomaterials such as porosity, size shape and crystallinity [63].

In the case of synthesis, bio-mediated can be classified as:

(a) Bacteria (b) fungi, (c) yeasts (d) biopolymers and (e) from plants and their extracts.

2.1 Bacteria

Synthesis of metallic and other novel nanoparticles by microbial classes broadly employed for viable biotechnological uses. Such as bioleaching, bioremediation, and genetic engineering [64]. The bacteriological straining choice that has been widely subjugated for the bio mediated preparation of reduced metal/ metal oxide nanoparticles and a suitable process is vital for their enormous scale manufacture [65]. In 1982, Pooley defined that bacteria aggregates silver on the bacterial cell wall. Suggested that the usage of bacteria toward technologically recuperate silver from its ore [66]. Klaus and coworkers in 1999 made AgNPs of 200 nm size employing the biomass of the bacteria *Pseudomonas stutzeri* AG259 Ag resistant [67]. For the research of nanoparticles, a wide range of bacterial species are used. Which includes *Aeromonas sp. SH10 Phaeocystisantarctica, Lactobacillus casei, Bacillus cereus, Escherichia coli Bacillus amyloliquefaciens, Pseudomonas proteolytica, Bacillus cecembensis, Bacillus indicus, Geobacter spp., Shewanella alga, Arthrobactergangotriensis, Enterobacter cloacae, Corynebacterium sp. SH09, Bacillus megaterium D01, Shewanellaoneidensis, Bacillus subtilis 168, Desulfovibriodesulfuricans, E. coli DH5a, Rhodopseudomonas capsulate, and Plectonemaboryanum* [68]. Kalimuthu and coworkers in 2008 described the production of AgNPs from the biomass of B. licheniformis with 50 nm size by the addition of silver nitrate aqueous solution [69]. Mokhtari coworkers in 2009 described the synthesis of AgNPs via photosynthesis with a size of 3 nm by the accumulation of silver nitrate solution to the culture supernatant of *Klebsiella pneumonia*, and prompted visible-light irradiation [70]. However, associated with conservative approaches, the exploitation of bacteria to reduce metal/ metal oxide leads to a slow development degree and a partial array of forms [71].

2.2 Fungi

The other excellent choice of microorganisms aimed at the biosynthesis of metal/metal oxide nanoparticles is fungi and an actual well-organized procedure for monodispersed nanoparticles' compeers through distinct morphology due to their massive kind of benefits over plants, yeast, bacteria, and physicochemical techniques [72, 73]. The possible mechanism aimed at forming nanoparticles (metallic) is secreting enzymes and proteins to reduce metal salts in the cell wall or inside the fungal cell. A significant measure is taken to develop metallic nanoparticles from discrete fungal strains inferred due to their t uniform in vitro activity [74]. Fungi-mediated (Rhizopusnigricans and Aspergillus flavus TFR7) synthesis of silver NPs [75] and TiO_2 [76] also one of the precise, effective procedures for the generation of monodispersed nanoparticles owing to the existence of enzymes, proteins, and reducing components on their cell surfaces. Xue B. et al. (2016) reported synthesizing AgNPs substrate concentration of 1.5 mM in basic pH, with 55 °C as reaction temperature using the fungal strain of Arthrodermafulvum. The silver nanoparticle synthesized brings into being crystallinity in nature with an optimized particle size of approx. 1.5 to 2.5 nm. Antifungal action was detected in contradiction of fungal strains, including Aspergillus, Candida Fusarium [77]. Honary and coworker in 2013 assessed a green synthesis process aimed at the extracellular production of silver NPs with a sphere shape having n regular diameter of 109 nm using Penicilliumcitrinum separated from soil [78].

2.3 Yeasts

Yeasts are single-cell microorganisms existing in eukaryotic cells. Many research groups describe the positive synthesis of Ag and Au NPs through yeast. Otari and coworkers in 2014 manufactured AgNPs using the culture supernatant of phencl (degraded) broth utilizing the reducing agent [79]. Eugenio M. et al. (2016) reported the biosynthesis of Ag NPs using yeast strains [80]. Ishida et al. in 2014 deliberate the combination of AgNPs exploiting the aqueous extract of the fungus Fusariumoxysporum and examined for antifungal activity [81]. Ag and Au nanoparticles' biosynthesis by a silver-tolerant yeast strain and Saccharomyces cerevisiae broth has been reported [82].

Bioinspired Nanomaterials
Materials Research Foundations **111** (2021) 224-263

Materials Research Forum LLC
https://doi.org/10.21741/9781644901571-8

2.4   Biopolymers

Green synthesis of metal/ metal oxide using biopolymers shows the dual character as stabilizing and reducing agent except intended for starch usage as a capping agent [83]. A.R. Futyra et al. in 2017 described that biopolymers improved the antimicrobial activity of AgNPs with a size smaller than 10 nm via ascorbic acid and chitosan utilized as the capping/reducing agent [84]. T.C. Leung et al. (2010) synthesized AgNPs employing carboxy methylated-curdlan/fucoidan as stabilizing as well as reducing agents [85]. Firm and constant starch-stabilized AgNPs produced with an average diameter 14.4+3.3 nm utilizing D-glucose employed as the reducing agent [86]. Ahmad MB et al. (2011) synthesized AgNPs using chitosan, and polyethylene glycol (PEG) used as the polymeric stabilizer and reliable support, respectively [87].

2.5   Plants

The existence of phytochemicals in plant-based synthesis consumes weird prospective to decrease metal ions in an extended smaller period than fungi and bacteria, which demands ample incubation time [88]. Additionally, extract from plant leaf plays a significant influence in nanoparticle production. Diverse plants include variable concentration levels of phytochemicals in plants such as sugars, carboxylic acids, amides, ketones, flavones, aldehydes, and terpenoids, accountable on behalf of bioreduction of nanoparticles [89]. The existence of various functional groups in flavonoids improved the capability to reduce metal ions. Due to tautomeric transformation, the reactive hydrogen atom is released from flavonoids through which the keto form is converted from enol form. This procedure is understood by the reduction process in the formation of metal ions into metal nanoparticles. In sweet basil Ocimumbasilicum extract, the transformation of enol to keto form is vital in producing biogenic Ag nanoparticles [90]. Gardea Torresdey and coworkers in 2003 offered the info on establishing AgNPs by a living plant system Alfalfa sprouts [91]. Sithara and coworkers created AgNPs through leaf extract of Acalyphahispida as the reducing/capping agent, and these AgNPs used to recognize Mn2+ ions [92]. Gavhane and coworkers in 2012 described the usage of the extract of Neem and Triphala to make AgNPs in which the size is in the range of 43 nm to 59 nm and they were spherical in shape [93]. B. Vellaichamy et al. (2014) used Simaroubaglauca, Crataevareligiosa, Bombaxceiba, and Madhucalongifolia (MLF) leaf extract as a reducing and capping agent to synthesize AgNPs and examined for decontamination   of   hazardous   pollutants,   4-hydroxynitrobenzene   and   4-nitrophenylamine, catalyzed dedying of industrial effluents and selective and sensitive sensing of Cd2+ and saturnism (lead poisoning) colorimetrically [94, 95, 96, 97]. Velmurugan et al. in 2015 made AgNPs utilizing peanut shell extract and associated their

Bioinspired Nanomaterials
Materials Research Foundations **111** (2021) 224-263

Materials Research Forum LLC
https://doi.org/10 21741/9781644901571-8

antifungal activity [98]. All the plant extracts act as both reducing agents and capping agents in various plant origin (Acalyphaindica, Citrus limon (lemon), Cycas sp. (cycas), Eucalyptus citriodora (neelagiri), Garcinia mangostana (mangosteen), Ludwigiaadscendens (ludwigia), Morus (mulberry), Nelumbonucifera (lotus), Ocimum sanctum (tulsi; root extract). Synthesis of Ag is also a very proficient procedure for the generation of monodispersed nanoparticles due to the presence of reducing components they have been preferably employed for Antibacterial activity against waterborne pathogens, Antibacterial, Antimicrobial activity against E. coli and S. aureus, Antimicrobial activity against E. coli, B. subtilis [99, 100, 101, 102, 103, 104, 105, 106, 107]. Iravani and coworkers [108] work for a complete impression of plant materials utilized for the biosynthesis of nanoparticles. The conceptual growth of methodologies for the eco-friendly bio-mediated synthesis of nanoparticles presented in Figure. 1.

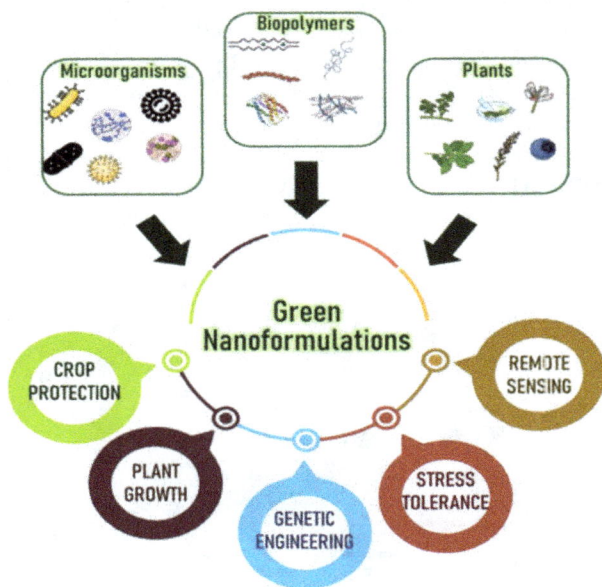

Figure 1. Illustration of bio-mediate routes for synthesis of nanoparticles. Reproduced permission from Ref. No. [34].

2.6   Mechanism for the bio-mediated synthesis

1. Electrostatic interaction between the functional groups of respective constituent of plant extract and metal ion by plants,

2. Intracellular and extracellular synthesis of metal nanoparticles by fungi

3. Electrostatic interface between metal ion and polymer with polar groups attached to biopolymers.

2.7   Bio-mediated synthesis of metal nanoparticles

For synthesis metal nanoparticles using bioactive components with water as a solvent, the phytochemicals act as capping, reduce metal ions, and stabilize nanoparticle formation agents. As proof, bacteria, algae, fungi, yeasts, actinomycetes, and plant extracts were utilized to reduce and stabilize the metallic nanoparticles in the "one-pot" synthesis process [109]. It has attracted the profound interest in crucial sensing materials for device development [110]. Hence, it is a large specific surface area, porous structure, high electron mobility, thermal, optical, and mechanical properties. Exhaustive research has been focused in recent years' on bio-mediated parts used to prepare various nanoparticles and nanocomposite, as shown in Figure 2. These synthesized metal nanoparticles have been applied in different fields like an electrochemical sensor, bio-sensing, and other various applications [111, 112, 113, 114]. Various bio-mediated nanoparticles were investigated that can be utilized for electrochemical determination of trace toxic chemicals.

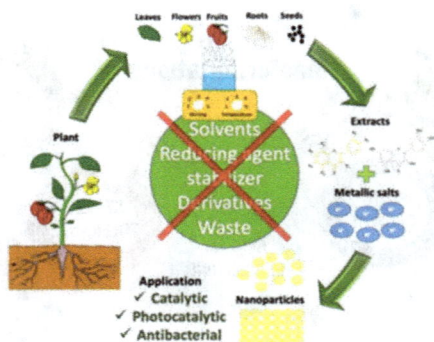

Figure 2. Shows the plants extracts mediated green synthesis of nanoparticles. Reproduced permission from Ref. No. [110].

Materials Research Forum LLC
https://doi.org/10.21741/9781644901571-8

3.   Electrochemical sensors for the determination of pollutants

Electroanalytical techniques are usually the inter-conversion between electricity and chemistry, which study an analyte by measuring the potential vs. current. Furthermore, nanostructured materials provide a unique physical and chemical property interfacing biological recognition events with electronic signal transduction [115]. A portable and straightforward electrochemical setup is used to measure the direct analysis of analyte information. Environmental detection consists of electrolyte (a solution double-layer phase ionic conductor) and electrode (an electronic conductor). The reactions occur at the interface between the working electrode and electrolyte solution, in the presence of reference and counter electrodes [116]. Finally, the sensing response and electrical excitation are connected to a portable electrochemical workstation with a power source, as shown in Figure 3. The sensing mode is mainly concerned with electrical quantities of current, potential, charge, and chemical parameters [117].   In the field of environment and medicine, electroanalytical chemistry is much fascinated and extended, and the design of molecular monolayer assembly, interfaces, the coupling of the electrochemical transducer with nanoscale materials. Lastly, chemical/biological signal into an electrical signal can be done by amperometry, voltammetry, potentiometry, or conductometry [118].

Used are many different electroanalytical methods to get various quantitative electrical signals, leading to a wide range of environmental applications based on nanomaterials. For pollutant detection, electrochemical techniques are classified into cyclic voltammetry, linear sweep voltammetry, differential pulse voltammetry, square wave voltammetry, amperometry, potentiometric techniques, and electrochemical impedance techniques are used in the sensor field [119]. All the methods, either potential or current, is controlled to determine the change of another parameter. For any electrochemical sensor, the essential parameters that play a significant role are sensitivity, selectivity, low detection limit, response time, dynamic range, linearity, and good stability and reproducibility [120].

Electroanalytical methods have spread to multiple research areas, including environmental analysis, diagnosis, enzyme kinetics, food sciences, and pharmacology [121]. Simple unmodified electrodes show less electrochemical activity, while the nanomaterials modification of bare electrodes shows better activity [122]. Different types of fabrication of electrode for electrochemical analyzer as shown in Figure. 4. One of the simple and fastest methods to fabricate electrodes is drop-casting. The nanomaterials were dispersing with water or ethanol to obtain the excellent suspension, and it was drop coated on the surface of bare electrodes. These modified electrodes have unique features, including selectivity, sensitivity, and short-time response towards various environmental

Materials Research Forum LLC
https://doi.org/10.21741/9781644901571-8

toxic hazards in detecting environmental chemicals that include heavy metals, pesticides, aromatic compounds, organic pollutants, and clinically relevant molecules. The functional nanomaterials are under extensive study, which has led to the environmental remediation by several analytical methods [123].

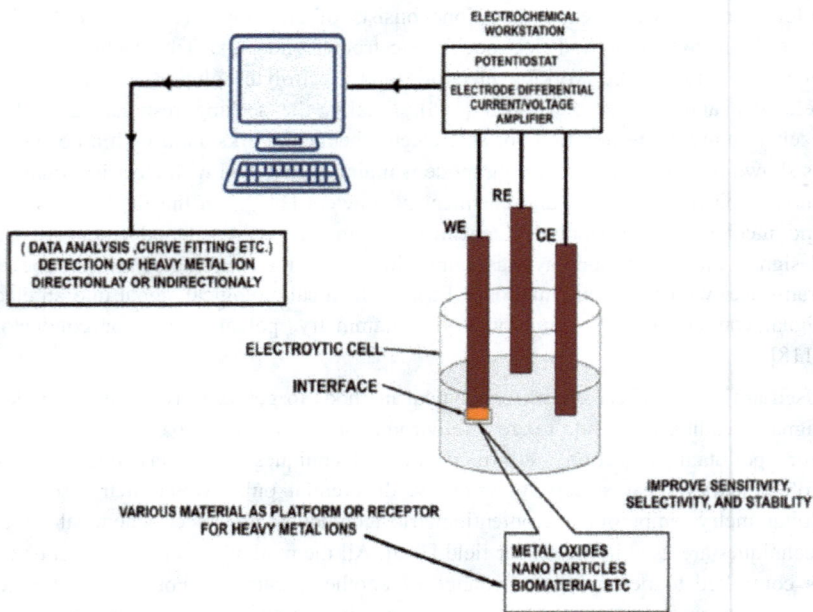

Figure 3 General electrochemical cell setup for sensors. Reproduced permission from Ref. No. [115].

Bioinspired Nanomaterials   Materials Research Forum LLC
Materials Research Foundations **111** (2021) 224-263   https://doi.org/10.21741/9781644901571-8

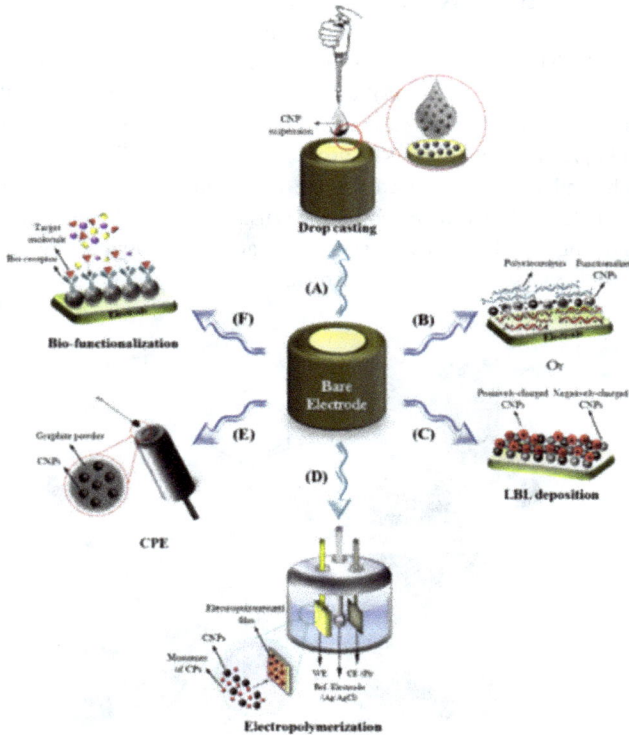

Figure. 4 Fabrication of electrode for electrochemical analyser. Reproduced permission from Ref. No. [121].

4.   Bio-mediated metal nanoparticles and its electrochemical sensing towards toxic chemicals

In sensor devices, nanometer-sized metal nanoparticles act as a crucial sensing material because of their high reactivity, surface area, and surface energy. For instance, A. Aravind et al. [124] reported that Cr(III) electrochemical sensing using Lycopersiconesculentum (LE) extract mediated AgNPs. The electrochemical detection of Cr(III) ions and 0.804 µM detection limit and the calibration range from 10 to 90 µM. M. Sebastian et al. [125] presented the silver nanoparticles from Agaricusbisporus extract (AgNP-AB) followed by modification of platinum electrode. Examined for electrochemical sensing behavior of Hg(II) ions. The LOD was found at 2.1×10^{-6} M. R. Emmanuel et al. [126] studied Acacia nilotica twig bark extract mediated green synthesis of gold nanoparticles (Au-NPs), and that can be utilized to trace level detection of nitrobenzene (NB). The developed Au-NPs/GCE displays a limit of detection of 0.016 μM with high sensitivity of $1.01 \mu A \, \mu M^{-1} cm^{-2}$ using a comprehensive 0.1 to 600 μM linear response [126]. The NB detection mechanism is shown in Figure. 5. Another study, C. Karuppiah et al., [127], reported Ag-NPs using Acacia nilotica Wild twig bark extract by green approach, and they are utilized for the detection of 4-nitrophenol (4-NP). The AgNPs modified electrode was assessed for the sensing of 4-NP. Finally, the 15 nM LOD was achieved using a wide linear response range from 100 nM-350 μM, and the $2.58 \pm 0.05 \, \mu A \mu M^{-1} \, cm^{-2}$ sensitivity was detected.

Figure. 5. Schematic electrochemical reduction of NB using green synthesis of Au-NPs. Reproduced permission from Ref. No. [126].

Justiciaglauca leaf extract mediated AgNPs were prepared and which can be used for the simple electrochemical detection of dopamine by using DPV. The LOD and high sensitivity is found as 0.017 μM and $7.85 \, \mu A \mu M^{-1} cm^{-2}$ by S. Palanisamy et al. [128]. Au-NPs were synthesized by fresh leaf extract of Justiciaglauca and the prepared nanoparticles utilized for the selective determination of Pb^{2+} via electrochemically at pH 5. The linear range and limit of detection were 0.005 to 800 μM L^{-1}, and 0.07 nM L^{-1}, respectively [129]. S.R. Dash et al. reported one-pot synthesis of AgNPs by

Psidiumguajava leaves and examined for electrochemical ascorbic acid sensor as shown in Figure. 6 [130].

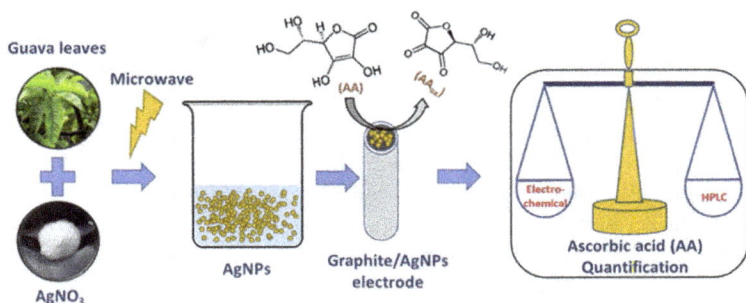

Figure. 6. Green synthesis of AgNPs from Psidiumguajava leaves and application for electrochemical reduction of ascorbic acid. Reproduced permission from Ref. No. [130].

Furthermore, G.D. Carlo et al. prepared Au−chitosan nanocomposites for electrochemical detection of caffeic acid and the limit of detection was 2.50×10^{-8} M [131].

4.1   Bio-mediated bimetallic nanoparticles and its electrochemical sensing applications

Bimetallic nanoparticles synthesis from bio-mediated is environmentally benign, less expensive, and less time-consuming. Which are of more interest than metal nanoparticles as they show better catalytic properties, optical, electrical, and medical applications due to their peculiar mixing of two metal elements and synergistic effects of two metal nanoparticles that form bimetallic [132]. The application of bimetallic nanoparticles as the sensor is the most promising due to the large surface area and small size, showing dramatic enhancement in activity, selectivity, and stability [133]. A. Pani et al. stated that electrochemical detection of Sudan dyes using peanut skin extract mediated Au-Ag bio-nano composite modified GCE. The oxidation and reduction Sudan IV concentration and detection limit were 10 to 80 μM and 4 μM, respectively [134]. P. Veerakumar et al. studied a new carbon material development by carbon precursor (Cassia fistula (golden shower)) fruit and decorated on Pt-Re NPs nanocomposite via a facile thermal reduction process. The prepared Pt-Re NPs/PAC modified GCE for furazolidone determination. The LOD and linear ranges were 75.5 nM and 1.0-299 μM, respectively [135].

4.2 Bio-mediated synthesis of various metal oxide nanoparticles and its electrochemical sensing applications

To enhance the sensitivity, selectivity, and catalytic activity via green synthesized the metal oxide nanoparticle (MO-NPs) because of high surface area. Further, MO-NPs showed an excellent candidate for the electrochemical detection of toxic chemicals because of their fast response and long lifetime [136]. A significant role of MO-NPs in electroanalysis involves direct electron transfer between the active sites of transducers and the electrode. The MO-NPs and the electrical contact of redox-centers in proteins with the transducers' surface enhance the catalytic properties [137]. N.S. Pavithra et al. [138] stated that simple and cost-effective Zinc oxide nanoparticles (ZnONps) were successfully prepared using chakkota fruit juice as novel fuel. Further, the synthesized ZnONps examined chronoamperometric quantification of dopamine and its linear range from 50-250 nM with a limit detection is 6 nM. S. Sukumar et al. [139] presented the CuONps were synthesized using Caesalpiniabonducella seed extract and used to detect riboflavin electrochemically. The modified electrode showed good stability and reproducibility and linear range of 3.13-56.3 nM with a LOD of 1.04 nM. S. Momeni et al. [140] had developed CuO/Cu_2O NPs through a simple synthesis route. The CuO/Cu_2O NPs modified carbon ionic liquid electrode was investigated towards formaldehyde oxidation. The results show a linear range from 0.1 $mmolL^{-1}$ - 110 $mmolL^{-1}$. The sensitivity, the low detection limit calculated to be 186.0 μAmM^{-1} cm^{-2}, 10 $\mu molL^{-1}$. Pechini modified method synthesized $DyMnO_3$-ZnO green-nanocomposites as shown in Figure. 7. The synthesized samples modified carbon paste electrode exhibits an excellent electrocatalytic effect for atenolol determination at pH 9.0 using voltammetry techniques. The linear ranges from 0.11 to 125.87 μM, and the LOD is 10.10 nM [141].

N. Matinise et al. reported MoringaOleifera to extract mediated synthesis of spinel $ZnFe_2O_4$ nanocomposites. The $ZnFe_2O_4$ modified GCE electrode shows good voltammetric response and high electro-activity. From EIS analysis, the time constant and exchange current of $ZnFe_2O_4$ nanocomposites were calculated, and the results are 5.2001×10^{-4} s/rad and 6.59432×10^{-4} A, respectively [142]. B.S. Surendra et al. studied that Jatropha extract was used to synthesize $ZnFe_2O_4$ nanoparticles used to detect Paracetamol, Dolo 650 Mifepristone-Misoprostol [143].

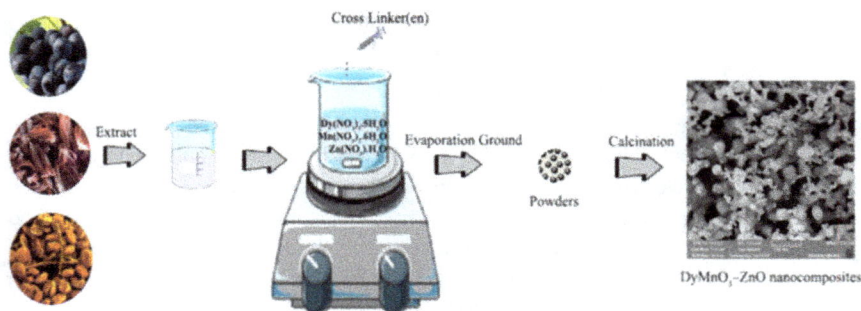

Figure. 7. Green synthesis of DyMnO₃-ZnO nanocomposite. Reproduced permission from Ref. No. [141].

4.3   Carbon based nanomaterials and its electrochemical sensing applications

Carbon-based nanomaterials (CBN) are becoming exceptional in science and technology because of their unique electronic, chemical, and mechanical properties. CBN are the most extensively used electrode materials; a decrease in over-potential can be observed with oxidized transducer owing to their catalytic property electronic conductivity [144]. Additional CBN's introduction to form the metallic-carbon bonds exhibits a robust synergistic effect due to the strong interactions between a metal oxide and CBN to enhance the electrochemical performance [145]. For example, T.V. Sathisha et al. [146] introduced CNPs based materials synthesized via a green and straightforward approach and castor oil as a natural precursor. The CNPs modified electrode was utilized for the electrochemical detection of DA, AA, and UA. Finally, the obtained results confirmed the sensitive DA determination. Furthermore, E. Vatandost and colleagues presented a simple and environmentally friendly method for synthesizing rGO using green tea extract. The electrocatalytic response of carbon paste electrodes modified rGO was also investigated toward 10 µM SY's electrochemical oxidation. The rGO/CPE shows high sensitivity (2328 µA mM^{-1}) with a linear range between 0.05-10 µM and 27 nM detection limit was achieved [147]. Recently, S. Nazarpour et al. described a novel method synthesis of reduced graphene oxide/gold nanoparticles (rGO/AuNPs) for the direct determination of L-tryptophan (Try). The green synthesized nanocomposite for selective analysis of tryptophan at 0.65 V at pH: 6.0. The linear between 0.5–500 µmol/L with LOD was calculated as 0.39 µmol/L [148]. In another study, a novel route has been introduced to the synthesis of Pd-rGO using Fenugreek seeds. The fabricated Pd-rGO based biosensor detects the triglycerides, and the result shows a wide linear range of 25 - 400 mgdL^{-1} with 1.3 µA mgdL^{-1}cm^{-1} sensitivity [149].

C. Karuppiah et al. reported that Justiciaglauca leaf extract mediated AgNPs were synthesized via the green method. The prepared AgNPs-RGO modified GCE was utilized for the detection of nitrobenzene (NB) electrochemically. The AgNPs-RGO modified electrode shows an electrocatalytic reduction of NB with good efficiency compared to other electrodes. The detection limit is 0.261 µM [150]. S. Palanisamy et al. investigated a green synthesized Au-NPs with reduced graphene oxide (RGO) using Terminalia chebula extract AuNPs-RGO showed good electrochemical behavior towards hydroquinone, catechol, and resorcinol [151].

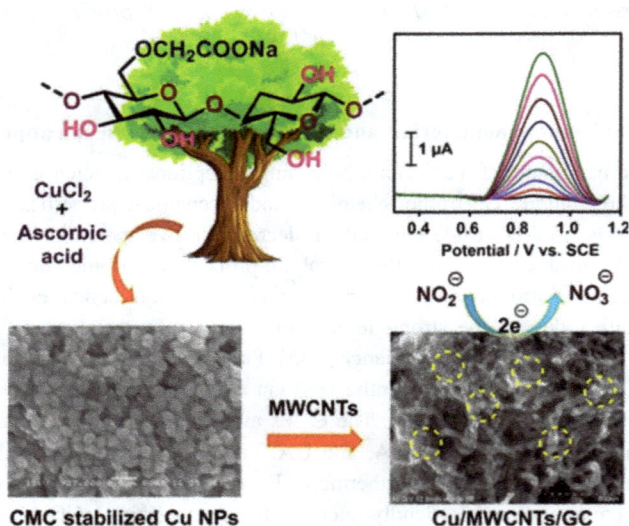

Figure. 8. Green synthesis of Cu/MWCNTs and application for electrochemical nitrite sensor. Reproduced permission from Ref. No. [111].

D. Sharma et al. documented an average size of 4-8 nm ZnO NPs using Carica papaya seed extract. To investigate the electrochemical application, ZnO NPs synthesized by integrating them with MWCNTs on GCE were further tested for silymarin sensing

Bioinspired Nanomaterials   Materials Research Forum LLC
Materials Research Foundations **111** (2021) 224-263   https://doi.org/10.21741/9781644901571-8

activities. The square wave voltammogram data was used for calculating the LOD of 0.08 mg L^{-1} [152]. P. Salazar et al. reported an electrochemical reduction of H_2O_2 using a nanocomposite of silver nanoparticles with reduced graphene oxide (rGox/AgNPs). This nanocomposite was synthesized using reduction of both silver ion and GO by the green method and a linear range of 0.002-20 mM and a LOD of 0.73 mM with a sensitivity of 236 μA $mM^{-1}cm^{-2}$ were achieved. In addition, without significant loss of sensitivity, reproducibility was achieved [153]. D. Manoj et al. had developed a carboxymethylcellulose (CMC) stabilized Cu nanoparticle was dispersed with MWCNTs. The resultant dispersion has been fabricated over GCE to obtain Cu/MWCNTs modified GCE and nitrite detection as shown in Figure. 8. The Cu/MWCNTs/GCE displays a good detection limit of 1.8 μM and sensitivity of 455.84 μA mM^{-1} cm^{-2} [111].

Green synthesis of multiwalled carbon nanotube-graphene hybrid with ZnO nanoparticle (ZnO-MWCNTs-sG) was investigated and its application towards organophosphorus biosensor. The detection limit is 1 pM [154]. R.K. Das et al. had developed a spherical AgNPs and rGO via bio-mediated route and examined for electrochemical sensing of H_2O_2. An amperometric method detection limit was found at 19.04 μM with high H_2O_2 concentration [155]. Further, T. Dodevska and coworkers have reported that Rosa damascena waste extracts synthesized AgNPs. Biosynthesized AgNPs were modified graphite electrode and investigated by detecting H_2O_2 and vanillin, as shown in Figure.9 [156].

R.K. Das and coworkers have reported bio-mediated (Sechiumedule fruit pieces extract) mixed spinel Co_3O_4 nanoparticles were synthesized. The Co_3O_4 NPs were immobilized on the carbon paste electrode and used to detect H_2O_2. The peak current was increased with the increase in H_2O_2 concentration, the limit of H_2O_2 detection, sensitivity was found as 1 to 1000 μM, 0.0217 μM and 65.32 $nA/\mu M/cm^2$ [157]. A Fe_3O_4@cellulose nanocrystals/Cu nanocomposite was prepared, and can be utilized for detection of venlafaxine. The dynamic linear range and the limit of venlafaxine detection were 0.05-600.0 μM, 0.01 μM [158]. S. Zhou et al. reported that the porous structure of nitrogen-doped ordered mesoporous carbons was synthesized via a hard-template method. The synthesized materials were used for the electrochemical detection of Catechol (CC) in the presence of HQ. The detection limit was estimated to be 0.9 μM [159]. The detailed comparison for the electrochemical determination of toxic pollutants is shown in Table 1.

Figure. 9. Green synthesis of AgNPs from extracts of Rosa damascena waste and application for electrochemical reduction of H_2O_2. Reproduced permission from Ref. No. [156].

Conclusions and future outlook

In this chapter, we briefly summarized the bio-mediated synthesis of various metal and metal oxide nanomaterials with electrochemical sensing properties towards toxic chemicals. However, the diverse bio-mediated nanomaterials with high sensitivity and specificity are especially important for obtaining better detection results in electrochemistry. Eventually still, the nanomaterials based electrochemical sensors face enormous challenges. In recent environmental problems, we still need further research work. Most importantly, critical challenges in the synthesis of nanomaterials and precise formation mechanism and electrochemical properties are mainly based on the results without deep scientific comprehension. It is difficult to control the nanocomposite size and morphology and challenge the electron transfer process to affect the electrochemical performance and reader clearance significantly. The nanomaterials' optimal conditions modified electrode surface, such as size, shape, concentration, and volume, will affect the detection performance. In the future, we expected bio mediated nanomaterials based metal oxide nanocomposites will make significant space and progress in these areas. Subsequently, we need several advancements, low-cost, and mass-produced nanomaterial-based catalysts that can offer new chances to make an invention for better electrodes research in this field. The novel electrochemical sensors serve better than conventional methods.

Table 1 *Electrochemical response parameters of bio mediated nanocomposite based modified electrodes for determination of toxic pollutants.*

Electrode material	Bio-Mediated material	Method	Sensitivity	Detection range	LOD	pH	Sensing analyte	Response time
AgNPs[124]	*Lycopersiconesculentum*	DPV	-	10 to 90 μM	0.804 μM	12.5	Cr(III)	-
AgNPs[125]	*Agaricusbisporus* (mushroom)	DPV	-	10-90 μM	2.1×10^{-6}		Hg(II)	-
Au-NPs[126]	*Acacia nilotica*Willd twig barka	DPV	$1.01 \mu A \mu M^{-1} cm^{-2}$	1to 600 μM	0.016 μM	7.0	nitrobenzene	-
Ag-NPs[127]	*Acacia nilotica*Willd twig barka	DPV	$2.57 \ \mu A \mu M^{-1} cm^{-2}$	0.1– 350 μM	0.015 μM	5.0	p-nitrophenol	-
AgNPs[128]	*Justiciaglauca* leaf extract	DPV	$7.85 \ \mu A \mu M^{-1} cm^{-2}$	0.05 to 45.35 μM	0.017 μM	7.0	dopamine	-
Au-NPs[129]	*Justiciaglauca* leaf extract	DPV	$2.93 \ \mu A \mu M^{-1} L^{-1} cm^{-2}$	0.005 to 800 μM L^{-1}	0.07 nM L^{-1}	5 0	lead poisoning	6 s
AgNPs[130]	*Psidiumguajava*leaves	Square wave voltammo gram	$0.719 \ \mu A/cm^{2} \cdot \mu$ M.	25 to 150	14.63 μM	9 5	ascorbic acid	-
AuNPs[131]	chitosan	DPV	-	5.00×10^{-8} M to 2.00 $\times \ 10^{-3}$ M	2.50×10^{-8} M	7 0	Caffeic Acid	-
Au-Ag bionanocomp osite[134]	*peanut skin* extract	Cyclic voltammo gram	-	10 to 80 μM	4 μM	-	Sudan IV	-
Pt-Re NPs/PAC[135]	*Cassia fistula* (golden shower) fruit	DPV	$19.20 \mu A \mu M^{-1} cm^{-2}$	0.2– 117.7μ M	20.8nM	7.0	furazolidone	-
ZnONps[138]	*chakkota*fruit juice	Amperogr am	-	50 to 250 nM	6 nM	7.4	dopamine	-
CuO[139]	*Caesalpiniabondu cella* seed extract	Square-Wave Voltamme try	-	3.13–5 6.3 nM	1.04 nM	5.0	riboflavin	-
CuO/Cu₂O[140]	Gum Arabic (GA)acacia tree	Cyclic voltammo	$186.0 \ \mu AmM^{-1}$	0.1 mmolL	10 $\mu molL^{-1}$	-	formaldehyd e	-

			grams	cm^{-2}	$^{-1}$–110 mmolL^{-1}			
DyMnO$_3$-ZnO[141]	*Vitisvinifera, Hibiscus sabdariffa* and *rhus* juices	DPV	-	0.11 and 125.87 µM	10.10 nM	9.0	atenolol	-
ZnFe$_2$O$_4$[142]	*MoringaOleiferae* xtract	EIS	-	-	-	-	-	-
ZnFe$_2$O$_4$[143]	*Jatropha extract*	Cyclic voltammograms	-	1-8 and 1-7 and 1-8 mM	-	-	Paracetamol, Dolo 650 and Mifepristone-Misoprostol	-
CNPs[146]	*Castor oil*	DPV	-	1×10^{-7}– 1000×10^{-7} M	0.12 µM	7.0	Dopamine	-
rGO[147]	*green tea extract*	DPV	2328 µAmM^{-1}	0.05-10µM	27 nM	4.0	sunset yellow	-
rGO/AuNPs[148]	*E. tereticornis leave*	DPV	-	0.5–500 µmol/L	0.39 µmol/L	6.0	L-tryptophan	-
Pd-rGO[149]	Fenugreek seeds	Hanes-Woolf plot	1.3 µA mgdL^{-1}cm^{-1}	25-400(mg/dL)	25(mg/dL)	7.4	Triglycerides	10s
AgNPs and RGO[150]	*Justiciaglauca leaf extract*	DPV	0.836 µA µM^{-1}cm^{-2}	0.5 to 900 µM	0.261 µM	7.0	nitrobenzene	-
RGO/Au-NPs[151]	*Terminalia chebula leaf extract*	DPV	5.12, 0.28 and 0.33 µA µM^{-1}cm^{-2}	3–90, 3–300 and 15–150 µM for HQ, CC and RC	0.15,0.12 and 0.78 µM for HQ, CC and RC	7.0	Hydroquinone (HQ), catechol (CC) and resorcinol (RC)	-
MWCNTs/ZnO NPs[152]	*Carica papaya* seed extract	square wave voltammograms	-	0.014 to 0.152 mg L^{-1}	0.08 mg L^{-1}	7.6	silymarin	-
rGox/AgNPs[153]	green *tea extract*	amperometry studies	236 µA mM^{-1}cm^{-2}	0.002 to 20mM	0.73 µM	8.0	H$_2$O$_2$	2s
Cu/MWCNTs[154]	carboxymethylcellulose (CMC)	Amperometric	455.84 µA mM^{-1}cm^{-2}	5 µM to 1260 µM	1.8 µM	7.0	nitrite	-
ZnO-MWCNT-sG[155]	Sunlight as the energy	Cyclic voltammogram	-	1 to 26 nM	1 pM	7.4	Paraoxon	-

Graphite/rGO /AgNPs[156]	*Sechiumedule* is an herbaceous perennial climber	Amperom etric	-	1–33mM	19.04 μM	6.74	H_2O_2	<3s
AgNPs/CS[157]	*Rosa damascena* extracts	amperome try studies	115.2μA mM^{-1} cm^{-2} for H_2O_2 56.8 for vanillin μAμM^{-1} cm^{-2}	0-6.6mM	8.4 μM	7.0	H_2O_2 and vanillin	2s
Co(OH)$_2$/Co$_3$O$_4$[158]	*Sechiumedule* fruit pieces extract	Cyclic voltammo grams	65.32nA/μ M/cm^2	1 to 1000 μM	0.0217 μM	7.2	H_2O_2	5s
Fe$_3$O$_4$@CNC/Cu[159]	*Petasiteshybridus* leaf	DPV	-	0.05-600.0 μM	0.01 μM	7.0	venlafaxine	-
NOMC-1000/GCE[160]	green biological dye (ethyl violet)	DPV	0.22 μA/μM	6-90 μM	0.9 μM	7.4	catechol and hydroquinon e	3s

References

[1] G. Maduraiveeran, M. Sasidharan, W. Jin, Earth-abundant transition metal and metal oxide nanomaterials: Synthesis and electrochemical applications, Prog. Mater Sci. 106 (2019) 100574-100624. https://doi.org/10.1016/j.pmatsci.2019.100574

[2] O.P. Bolade, A.B. Williams, N.U. Benson, Green synthesis of iron-based nanomaterials for environmental remediation: A review, Environ. Nanotechnol. Monit. Manage. 13(2020)100279-100346. https://doi.org/10.1016/j.enmm.2019.100279

[3] S. Kargozar, M. Mozafari,Nanotechnology and nanomedicine: Start small, think big, Mater. Today:. Proc.5(7) (2018) 15492-15500. https://doi.org/10.1016/j.matpr.2018.04.155

[4] R. Purohit, A.Mittal, S. Dalela, V. Warudkar, K. Purohit, S. Purohit, Social, Environmental and ethical impacts of nanotechnology, Mater. Today:. Proc. 4(4) (2017) 5461-5467. https://doi.org/10.1016/j.matpr.2017.05.058

[5] P.D. Sia, Nanotechnology among innovation, health and risks, ProcediaSocial and Behav.Sci.237 (2017) 1076-1080. https://doi.org/10.1016/j.sbspro.2017.02.158

[6] J.C. Glenn, Nanotechnology: Future military environmental health considerations, Technol. Forecasting Social Change73(2) (2006) 128-137. https://doi.org/10.1016/j.techfore.2005.06.010

[7] P.S. Reddy, A.J. Chamkha, Influence of size, shape, type of nanoparticles, type and temperature of the base fluid on natural convection MHD of nanofluids, Alexandria Eng. J. 55(1)(2016) 331-341. https://doi.org/10.1016/j.aej.2016.01.027

[8] Y. Liu, Z. Liu, D. Huang, M. Cheng, G. Zeng, C. Lai, C. Zhang, C. Zhou, W. Wang, D. Jiang, H. Wang, B. Shao, Metal or metal-containing nanoparticle@MOF nanocomposites as a promising type of photocatalyst, Coord. Chem. Rev. 388(2019) 63-78. https://doi.org/10.1016/j.ccr.2019.02.031

[9] L. Chen, J. Liang, An overview of functional nanoparticles as novel emerging antiviral therapeutic agents, Mater. Sci. Eng., C (2020) 110924-110969. https://doi.org/10.1016/j.msec.2020.110924

[10] P. Nayar, S. Waghmare, P. Singh, M. Najar, S. Puttewar, A. Agnihotri, Comparative study of phase transformation of Al_2O_3 nanoparticles prepared by chemical precipitation and sol-gel auto combustion methods, Mater. Today:. Proc. (2019). https://doi.org/10.1016/j.matpr.2019.05.450

[11] F. Huang, Y. Guo, S. Wang, S. Zhang, M. Cui, Solgel-hydrothermal synthesis of Tb/Tourmaline/TiO_2nano tubes and enhanced photocatalytic activity, Solid State Sci. 64(2017) 62-68. https://doi.org/10.1016/j.solidstatesciences.2016.12.013

[12] T. Jahanbin, M. Hashim, K.A. Mantori, Comparative studies on the structure and electromagnetic properties of Ni−Zn ferrites prepared via co-precipitation and conventional ceramic processing routes, J. Magn. Magn. Mater. 322(18)(2010) 2684-2689. https://doi.org/10.1016/j.jmmm.2010.04.008

[13] Q. He, Y. Zhang, X. Chen, Z. Wang, H. Ji, M. Ding, B. Xie, P. Yu, The effects of substrate temperatures on the electrical properties of $CaZrO_3$ thin films prepared by RF magnetron sputtering, Curr. Appl Phys. 20(4) (2020) 557-561. https://doi.org/10.1016/j.cap.2020.02.004

[14] Y. Liu, H. Wan, N. Jiang, W. Zhang, H. Zhang, B. Chang, Q. Wang, Y. Zhang, Z. Wang, S. Luo, H. Sun, Chemical reduction-induced oxygen deficiency in Co_3O_4nanocubes as advanced anodes for lithium ion batteries, Solid State Ionics 334(2019) 117-124. https://doi.org/10.1016/j.ssi.2019.02.014

[15] M. Akbari, A.A. Mirzaei, M. Arsalanfar, Microemulsion based synthesis of promoted Fe−Co/MgOnanocatalyst: Influence of calcination atmosphere on the physicochemical properties, activity and light olefins selectivity for hydrogenation of carbon monoxide, Mater. Chem. Phys. 249 (2020) 123003. https://doi.org/10.1016/j.matchemphys.2020.123003

[16]   J. Jeoung-Ho, C. Min-Cheol, C. Seong-Jai, B. Dong-Sik, Synthesis and characterization of metallic Pd embedded TiO_2 nanoparticles by reverse micelle and sol-gel processing, Trans. Nonferrous Met. Soc. China 19(2009) s96-s99. https://doi.org/10.1016/S1003-6326(10)60253-1

[17]   M. Lüsi, H. Erikson, M. Merisalu, M. Rähn, V. Sammelselg, K. Tammeveski, Electrochemical reduction of oxygen in alkaline solution on Pd/C catalysts prepared by electrodeposition on various carbon nanomaterials, J. Electroanal. Chem. 834 (2019) 223-232. https://doi.org/10.1016/j.jelechem.2018.12.061

[18]   L. Kang, H.L. An, S. Jung, S. Kim, S. Nahm, D.-G. Kim, C.G. Lee, Low-voltage operating solution-processed CdS thin-film transistor with $Ca_2Nb_3O_{10}$ nanosheets deposited using Langmuir–Blodgett method for a gate insulator, Appl. Surf. Sci. 476 (2019) 374-377. https://doi.org/10.1016/j.apsusc.2019.01.132

[19]   S. Kundu, L. Ma, Y. Chen, H. Liang, Microwave assisted swift synthesis of $ZnWO_4$ nanomaterials: material for enhanced photo-catalytic activity, J. Photochem. Photobiol. A 346 (2017) 249-264. https://doi.org/10.1016/j.jphotochem.2017.05.004

[20]   X.-P. Li, Y.-L. Sun, C.-W. Luo, Z.-S. Chao, UV-resistant hydrophobic CeO_2 nanomaterial with photocatalytic depollution performance, Ceram. Int. 44(11) (2018) 13439-13443. https://doi.org/10.1016/j.ceramint.2018.04.132

[21]   W. Shang, T. Cai, Y. Zhang, D. Liu, S. Liu, Facile one pot pyrolysis synthesis of carbon quantum dots and graphene oxide nanomaterials: All carbon hybrids as eco-environmental lubricants for low friction and remarkable wear-resistance, Tribol. Int. 118 (2018) 373-380. https://doi.org/10.1016/j.triboint.2017.09.029

[22]   M. Mahmoodian, H. Hajihoseini, S. Mohajerzadeh, M. Fathipour, Nano patterning and fabrication of single polypyrrole nanowires by electron beam Lthography, Synth. Met. 249(2019) 14-24. https://doi.org/10.1016/j.synthmet.2019.01.013

[23]   A.V. Kabashin, M. Meunier, Laser ablation-based synthesis of functionalized colloidal nanomaterials in biocompatible solutions, J. Photochem. Photobiol. A 182(3) (2006) 330-334. https://doi.org/10.1016/j.jphotochem.2006.06.008

[24]   G. Shruthi, K.S. Prasad, T. P. Vinod, V. Balamurugan, and C. Shivamallu, Green synthesis of biologically active silver nanoparticles through a phyto-mediated approach using areca catechu leaf extract, Chem.Select 2 (2017) 10354 –10359. https://doi.org/10.1002/slct.201702257

[25] R.M. Tripathi, S.J. Chung, Biogenic nanomaterials: Synthesis, characterization, growth mechanism, and biomedical applications, J. Microbiol. Methods 157 (2019) 65–80. https://doi.org/10.1016/j.mimet.2018.12.008

[26] R.K. Das, S.K. Brar, Plant mediated green synthesis: modified approaches, Nanoscale 5 (2013) 10155–10162. https://doi.org/10.1039/c3nr02548a

[27] S. Ullah, A. Ahmad, H. Ri, A.U. Khan, U.A. Khan, Q. Yuan, Green synthesis of catalytic zinc oxide nano-flowers and their bacterial infection therapy, Appl. Organometal. Chem. 34(1)(2020) e5298-e5309. https://doi.org/10.1002/aoc.5298

[28] S.G. Patra, K. Sathiyan, M. Meistelman, and T. Zidki, Green synthesis of M^0 nanoparticles (M=Pd, Pt, and Ru) for electrocatalytic hydrogen evolution, Isr. J. Chem. 60 (2020) 1 – 9. https://doi.org/10.1002/ijch.201900175

[29] X. Yang, K. Fu, L. Mao, W. Peng, J. Jin, S. Yang, G. Li,Bio-mediated synthesis of a-Ni(OH)$_2$nanobristles on hollow porous carbon nanofibers for rechargeable alkaline batteries, Chem. Eng. Sci. 205 (2019) 269–277. https://doi.org/10.1016/j.ces.2019.04.039

[30] Y. Zhang, W. Xu, X.Wang, S. Ni, E. Rosqvist, J.-H. Smått, J. Peltonen, Q. Hou, M. Qin, S. Willför, C. Xu,From biomass to nanomaterials: A green procedure for preparation of holistic bamboo multifunctional nanocomposites based on formic acid rapid fractionation, ACS Sustainable Chem. Eng. 7(7)(2019) 6592-6600. https://doi.org/10.1021/acssuschemeng.8b05502

[31] J. Iqbal, B.A. Abbasi, A. Munir, S. Uddin, S. Kanwal, T. Mahmood, Facile green synthesis approach for the production of chromium oxide nanoparticles and their different in vitro biological activities, Microsc. Res. Tech. (2020) 1–14. https://doi.org/10.1002/jemt.23460

[32] M. Qasem, R.E. Kurdi, and D. Patra, Green synthesis of curcumin conjugated CuO nanoparticles for catalytic reduction of methylene blue, Chem.Select 5 (2020) 1694 – 1704. https://doi.org/10.1002/slct.201904135

[33] A. Chandra, A. Bhattarai, A.K. Yadav, J. Adhikari, M. Singh, and B. Giri, Green Synthesis of Silver Nanoparticles Using Tea Leaves from Three Different Elevations, Chem.Select 15 (2000) 1–9

[34] C. Bartolucci, A. Antonacci, F. Arduini, D. Moscone, L. Fraceto, E. Campos, R. Attaallah, A. Amine, C. Zanardi, L. Cubillana, J.M. Palacios Santander, V. Scognamiglio, Green nanomaterials fostering agrifood sustainability, Trends Anal. Chem. 125 (2020)115840–115884. https://doi.org/10.1016/j.trac.2020.115840

[35] R.M. Tripathi, S.J. Chung,Biogenic nanomaterials: Synthesis, characterization, growth mechanism, and biomedical applications, J. Microbiol. Methods 157 (2019) 65–80. https://doi.org/10.1016/j.mimet.2018.12.008

[36] J. Han, L. Xiong, X. Jiang, X. Yuan, Y. Zhao, D. Yang, Bio-functional electrospun nanomaterials: From topology design to biological applications, Prog. Polym. Sci. 91 (2019) 1-28. https://doi.org/10.1016/j.progpolymsci.2019.02.006

[37] L.Zeng, J. Gao, Y. Liu, J. Gao, L.Yao, X. Yang, X. Liu, B. He, L. Hu, J. Shi, M. Song G. Qu, G. Jiang, Role of protein corona in the biological effect of nanomaterials: Investigating methods, Trends Anal. Chem. 118 (2019) 303-314. https://doi.org/10.1016/j.trac.2019.05.039

[38] L. García-Carmona, M. C. González, A. Escarpa, Nanomaterial-based electrochemical (bio)-sensing: One step ahead in diagnostic and monitoring of metabolic rare diseases, Trends Anal. Chem.118 (2019) 29-42. https://doi.org/10.1016/j.trac.2019.05.020

[39] R. Eivazzadeh-Keihan, P. Pashazadeh, M. Hejazi, M.D.L. Guardia, A. Mokhtarzadeh, Recent advances in Nanomaterial-mediated Bio and immune sensors for detection of aflatoxin in food products, Trends Anal. Chem. 87 (2017) 112-128. https://doi.org/10.1016/j.trac.2016.12.003

[40] W. Dudefoi, A. Villares, S. Peyron, C. Moreau, M.-H. Ropers, N. Gontard, B. Cathala , Nanoscience and nanotechnologies for biobased materials, packaging and food applications: Newopportunities and concerns, Innovative Food Sci. Emerg. Technol. 46 (2018) 107-121. https://doi.org/10.1016/j.ifset.2017.09.007

[41] T.A. Saleh, G. Fadillah, Recent trends in the design of chemical sensors based on graphene–metal oxide nanocomposites for the analysis of toxic species and biomolecules, Trends Anal. Chem.120 (2019) 115660. https://doi.org/10.1016/j.trac.2019.115660

[42] A. Sabarwal, K. Kumar, R. P.Singh, Hazardous effects of chemical pesticides on human health–Cancer and other associated disorders, Environ. Toxicol. Pharmacol. 63 (2018) 103-114. https://doi.org/10.1016/j.etap.2018.08.018

[43] P. Chowdhary, A. Raj, R.N. Bharagava, Environmental pollution and health hazards from distillery wastewater and treatment approaches to combat the environmental threats: A review, Chemosphere 194(2018) 229-246. https://doi.org/10.1016/j.chemosphere.2017.11.163

[44] S.S. Chandel, T. Agarwal, Review of current state of research on energy storage, toxicity, health hazards and commercialization of phase changing materials, Renewable Sustainable Energy Rev. 67(2017) 581-596. https://doi.org/10.1016/j.rser.2016.09.070

[45] Z. Peng, X. Liu, W. Zhang, Z. Zeng, Z. Liu, C. Zhang, Y. Liu, B. Shao, Q. Liang, W. Tang, X. Yuan, Advances in the application, toxicity and degradation of carbon nanomaterials in environment: A review, Environ. Int.134 (2020) 105298. https://doi.org/10.1016/j.envint.2019.105298

[46] E. Mahmoud, M. Ibrahim, N. Ali, H. Ali, Spectroscopic analyses to study the effect of biochar and compost on dry mass of canola and heavy metal immobilization in soil, Commun. Soil Sci. Plant Anal. 49 (2018) 1990-2001. https://doi.org/10.1080/00103624.2018.1492601

[47] J.F. Ping, J. Wu, Y.B. Ying, M.H. Wang, G. Liu, M. Zhang, Evaluation of trace heavy metal levels in soil samples using an ionic liquid modified carbon paste electrode, J. Agric. Food Chem., 59 (2011) 4418-4423. https://doi.org/10.1021/jf200288e

[48] S. Zhang, J. Li, M. Zeng, J. Xu, X. Wang, W. Hu, Polymer nanodots of graphitic carbon nitride as effective fluorescent probes for the detection of Fe^{3+} and Cu^{2+} ions, Nanoscale 6 (2014) 4157-4162. https://doi.org/10.1039/c3nr06744k

[49] M. Rong, L. Lin, X. Song, Y. Wang, Y. Zhong, J. Yan, Y. Feng, X. Zeng, X. Chen, Fluorescence sensing of chromium (VI) and ascorbic acid using graphitic carbon nitride nanosheets as a fluorescent switch, Biosens. Bioelectron. 68 (2015) 210-217. https://doi.org/10.1016/j.bios.2014.12.024

[50] Z. Koudelkova, T. Syrovy, P. Ambrozova, Z. Moravec, L. Kubac, D. Hynek, L. Richtera, V. Adam, Determination of zinc, cadmium, lead, copper and silver using a carbon paste electrode and a screen printed electrode modified with chromium(III) oxide, Sensors 17 (2017) 1832-1846. https://doi.org/10.3390/s17081832

[51] T.A. Ali, G.G. Mohamed, A.R. Othman, Design and construction of new potentiometric sensors for determination of copper(II) ion based on copper oxide nanoparticles, Int. J. Electrochem. Sci., 10 (2015) 8041-8057.

[52] R. Ramachandran, T.W. Chen, S.M. Chen, T. Baskar, R. Kannan, P. Elumalai, P. Raja, T. Jeyapragasam, K. Dinakaran, G. Gnanakumar, A review of the advanced developments of electrochemical sensors for the detection of toxic and bioactive

Bioinspired Nanomaterials
Materials Research Foundations **111** (2021) 224-263

Materials Research Forum LLC
https://doi.org/10.21741/9781644901571-8

molecules, Inorg. Chem. Front. 6 (2019) 3418-3439.
https://doi.org/10.1039/C9QI00602H

[53] K.D. Roy, M. Debiprosad, Review on nanomaterials-enabled electrochemical sensors for ascorbic acid detection, Anal. Biochem. 586 (2019) 113415-113432. https://doi.org/10.1016/j.ab.2019.113415

[54] F. Laghrib, M. Bakasse, S. Lahrich, M.A. El Mhammedi, Electrochemical sensors for improved detection of paraquat in food samples: A review, Mater. Sci.Eng. C, 107 (2020) 110349-110399. https://doi.org/10.1016/j.msec.2019.110349

[55] Y.M. Díaz-González, M. Gutiérrez-Capitán, P. Niu, A. Baldi, C. Jiménez-Jorquera, C. Fernández-Sánchez, Electrochemical devices for the detection of priority pollutants listed in the EU water framework directive, Trends Anal. Chem., 77 (2016) 186-202. https://doi.org/10.1016/j.trac.2015.11.023

[56] S. Cinti, F. Arduini, Graphene-based screen-printed electrochemical (bio)sensorsand their applications: efforts and criticisms, Biosens. Bioelectron. 89 (2016) 107-122. https://doi.org/10.1016/j.bios.2016.07.005

[57] L. Rassaei, F. Marken, M. Sillanpää, M. Amiri, C.M. Cirtiu, M. Sillanpää, Nanoparticles in electrochemical sensors for environmental monitoring, Trends Anal. Chem., 30 (2011) 1704-1715. https://doi.org/10.1016/j.trac.2011.05.009

[58] K. Murtada, V. Moreno, Nanomaterials-based electrochemical sensors for the detection of aroma compounds - towards analytical approach, J. Electroanal. Chem.861 (2020) 113988-114036. https://doi.org/10.1016/j.jelechem.2020.113988

[59] M.A. Beluomini, J.L. Silva, A.C.D Sá, E. Buffon, T.C. Pereira, N.R. Stradiotto, Electrochemical sensors based on molecularly imprinted polymer on nanostructured carbon materials: A review, J. Electroanal. Chem., 840(2019) 343-366. https://doi.org/10.1016/j.jelechem.2019.04.005

[60] L. Shang, J. Xu, G.U. Nienhaus, Recent advances in synthesizing metal nanocluster-based nanocomposites for application in sensing, imaging and catalysis, Nano Today 28(2019) 100767. https://doi.org/10.1016/j.nantod.2019.100767

[61] F.C. Adams, C. Barbante, Nanoscience, nanotechnology and spectrometry, Spectrochim. Acta, Part B 86 (2013) 3–13. https://doi.org/10.1016/j.sab.2013.04.008

[62] M. Bandeira, M. Giovanela, M. Roesch-Ely, D.M. Devine, J.D.S. Crespo,Green synthesis of zinc oxide nanoparticles: A review of the synthesis methodology and

mechanism of formation, Sustainable Chem. Pharm. 15 (2020) 100223-1002333. https://doi.org/10.1016/j.scp.2020.100223

[63]   A. Roy, O. Bulut, S. Some, A.K. Mandal and M. D. Yilmaz, Green synthesis of silver nanoparticles: biomolecule-nanoparticle organizations targetingantimicrobial activity, RSC Adv. 9 (2019) 2673–2702. https://doi.org/10.1039/C8RA08982E

[64]   M. Gericke,A. Pinches, Microbial production of gold nanoparticles. Gold Bull. 39 (2006) 22–28. https://doi.org/10.1007/BF03215529

[65]   S. Iravani Bacteria in nanoparticle synthesis: current status and future prospects. IntSch Res Not. 2014 (2014) 1–18. https://doi.org/10.1155/2014/359316

[66]   F. D. Pooley, Bacteria accumulate silver during leaching of sulphide ore minerals, Nature 296 (1982) 642–643. https://doi.org/10.1038/296642a0

[67]   T. Klaus, R. Joerger, E. Olsson and C.-G. Granqvist, Silverbased crystalline NPs, microbially fabricated, Proc. Natl. Acad. Sci. U. S. A. 96 (1999) 13611–13614. https://doi.org/10.1073/pnas.96.24.13611

[68]   J. Singh, T. Dutta, K.-H. Kim, M. Rawat, P. Samddar, P. Kumar, 'Green' synthesis of metals and their oxide nanoparticles: applications for environmental remediation, J.Nanobiotechnol. 16 (2018) 84-108. https://doi.org/10.1186/s12951-018-0408-4

[69]   K. Kalimuthu, R. S. Babu, D. Venkataraman, M. Bilal and S. Gurunathan, Biosynthesis of silver nanocrystals by Bacillus licheniformis, Colloids Surf. B 65(1)(2008) 150–153. https://doi.org/10.1016/j.colsurfb.2008.02.018

[70]   N. Mokhtari, S. Daneshpajouh, S. Seyedbagheri, R. Atashdehghan, K. Abdi, S. Sarkar, S. Minaian, H. R. Shahverdi, A. R. Shahverdi, Biological synthesis of very small silver NPs by culture supernatant of Klebsiella pneumonia: The effects of visible-light irradiation and the liquid mixing process, Mater. Res. Bull.44(6)(2009) 1415–1421. https://doi.org/10.1016/j.materresbull.2008.11.021

[71]   O.V. Kharissova, H.R. Dias, B.I. Kharisov, B.O. P´erez, V.M. P´erez, The greener synthesis of NPs, Trends Biotechnol. 31(4)(2013) 240–248. https://doi.org/10.1016/j.tibtech.2013.01.003

[72]   Y-L. Chen, H-Y. Tuan, C-W. Tien, W.H. Lo, H.C. Liang, Y.C. Hu, Augmented biosynthesis of cadmium sulfide nanoparticles by genetically engineered Escherichia coli. BiotechnolProg. 25(2009) 1260–1266. https://doi.org/10.1002/btpr.199

[73]   P.Mohanpuria, N.K. Rana, S.K. Yadav. Biosynthesis of nanoparticles: technological concepts and future applications. J Nanoparticle Res. 10 (2008) 507–517. https://doi.org/10.1007/s11051-007-9275-x

[74]   K.S. Siddiqi, A. Husen, Fabrication of metal NPs from fungi and metal salts: scope and application, Nanoscale Res. Lett., 11(1)(2016) 98-. https://doi.org/10.1186/s11671-016-1311-2

[75]   B.K. Ravindra, A.H. Rajasab, A comparative study on biosynthesis of silver nanoparticles using four different fungal species, Int J Pharm Pharm Sci. 6(1) (2014) 372–376.

[76]   R.Raliya, P. Biswas, J.C.Tarafdar, TiO$_2$ nanoparticle biosynthesis and its physiological effect on mung bean (Vignaradiata L.). Biotechnol Rep. 5 (2015) 22–26. https://doi.org/10.1016/j.btre.2014.10.009

[77]   K. S. Siddiqi, A. Husen, Fabrication of metal NPs from fungi and metal salts: scope and application, Nanoscale Res. Lett., 11(1) (2016) 98-113. https://doi.org/10.1186/s11671-016-1311-2

[78]   B. Xue, D. He, S. Gao, D. Wang, K. Yokoyama, L. Wang, Biosynthesis of silver NPs by the fungus Arthrodermafulvum and its antifungal activity against genera of Candida, Aspergillus and Fusarium, Int. J. Nanomed., 11 (2016) 1899-1906. https://doi.org/10.2147/IJN.S98339

[79]   S.V. Otari, R.M. Patil, N.H. Nadaf, S.J. Ghosh, S.H. Pawar, Green synthesis of silver NPs by microorganism using organic pollutant: its antimicrobial and catalytic application, Environ. Sci. Pollut. Res. 21(2) (2014) 1503–1513. https://doi.org/10.1007/s11356-013-1764-0

[80]   M. Eugenio, N. M"uller, S. Fras´es, R. Almeida-Paes, L.M. Lima, L. Lemgruber, M. Farina, W. de Souza, C. Sant'Anna, Yeast-derived biosynthesis of silver/silver chloride NPs and their antiproliferative activity against bacteria, RSC Adv. 6(12) (2016) 9893–9904. https://doi.org/10.1039/C5RA22727E

[81]   K. Ishida, T.F. Cipriano, G.M. Rocha, G. Weissm"uller, F. Gomes, K. Miranda, S. Rozental, Silver nanoparticle production by the fungus Fusariumoxysporum: nanoparticle characterisation and analysis of antifungal activity against pathogenic yeasts, Mem. Inst. Oswaldo Cruz 109(2)(2014) 220–228. https://doi.org/10.1590/0074-0276130269

[82]   Mourato A, Gadanho M, Lino AR, Tenreiro R. Biosynthesis of crystalline silver and gold nanoparticles by extremophilic yeasts. BioinorgChem Appl. 1 (2011) 1-9. https://doi.org/10.1155/2011/546074

[83]   A. M. Elgorban, A. N. Al-Rahman, S. R. Sayed, A. Hirad, A. A.-F. Mostafa and A. H. Bahkali, Antimicrobial activity and green synthesis of AgNPs using Trochodermaviride, Biotechnol. Biotechnol. Equip. 30(2)(2016) 299–304. https://doi.org/10.1080/13102818.2015.1133255

[84]   A. Regiel-Futyra, M. Kus-Li´skiewicz, V. Sebastian, S. Irusta, M. Arruebo, A. Kyzioł, G. Stochel, Development of noncytotoxic silver–chitosan nanocomposites for efficient control of biofilm forming microbes, RSC Adv. 7(83)(2017) 52398-52413. https://doi.org/10.1039/C7RA08359A

[85]   T.C. Leung, C. K. Wong,Y. Xie, Green synthesis of silver NPs using biopolymers, carboxymethylated-curdlan and fucoidan, Mater. Chem. Phys. 121(3) (2010) 402–405. https://doi.org/10.1016/j.matchemphys.2010.02.026

[86]   P. Vasileva, B. Donkova, I. Karadjova, C. Dushkin, Synthesis of starch-stabilized silver NPs and their application as a surface plasmon resonance-based sensor of hydrogen peroxide, Colloids Surf. A 382 (2011) 203–210. https://doi.org/10.1016/j.colsurfa.2010.11.060

[87]   M.B. Ahmad, M.Y. Tay, K. Shameli, M.Z. Hussein, J.J. Lim, Green synthesis and characterization of silver/ chitosan/polyethylene glycol nanocomposites without any reducing agent, Int. J. Mol. Sci. 12(8)(2011) 4872–4884. https://doi.org/10.3390/ijms12084872

[88]   P. Malik, R. Shankar, V. Malik, N. Sharma, T.K. Mukherjee,Green chemistry based benign routes for nanoparticle synthesis. J. Nanoparticles 2014 (2014) 1–14. https://doi.org/10.1155/2014/302429

[89]   K.S.Mukunthan, S.Balaji, Cashew apple juice (Anacardiumoccidentale L.) speeds up the synthesis of silver nanoparticles. Int. J. Green Nanotechnol. 4 (2012) 71–85. https://doi.org/10.1080/19430892.2012.676900

[90]   N. Ahmad, S. Sharma, M.K.Alam, V.N. Singh, S.F.Shamsi, B.R. Mehta, A.Fatma,Rapid synthesis of silver nanoparticles using dried medicinal plant of basil. Colloids Surf B Biointerfaces. 81 (2010) 81–86. https://doi.org/10.1016/j.colsurfb.2010.06.029

[91]   J.L. Gardea-Torresdey, E. Gomez, J.R. Peralta-Videa, J.G. Parsons, H. Troiani, M. Jose-Yacaman, Alfalfa sprouts: a natural source for the synthesis of silver NPs, Langmuir 19(4)(2003) 1357–1361. https://doi.org/10.1021/la020835i

[92]   R. Sithara, P. Selvakumar, C. Arun, S. Anandan, P. Sivashanmugam, Economical synthesis of silver nanoparticles using leaf extract of Acalyphahispida and its application in the detection of Mn (II) ions, J. Adv. Res. 8(6) (2017) 561–568. https://doi.org/10.1016/j.jare.2017.07.001

[93]   A.J. Gavhane, P. Padmanabhan, S.P. Kamble, S.N. Jangle, Synthesis of silver NPs using extract of neem leaf and triphala and evaluation of their antimicrobial activities, Int. J. Pharma Bio Sci. 3(3)(2012) 88–100.

[94]   B. Vellaichamy, P. Periakaruppan, Silver-nanospheres as a green catalyst for the decontamination of hazardous pollutants, RSC Adv. 5 (2015) 105917–105924. https://doi.org/10.1039/C5RA21599D

[95]   B. Vellaichamy, P. Periakaruppan, Ag nanoshell catalyzed dedying of industrial effluents,RSC Adv. 6 (2016) 31653–31660. https://doi.org/10.1039/C6RA02937J

[96]   B. Vellaichamy, P. Periakaruppan, Green synthesized nanospherical silver for selective and sensitive sensing of Cd^{2+}colorimetrically, RSC Adv. 6 (2016) 35778–35784. https://doi.org/10.1039/C6RA04381J

[97]   V. Balakumar, P. Prakash, K. Muthupandi, A. Rajan, Nanosilver for selective and sensitive sensing of saturnism,Sens. Actuators, B 241 (2017) 814–820. https://doi.org/10.1016/j.snb.2016.10.142

[98]   P. Velmurugan, S. Sivakumar, S. Young-Chae, J. Seong-Ho, Y. Pyoung-In, H. Sung-Chul, Synthesis and characterization comparison of peanut shell extract silver NPs with commercial silver NPs and their antifungal activity, J. Ind. Eng. Chem. 31 (2015) 51–54. https://doi.org/10.1016/j.jiec.2015.06.031

[99]   C.Krishnaraj, E.G.Jagan, S.Rajasekar, P.Selvakumar, P.T.Kalaichelvan, N. Mohan,Synthesis of silver nanoparticles using Acalyphaindica leaf extracts and its antibacterial activity against water borne pathogens. Colloids Surf. B76(1) (2010) 50-56. https://doi.org/10.1016/j.colsurfb.2009.10.008

[100] T.C.Prathna, N.Chandrasekaran, A.M.Raichur, A.Mukherjee Biomimetic synthesis of silver nanoparticles by Citrus limon (lemon) aqueous extract and theoretical prediction of particle size. Colloids Surf. B 82 (2011) 152–159. https://doi.org/10.1016/j.colsurfb.2010.08.036

255

[101] A.K.Jha, K. Prasad, Green synthesis of silver nanoparticles using cycas leaf. Int J Green Nanotechnol. Phys. Chem. 1 (2010) 110–117. https://doi.org/10.1080/19430871003684572

[102] S.Ravindra, Y.M. Mohan, N.R. Narayana, K.M. Raju, Fabrication of antibacterial cotton fibres loaded with silver nanoparticles via "green approach". Colloids Surf. A Physicochem. Eng. Asp. 367 (2010) 31– 40. https://doi.org/10.1016/j.colsurfa.2010.06.013

[103] R. Veerasamy, T.Z. Xin, S.Gunasagaran, T.F.W. Xiang, E. Fang, C. Yang, N. Jeyakumar, S.A. Dhanaraj, Biosynthesis of silver nanoparticles using mangosteen leaf extract and evaluation of their antimicrobial activities. J. Saudi. Chem. Soc. 15(2) (2011) 113-120. https://doi.org/10.1016/j.jscs.2010.06.004

[104] T.Mochochoko, O.S.Oluwafemi, D.N.Jumbam, S.P.Songca. Green synthesis of silver nanoparticles using cellulose extracted from an aquatic weed; water hyacinth. Carbohydr.Polym. 98 (2013) 290–294. https://doi.org/10.1016/j.carbpol.2013.05.038

[105] J. Singh, N. Singh, A. Rathi, D.Kukkar, M.Rawat,Facile approach to synthesize and characterization of silver nanoparticles by using mulberry leaves extract in aqueous medium and its application in antimicrobial activity. J Nanostructures. 7 (2017) 134–40.

[106] T.Santhoshkumar, A.A.Rahuman, G.Rajakumar, S.Marimuthu, A.Bagavan, C.Jayaseelan, A.A.Zahir, G.Elango, C. Kamaraj,Synthesis of silver nanoparticles using Nelumbonucifera leaf extract and its larvicidal activity against malaria and filariasis vectors. Parasitol Res. 108 (2011) 693–702. https://doi.org/10.1007/s00436-010-2115-4

[107] J. Singh, A. Mehta, M.Rawat, S. Basu, Green synthesis of silver nanoparticles using sun dried tulsi leaves and its catalytic application for 4-nitrophenol reduction. J Environ. Chem. Eng. 6 (2018) 1468–1474. https://doi.org/10.1016/j.jece.2018.01.054

[108] S. Iravani, Green synthesis of metal nanoparticles using plants. Green Chem. 13 (2011) 2638-2650. https://doi.org/10.1039/c1gc15386b

[109] C.V. Rao, A.K. Golder, Development of a bio-mediated technique of silver-doping on titania, Colloids and Surfaces A: Physicochem. Eng. Aspects 506 (2016) 557–565. https://doi.org/10.1016/j.colsurfa.2016.07.031

[110] P. Gómez-López, A. Puente-Santiago, A. Castro-Beltrán, L.A. Santos do Nacimiento, A.M. Balu, R. Luque, C.G. Alvarado-Beltrán, Nanomaterials and

Catalysis for Green Chemistry, Current Opinion in Green and Sustainable Chemistry, 24 (2020) 48-55. https://doi.org/10.1016/j.cogsc.2020.03.001

[111] D. Manoj, R. Saravanan, J. Santhanalakshmi, S. Agarwal, V.K. Gupta, R. Boukherroub, Towards green synthesis of monodisperse Cu nanoparticles: An efficient and high sensitive electrochemical nitrite sensor, Sens. Actuators, B 266(2018) 873-882. https://doi.org/10.1016/j.snb.2018.03.141

[112] B.K.Ravindra, A.H.Rajasab. A comparative study on biosynthesis of silver nanoparticles using four different fungal species. Int. J. Pharm. Pharm. Sci. 6(1) (2014) 372–376.

[113] A.M.Fayaz, K.Balaji, M.Girilal, R. Yadav, P.T. Kalaichelvan, R. Venketesan, Biogenic synthesis of silver nanoparticles and their synergistic effect with antibiotics: a study against grampositive and gram-negative bacteria. Nanomed. Nanotechnol. Biol. Med. Sci. 6(1) (2010) 103–109. https://doi.org/10.1016/j.nano.2009.04.006

[114] R.Raliya, J.C.Tarafdar, Biosynthesis and characterization of zinc, magnesium and titanium nanoparticles: an eco-friendly approach. Int. Nano Lett. (2014) 493-103. https://doi.org/10.1007/s40089-014-0093-8

[115] BK.Bansod,T. Kumar,R. Thakur,S. Rana, I. Singh, A review on various electrochemical techniques for heavy metal ions detection with different sensing platforms. Biosensors and Bioelectronics. 94(2017) 443–55. https://doi.org/10.1016/j.bios.2017.03.031

[116] B. Bansod, T. Kumar, R. Thakur, S. Rana, I. Singh, A review on various electrochemical techniques for heavy metal ions detection with different sensing platforms, Biosens. Bioelectron. 94 (2017) 443-455. https://doi.org/10.1016/j.bios.2017.03.031

[117] A. Joshi, K.-H. Kim, Recent advances in nanomaterial-based electrochemical detection of antibiotics: Challenges and future perspectives, Biosens. Bioelectron. 153 (2020) 112046-112121. https://doi.org/10.1016/j.bios.2020.112046

[118] M. Labib, E.H. Sargent, S.O. Kelley, Electrochemical methods for the analysis of clinically relevant biomolecules. Chem. Rev. 116 (2016), 9001-9090. https://doi.org/10.1021/acs.chemrev.6b00220

[119] Y. Shao, J. Wang, H. Wu, J. Liu, I.A. Aksay, Y. Lin, Graphene Based Electrochemical Sensors and Biosensors: A Review, Electroanalysis 22(10)(2010) 1027-1036. https://doi.org/10.1002/elan.200900571

[120] X. Gan, H. Zhao, Understanding signal amplification strategies of nanostructured electrochemical sensors for environmental pollutants, Curr.Opin.Electrochem. 17(2019) 56-64. https://doi.org/10.1016/j.coelec.2019.04.016

[121] E. Asadian, M. Ghalkhani, S. Shahrokhian, Electrochemical sensing based on carbon nanoparticles: A review, Sensors & Actuators: B. Chemical 293 (2019) 183–209. https://doi.org/10.1016/j.snb.2019.04.075

[122] F. Arduini, S.Cinti, V. Mazzaracchio, V. Scognamiglio, A. Amine, D. Moscone,Carbon black as an outstanding and affordable nanomaterial for electrochemical (bio)sensor design, Biosens. Bioelectron.156 (2020) 112033-112085. https://doi.org/10.1016/j.bios.2020.112033

[123] B. Vellaichamy, P.Periakaruppan, S.K.Ponnaiah, A new in-situ synthesized ternary CuNPs-PANI-GO nano composite forselective detection of carcinogenic hydrazine, Sens. Actuators, B245 (2017) 156–165. https://doi.org/10.1016/j.snb.2017.01.117

[124] A. Aravind, M. Sebastiana, B. Mathew,Green synthesized unmodified silver nanoparticles as a multi-sensor for Cr(III) ions, Environ. Sci.: Water Res. Technol. 4 (2018) 1531-1542. https://doi.org/10.1039/C8EW00374B

[125] M. Sebastian, A. Aravind and B. Mathew, Green silver-nanoparticle-based dual sensor for toxic Hg(II) ions, Nanotechnology 29(35) (2018)355502-355530. https://doi.org/10.1088/1361-6528/aacb9a

[126] R. Emmanuel, C.Karuppiaha, S.-M. Chena, S. Palanisamya,S. Padmavathy, P. Prakash, Green synthesis of gold nanoparticles for trace level detection of ahazardous pollutant (nitrobenzene) causing Methemoglobinaemia, J. Hazard. Mater. 279 (2014) 117–124. https://doi.org/10.1016/j.jhazmat.2014.06.066

[127] C. Karuppiah, S. Palanisamy, S.-M. Chen, R. Emmanuel, M.A. Ali, P. Muthukrishnan, P. Prakash, F.M.A. Al-Hemaid, Green biosynthesis of silver nanoparticles and nanomolardetection of p-nitrophenol, J. Solid State Electrochem. 18 (2014) 1847–1854. https://doi.org/10.1007/s10008-014-2425-z

[128] S. Palanisamy, B. Thirumalraj, S.-M. Chen, M.A. Ali, K. Muthupandi, R. Emmanuel, P. Prakash, and Fahad M.A. Al-Hemaid,Fabrication of silver nanoparticles decorated on activated screen printed carbon electrode and its application for ultrasensitive detection of dopamine, Electroanalysis 27 (2015) 1998 – 2006. https://doi.org/10.1002/elan.201500079

[129] C. Karuppiah, S. Palanisamy, S.-M. Chen, R. Emmanuel, K. Muthupandi and P. Prakash,Green synthesis of gold nanoparticles and its application for the trace level determination of painter's colic, RSC Adv. 5 (2015) 16284–16291. https://doi.org/10.1039/C4RA14988B

[130] S.R. Dash, S.S. Bag, A.K. Golder, Synergized AgNPs formation using microwave in a bio-mediated route: Studies on particle aggregation and electrocatalytic sensing of ascorbic acidfrom biological entities, J. Electroanal. Chem.827 (2018) 181–192. https://doi.org/10.1016/j.jelechem.2018.09.023

[131] G.D. Carlo, A. Curulli, R.G. Toro, C. Bianchini, T.D. Caro, G. Padeletti, D. Zane, G M. Ingo,Green synthesis of gold–chitosan nanocomposites for caffeic acidsensing, Langmuir 28 (2012) 5471–5479. https://doi.org/10.1021/la204924d

[132] K. Gopinath, S. Kumaraguru, K. Bhakyaraj, S. Mohan, K.S. Venkatesh, M. Esakkirajan, P. Kaleeswarran, N.S. Alharbi, S. Kadaikunnan, M. Govindarajan, G. Benelli, A. Arumugam, Green synthesis of silver, gold and silver/gold bimetallic nanoparticles using the Gloriosasuperba leaf extract and their antibacterial and antibiofilmactivities. Microb. Pathog. 101 (2016) 1-11. https://doi.org/10.1016/j.micpath.2016.10.011

[133] J. Rick, M.-C. Tsai, and B.J. Hwang, Biosensors incorporating bimetallic nanoparticles, Nanomaterials 6 (2016) 5-35. https://doi.org/10.3390/nano6010005

[134] A. Pani, T.D.Thanh, N. H. Kim, J. H. Lee, S.-I.Yun, Peanut Skin Extract Mediated Synthesis of AuNPs, AgNPs and Au-Ag Bionanocomposite forElectrochemical Sudan IV Sensing,IET Nanobiotechnol. 10(6) (2016) 431-437. https://doi.org/10.1049/iet-nbt.2016.0017

[135] P. Veerakumar, A. Sangili, S.-M. Chen, A. Pandikumar, and K.-C. Lin, Fabrication of platinum-rhenium nanoparticles-decoratedporous carbons: voltammetric sensing of furazolidone, ACS Sustainable Chem. Eng. 8(9)(2020) 3591-3605. https://doi.org/10.1021/acssuschemeng.9b06058

[136] J.M. George, A. Antony, B.Mathew, Metal oxide nanoparticles in electrochemical sensing and biosensing: a review, Microchim. Acta185(7) (2018) 358-384. https://doi.org/10.1007/s00604-018-2894-3

[137] M.U.A. Prathap, B. Kaur, R. Srivastava, Electrochemical Sensor Platforms Based on Nanostructured Metal Oxides, and Zeolite-Based Materials, Chem. Rec. 19(5) (2019) 883-907. https://doi.org/10.1002/tcr.201800068

[138] N.S. Pavithra, K. Lingaraju, G.K. Raghu, G. Nagaraju,Citrus maxima (Pomelo) juice mediated eco-friendly synthesis of ZnO nanoparticles: Applications to photocatalytic, electrochemical sensor and antibacterial activities,Spectrochim. Acta, Part A   185 (2017) 11–19. https://doi.org/10.1016/j.saa.2017.05.032

[139] S. Sukumar, A. Rudrasenan, D.P. Nambiar,Green-Synthesized Rice-Shaped Copper Oxide Nanoparticles Using Caesalpiniabonducella Seed Extract and Their Applications, ACS Omega 5(2)(2020) 1040-1051. https://doi.org/10.1021/acsomega.9b02857

[140] S. Momeni, F. Sedaghati,CuO/Cu_2O nanoparticles: A simple and green synthesis, characterization and their electrocatalytic performance toward formaldehyde oxidation,Microchem. J. 143 (2018) 64–71. https://doi.org/10.1016/j.microc.2018.07.035

[141] M. Valian, A. Khoobi, M. Salavati-Niasari, Green synthesis and characterization of $DyMnO_3$-ZnO ceramicnanocomposites for the electrochemical ultratrace detection of atenolol, Mater. Sci. Eng., C 111 (2020) 110854-110865. https://doi.org/10.1016/j.msec.2020.110854

[142] N. Matinise, K. Kaviyarasu, N. Mongwaketsi, S. Khamlich, L. Kotsedi, N. Mayedwa, M. Maaza,Green synthesis of novel zinc iron oxide ($ZnFe_2O_4$) nanocomposite via MoringaOleifera natural extract for electrochemical applications, Appl. Surf. Sci.446 (2018) 66–73. https://doi.org/10.1016/j.apsusc.2018.02.187

[143] B.S. Surendra, H.P. Nagaswarupa, M.U. Hemashree, J. Khanum, Jatropha extract mediated synthesis of $ZnFe_2O_4$nanopowder: Excellent performance as an electrochemical sensor, UV photocatalyst and an antibacterial activity, Chem. Phys. Lett. 739 (2020) 136980-137012. https://doi.org/10.1016/j.cplett.2019.136980

[144] T. Laurila, S. Sainio, M.A. Caro, Hybrid carbon based nanomaterials for electrochemical detection of biomolecules, Prog. Mater Sci. 88(2017) 499-594. https://doi.org/10.1016/j.pmatsci.2017.04.012

[145] J.N. Tiwari, V. Vij, K.C. Kemp, K.S. Kim, Engineered Carbon-Nanomaterial-Based Electrochemical Sensors for Biomolecules, ACS Nano 10(1) (2016) 46-80. https://doi.org/10.1021/acsnano.5b05690

[146] T.V. Sathisha, B.K. Swamy, M. Schell, B. Eswarappa, Synthesis and characterization of carbon nanoparticles and their modified carbon paste electrode for the determination of dopamine, J. Electroanal. Chem. 720 (2014) 1–8. https://doi.org/10.1016/j.jelechem.2014.02.020

Bioinspired Nanomaterials
Materials Research Foundations **111** (2021) 224-263

Materials Research Forum LLC
https://doi.org/10.21741/9781644901571-8

[147] E. Vatandost, A. Ghorbani-HasanSaraei, F. Chekin, S. NaghizadehRaeisi, S-A. Shahidi, Green tea extract assisted green synthesis of reduced graphene oxide: Application for highly sensitive electrochemical detection of sunset yellow in food products, Food Chem. (2020) 100085 In Press. https://doi.org/10.1016/j.fochx.2020.100085

[148] S. Nazarpour, R. Hajian, M.H. Sabzvari, A novel nanocomposite electrochemical sensor based on green synthesis of reduced graphene oxide/gold nanoparticles modified screen printed electrode for determination of tryptophan using response surface methodology approach, Microchemical Journal 154 (2020) 104634-104641. https://doi.org/10.1016/j.microc.2020.104634

[149] C. Singh, A. Ali, and G. Sumana, Green Synthesis of Graphene Based Biomaterial using Fenugreek Seeds for Lipid Detection, ACS Sustainable Chem. Eng. 4(3)(2016) 871-880. https://doi.org/10.1021/acssuschemeng.5b00923

[150] C. Karuppiah, K. Muthupandi, S.-M. Chen, M.A. Ali, S. Palanisamy, A. Rajan, P. Prakash, F.M.A. Al-Hemaid, B.-S. Lou, Green synthesized silver nanoparticles decorated on reduced graphene oxide for enhanced electrochemical sensing of nitrobenzene in waste water samples, RSC Adv. 5 (2015) 31139–31146. https://doi.org/10.1039/C5RA00992H

[151] S. Palanisamy, C. Karuppiah, S.-M. Chen, K. Muthupandi, R. Emmanuel, P. Prakash, M.S. Elshikh, M.A. Ali, F.M. A. Al-Hemaid, Selective and Simultaneous Determination of Dihydroxybenzene Isomers Based on Green Synthesized Gold Nanoparticles Decorated Reduced Graphene Oxide, Electroanalysis 27(5)(2015) 1144-1151. https://doi.org/10.1002/elan.201400657

[152] D. Sharma, M.I. Sabela, S. Kanchi, K. Bisetty, A.A. Skelton, B. Honarparvar,Green synthesis, characterization and electrochemical sensing of silymarin by ZnO nanoparticles: Experimental and DFT studies, J. Electroanal. Chem. 808 (2018) 160–172. https://doi.org/10.1016/j.jelechem.2017.11.039

[153] P. Salazar, I. Fernandez, M.C. Rodríguez, A. Hernandez-Creus, J.L. Gonzalez-Mora, One-step green synthesis of silver nanoparticle-modified reduced graphene oxide nanocomposite for H_2O_2 sensing applications, J. Electroanal. Chem. 855 (2019) 113638-113647. https://doi.org/10.1016/j.jelechem.2019.113638

[154] P. Nayak, B. Anbarasan, and S. Ramaprabhu, Fabrication of Organophosphorus Biosensor Using ZnO Nanoparticle-Decorated Carbon Nanotube−Graphene Hybrid Composite Prepared by a Novel Green Technique,J. Phys. Chem. C 117(25)(2013) 13202-13209. https://doi.org/10.1021/jp312824b

Materials Research Forum LLC
https://doi.org/10.21741/9781644901571-8

[155] R.K. Das, S. Saha, V.R. Chelli, A.K. Golder, Bio-inspired AgNPs, multilayers-reduced graphene oxide andgraphite nanocomposite for electrochemical H_2O_2 sensing, Bull. Mater. Sci. 41 (2018) 86-97. https://doi.org/10.1007/s12034-018-1592-4

[156] T. Dodevska, I. Vasileva, P. Denev, D. Karashanova, B. Georgieva, D. Kovacheva, N. Yantcheva, A. Slavov,Rosa damascena waste mediated synthesis of silver nanoparticles: Characteristics and application for an electrochemical sensing of hydrogen peroxide and vanillin, Materials Chemistry and Physics 231 (2019) 335–343. https://doi.org/10.1016/j.matchemphys.2019.04.030

[157] R.K. Das, A.K.Golder, Co_3O_4 spinel nanoparticles decorated graphite electrode: Bio-mediated synthesis and electrochemical H_2O_2 sensing, Electrochim.Acta 251 (2017) 415-426. https://doi.org/10.1016/j.electacta.2017.08.122

[158] M.A. Khalilzadeh, S. Tajik, H. Beitollahi, R.A. Venditti, Green synthesis of magnetic nanocomposite with iron oxide deposited on cellulose nanocrystals with copper ($Fe_3O_4@CNC/Cu$): investigation ofcatalytic activity for the development of a venlafaxine electrochemical sensor, Ind. Eng. Chem. Res. 59(10)(2020) 4219-4228. https://doi.org/10.1021/acs.iecr.9b06214

[159] S. Zhou, H. Xu, Q. Yuan, H. Shen, X. Zhu, Y. Liu, W. Gan, N-doped ordered mesoporous carbon originated from a green biological dye for electrochemical sensing and high pressure CO_2 storage, ACS Appl. Mater. Interfaces 8(1) (2016) 918-926. https://doi.org/10.1021/acsami.5b10502

Keyword Index

About the Editors

Dr. Alagarsamy Pandikumar is currently working as Scientist in Functional Materials Division, CSIR-Central Electrochemical Research Institute, Karaikudi, India. He obtained his Ph.D. in Chemistry (2014) from the Madurai Kamaraj University, Madurai and then successfully completed his post-doctoral fellowship tenure (2014-2016) at the University of Malaya, Malaysia under High Impact Research Grant. His current research involves development of novel materials with graphene, graphitic carbon nitride, in combination to metals, metal oxides, polymers and carbon nanotubes for energy conversion and storage and dye-sensitized solar cells applications. His results outcomes were documented in 134 in peer-reviewed journals including 10 review articles and also have more than 3300 citations with the h−index of 39. On other side, he served as Guest Editor for a special issue in Materials Focus journal and edited 12 books for reputed publishers.

Dr. Perumal Rameshkumar is currently working as an Assistant Professor of Chemistry at Kalasalingam Academy of Research and Education, India. He obtained his M.Sc. (chemistry) (2009) from Madurai Kamaraj University. He joined as Junior Research Fellow (2010) at the same University and subsequently promoted as Senior Research Fellow (2012). His doctoral thesis focused on 'polymer encapsulated metal nanoparticles for sensor and energy conversion applications'. He worked as Post-Doctoral Research Fellow (2014) at University of Malaya, Malaysia in the field of 'graphene-inorganic nanocomposite materials for electrochemical sensor and energy conversion'. His current research interests include synthesis of functionalized nanomaterials, electrochemical sensors, energy-related electrocatalysis and photoelectrocatalysis. His research findings were documented in 34 peer reviewed journals including 01 review article. For his credit, he edited 02 books under Elsevier publications.

www.ingramcontent.com/pod-product-compliance
Lightning Source LLC
Chambersburg PA
CBHW061205220326
41597CB00015BA/1490